Springer Proceedings in Physics

Volume 198

The series Springer Proceedings in Physics, founded in 1984, is devoted to timely reports of state-of-the-art developments in physics and related sciences. Typically based on material presented at conferences, workshops and similar scientific meetings, volumes published in this series will constitute a comprehensive up-to-date source of reference on a field or subfield of relevance in contemporary physics. Proposals must include the following:

- name, place and date of the scientific meeting
- a link to the committees (local organization, international advisors etc.)
- scientific description of the meeting
- list of invited/plenary speakers
- an estimate of the planned proceedings book parameters (number of pages/ articles, requested number of bulk copies, submission deadline).

More information about this series at http://www.springer.com/series/361

Nicolae Herisanu · Vasile Marinca
Editors

Acoustics and Vibration of Mechanical Structures—AVMS-2017

Proceedings of the 14th AVMS Conference, Timisoara, Romania, May 25–26, 2017

Springer

Editors
Nicolae Herisanu
Department of Mechanics and Strength of Materials
University Politehnica Timişoara
Timişoara
Romania

Vasile Marinca
Centre for Fundamental and Advanced Technical Research, Romanian Academy
Branch of Timişoara
Timişoara
Romania

ISSN 0930-8989 ISSN 1867-4941 (electronic)
Springer Proceedings in Physics
ISBN 978-3-319-69822-9 ISBN 978-3-319-69823-6 (eBook)
https://doi.org/10.1007/978-3-319-69823-6

Library of Congress Control Number: 2017956331

Printed on acid-free paper

This Springer imprint is published by Springer Nature
The registered company is Springer International Publishing AG
The registered company address is: Gewerbestrasse 11, 6330 Cham, Switzerland

Preface

The Proceedings of the 14th International Conference on Acoustics and Vibration of Mechanical Structures—AVMS-2017—contains selected papers contributed to the conference held during 25–26 May 2017 in Timişoara (Romania).

This book is organised into five chapters covering a broad range of topics related to acoustics and vibration problems, such as:

- Noise and vibration control;
- Noise and vibration generation and propagation;
- Effects of noise and vibration;
- Condition monitoring and vibration testing;
- Nonlinear acoustics and vibration;
- Analytical, numerical and experimental techniques for noise and vibration;
- Modelling, prediction and simulations of noise and vibration;
- Environmental and occupational noise and vibration;
- Noise and vibration attenuators;
- Biomechanics and bioacoustics.

There are presented some analytical, numerical and experimental techniques applicable to analyse linear and nonlinear noise and vibration problems.

Each paper went through a rigorous review process performed by the members of the International Scientific Committee and specialised external reviewers, and the accepted papers are reported in this volume.

We would like to express our sincere appreciation to keynote speakers and all contributors of the presented papers for sharing their knowledge and experiences with all the participants. We are also expressing our sincere thanks to the members of the International Scientific Committee for spending their valuable time to review the papers and also to the members of the Organising Committee for ensuring the success of this conference which would not have been possible without their efforts.

Finally, special thanks are given to Springer for producing this volume.

Timişoara, Romania

Nicolae Herisanu
Vasile Marinca

Organising Committee

Organised By

- University Politehnica Timisoara—Acoustics and Vibration Laboratory;
- University of Nis—Noise and Vibration Laboratory;
- Romanian Academy—Centre for Fundamental and Advanced Technical Research;
- Romanian Acoustical Society;
- The General Association of the Engineers in Romania—AGIR

Organization

Nicolae Herisanu—Chairman, University Politehnica Timişoara, Romania
Momir Praščevič—Co-chairman, University of Nis, Serbia
Members
Viorel-Aurel Şerban, Rector—University Politehnica Timişoara, Romania
Vasile Marinca, University Politehnica Timişoara, Romania
Darko Mihajlov, University of Nis, Serbia
Ramona Nagy, University Politehnica Timişoara, Romania
Karoly Menyhardt, University Politehnica Timişoara, Romania
Cosmina Vigaru, University Politehnica Timişoara, Romania

International Scientific Committee

Jan Awrejcewicz, Lodz University of Technology, Poland
Vasile Bacria, University Politehnica Timişoara, Romania
Malvina Baica, University of Wisconsin, USA

Contents

Part I
Analytical Approaches to Vibration Problems

Chapter 1
Fractional Calculus: Fundamentals and Applications

J. A. Tenreiro Machado

Abstract This paper presents the fundamental aspects of the theory of Fractional Calculus. Several approximation methods for the calculation of fractional-order derivatives are discussed. The application of Fractional Calculus in automatic control systems and their main properties are also analyzed.

1.1 Introduction

The generalization of the concept of derivative $D^{\alpha} f(x)$ to non-integer values of α started with the theory of differential calculus. Leibniz, in his correspondence with L'Hopital (1695), had several notes about the calculation of $D^{1/2} f(x)$. The development of the theory of Fractional Calculus (FC) received the contributions of many important mathematicians such as Euler, Liouville, Riemann and Letnikov [11, 13, 17]. In the fields of physics and engineering, FC is presently associated with the modelling of electro-chemical reactions, irreversibility and electromagnetism [4, 10]. The adoption of the FC in control algorithms has been studied [1, 2, 6, 8, 18, 20] using the frequency and discrete-time domains.

The chapter is organized as follows. Section 1.2 outlines the main mathematical aspects of the theory of FC. Section 1.3 introduces several algorithms to approximate fractional-order derivatives. Section 1.4 present controllers using FC. Finally, Sect. 1.5 draws the main conclusions.

J. A. T. Machado (✉)
Department of Electrical Engineering, Institute of Engineering, Polytechnic of Porto, R. Dr. António Bernardino de Almeida, 431, 4249-015 Porto, Portugal
e-mail: jtm@isep.ipp.pt

N. Herisanu and V. Marinca (eds.), *Acoustics and Vibration of Mechanical Structures—AVMS-2017*, Springer Proceedings in Physics 198,
https://doi.org/10.1007/978-3-319-69823-6_1

1.2 Mathematical Fundamentals of Fractional Calculus

Can the order of derivatives and integrals be extended to have meaning with any number irrational, fractional or complex? The concept motivated researchers to develop the concept of FC both in theoretical aspects and practical implementations. During the last decades FC was recognized to be a solid tool to model and to analyze complex dynamical systems and the scientific community witnessed the emergence of this scientific area.

There are several definitions of fractional derivatives, their comparison and discussion of pros and cons is outside the scope of this work. The most used definitions of a fractional derivative of order α are the Riemann-Liouville (RL), Grunwald-Letnikov (GL) and Caputo (C) formulations [5] given by the expressions:

$$ {}_{a}^{RL}D_{t}^{\alpha}f(t) = \frac{1}{\Gamma(n-\alpha)}\frac{d^{n}}{dt^{n}}\int_{a}^{t}\frac{f(\tau)}{(t-\tau)^{\alpha-n+1}}d\tau, \quad t>a, \quad \mathrm{Re}(\alpha)\in[n-1,n] \tag{1.1}$$

$$ {}_{a}^{GL}D_{t}^{\alpha}f(t) = \lim_{h\to\infty}\frac{1}{h^{\alpha}}\sum_{k=0}^{\frac{t-a}{h}}(-1^{k})\binom{\alpha}{k}f(t-kh), \quad t>a, \quad \alpha>0 \tag{1.2}$$

$$ {}_{a}^{C}D_{t}^{\alpha}f(t) = \frac{1}{\Gamma(n-\alpha)}\int_{a}^{t}\frac{f^{(n)}(\tau)}{(t-\tau)^{\alpha-n+1}}d\tau, \quad t>a, \quad n-1<\alpha<n \tag{1.3}$$

where $\Gamma(\cdot)$ is Euler's gamma function, $[x]$ means the integer part of x, and h is the step time increment.

These operators capture the history of all past events, in opposition to integer derivatives that are merely 'local' operators. This means that fractional order systems have a memory of the dynamical evolution. This property was recognized in several natural and artificial phenomena and their modeling becomes much simpler using the tools of FC.

Adopting the Laplace transform we have the expressions:

$$ \mathcal{L}\left\{{}_{0}^{RL}D_{t}^{\alpha}f(t)\right\} = s^{\alpha}\mathcal{L}\{f(t)\} - \sum_{k=0}^{n-1}s^{k}{}_{0}^{RL}D_{t}^{\alpha-k-1}f(0^{+}) \tag{1.4}$$

$$ \mathcal{L}\left\{{}_{0}^{C}D_{t}^{\alpha}f(t)\right\} = s^{\alpha}\mathcal{L}\{f(t)\} - \sum_{k=0}^{n-1}s^{a-k-1}f^{(k)}(0) \tag{1.5}$$

where s and L denote the Laplace variable and operator, respectively.

The Mittag-Leffler (ML) function $E_\alpha(t)$ is defined as:

$$\mathrm{E}_\alpha(t) = \sum_{k=0}^{\infty} \frac{t^k}{\Gamma(\alpha k + 1)}, \quad \alpha \in C, \;\; \mathrm{Re}(\alpha) > 0. \tag{1.6}$$

Expression (1.6) represents a generalization of the exponential series. The ML function forms a bridge between the exponential and the power laws. The first occurs in phenomena governed by integer order and the second in fractional order dynamics. In particular, when $\alpha = 1$ the ML function simplifies and we have $\mathrm{E}_1(t) = \mathrm{e}^t$, while, for large values of t, the asymptotic behavior yields:

$$E_\alpha(-t) \approx \frac{1}{\Gamma(1-\alpha)} \frac{1}{t}, \quad \alpha \neq 1, \;\; 0 < \alpha < 2 \tag{1.7}$$

The Laplace transform leads to the formula:

$$\mathcal{L}\{E_\alpha(\pm a t^\alpha)\} = \frac{s^{\alpha-1}}{s^\alpha \mp a} \tag{1.8}$$

and, therefore, we verify a generalization of the Laplace transform pairs from the exponential towards the ML, that is, from integer up to fractional powers of s.

The two-parameter ML function is given by:

$$\mathrm{E}_{\alpha,\beta}(t) = \sum_{k=0}^{\infty} \frac{t^k}{\Gamma(\alpha k + 1)}, \alpha, \quad \beta \in C, \;\; \mathrm{Re}(\alpha), \;\; \mathrm{Re}(\beta) > 0 \tag{1.9}$$

Function (9) is a generalization of (6), since $E_\alpha(t) = E_{\alpha,1}(t)$.

1.3 Approximations to Fractional-Order Derivatives

We now discuss two methods for implementing fractional-order derivatives, namely the frequency-based and the discrete-time approaches, and their implication in control algorithms.

In order to analyze a frequency-based approach to $D^\alpha, 0 < \alpha < 1$, we consider the recursive circuit [3, 14] represented on Fig. 1.1:

$$i = \sum_{k=0}^{n} i_k \tag{1.10}$$

$$R_{k+1} = \frac{1}{\varepsilon} R_k \tag{1.11}$$

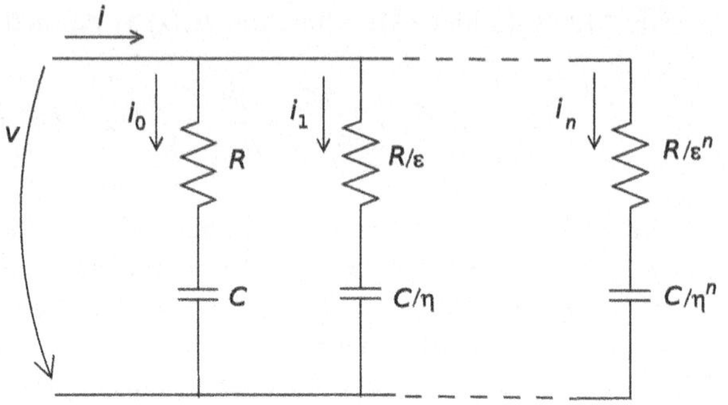

Fig. 1.1 Electrical circuit with a recursive association of resistance and capacitance elements

$$C_{k+1} = \frac{1}{\eta} C_k \tag{1.12}$$

where ε and η are scale factors, i is the current due to an applied voltage v and R_k and C_k are the resistance and capacitance elements of the kth branch of the circuit. The admittance $Y(j\omega)$ is given by:

$$Y(j\omega) = \frac{I(j\omega)}{V(j\omega)} = \sum_{k=0}^{n} \frac{j\omega C\varepsilon^k}{j\omega CR + (\varepsilon\eta)^k} \tag{1.13}$$

where F denotes the Fourier transform operator, ω represents the frequency, $\mathcal{F}\{i(t)\} = I(j\omega), \mathcal{F}\{v(t)\} = V(j\omega)$ and $j = \sqrt{-1}$.

Figure 1.2 shows the asymptotic Bode diagrams of amplitude and phase of Y (jω). The frequencies of the poles ω_k and zeros $\omega_k{}'$ obey the recursive relationships:

$$\frac{\omega'_{k+1}}{\omega'_k} = \frac{\omega_{k+1}}{\omega_k} = \varepsilon\eta \tag{1.14}$$

$$\frac{\omega_{k+1}}{\omega'_k} = \varepsilon \tag{1.15}$$

$$\frac{\omega'_k}{\omega_k} = \eta \tag{1.16}$$

From the Bode diagram of amplitude or of phase, the average slope m' can be calculated as:

$$m' = \frac{\ln \varepsilon}{\ln \varepsilon + \ln \eta} \tag{1.17}$$

In this perspective, the circuit of Fig. 1.1 implements an approximation to D^{α}, $0 < \alpha < 1$, with $m' = \alpha$, based on a recursive pole/zero placement in the frequency domain. As mentioned in Sect. 1.2, the Laplace definition for a derivative of order

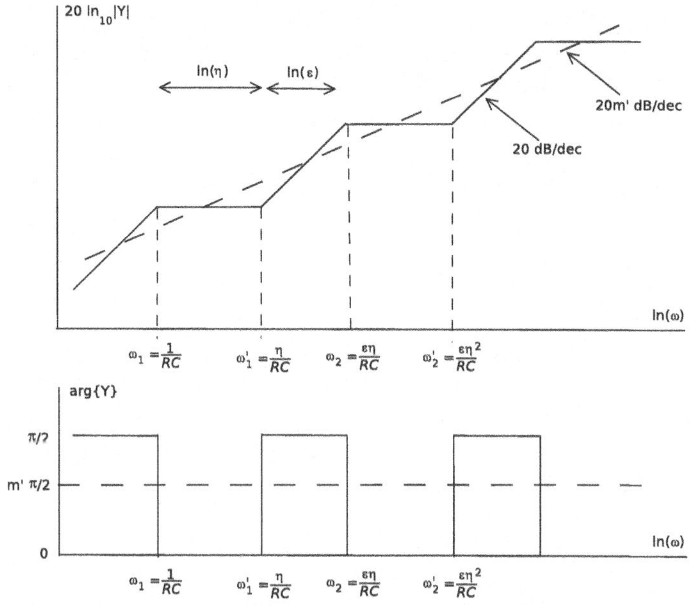

Fig. 1.2 Bode diagrams of amplitude and phase of $Y(j\omega)$

$\alpha \in C$ generalizes the classical integer-order property with the multiplication of the signal transform by the s operator. Therefore, in what concerns automatic control theory this means that frequency-based analysis methods have a straightforward adaptation to their fractional-order counterparts. Nevertheless, the implementation based on the Laplace definition (adopting the frequency domain) requires an infinite number of poles and zeros obeying a recursive relationship. In a real approximation the finite number of poles and zeros yields a ripple in the frequency response and a limited bandwidth. The Grunwald-Letnikov definition (2) of a derivative of fractional order α of the signal $f(t)$, $D^{\alpha}f(t)$, suggests a discrete-time algorithm with the approximation of the time increment h by means of the sampling period T, yielding the equation in the Z-transform:

$$\frac{Z\{D^{\alpha}f(t)\}}{Z\{f(t)\}} = \frac{1}{T^{\alpha}} \sum_{k=0}^{\infty} \frac{(-1)^k \Gamma(\alpha+1)}{k!\Gamma(\alpha-k+1)} z^{-k} = \left(\frac{1-z^{-1}}{T}\right)^{\alpha} \tag{1.18}$$

An implementation of (1.18) corresponds to a r-term truncated series given by:

$$\frac{Z\{D^{\alpha}f(t)\}}{Z\{f(t)\}} = \frac{1}{T^{\alpha}} \sum_{k=0}^{r} \frac{(-1)^k \Gamma(\alpha+1)}{k!\Gamma(\alpha-k+1)} z^{-k}. \tag{1.19}$$

In order to have good approximations, we must have a large r and a small T.

Expression (1.18) represents the Euler, or first backward difference, approximation in the so-called $s \to z$ conversion scheme. Another possibility, often adopted in control system design, consists in the Tustin (or bilinear) rule. The Euler and Tustin rational expressions, $\psi_0(z^{-1}) = \frac{1-z^{-1}}{T}$ and $\psi_1(z^{-1}) = \frac{2}{T}\frac{1-z^{-1}}{1+z^{-1}}$, denoted as generating approximants of zero and first order, respectively. Therefore, the generalization [7] of these conversion methods to the non-integer order a can be expressed as:

$$s^\alpha \approx \left(\frac{1-z^{-1}}{T}\right)^\alpha \tag{1.20}$$

$$s^\alpha \approx \left(\frac{2}{T}\frac{1-z^{-1}}{1+z^{-1}}\right)^\alpha \tag{1.21}$$

We obtain a family of fractional differentiators generated by $\psi_0^\alpha(z^{-1}) = [\psi_0(z^{-1})]^\alpha$ and $\psi_1^\alpha(z^{-1}) = [\psi_1(z^{-1})]^\alpha$ weighted by the factors p and $1-p$, yielding:

$$\psi_{av}^\alpha(z^{-1}) = p\psi_0^\alpha(z^{-1}) + (1-p)\psi_1^\alpha(z^{-1}) \tag{1.22}$$

For obtaining a rational approximation, the final expression is the truncated Taylor series or the expansion by means of a rational fraction. Due to its superior performance, a simpler initialization is often used a fraction of order r:

$$\psi(z^{-1}) = \frac{\sum_{k=0}^{r} a_k z^{-k}}{\sum_{k=0}^{r} b_k z^{-k}} \tag{1.23}$$

In many cases it is adopted a Padé expansion in the neighborhood of $z = 0$ and, since one coefficients in the polynomials is linearly dependent, it is usually considered $b_0 = 1$. The arithmetic mean (1.22) motivates the study of an averaging method [9] based on the generalized formula of averages (often called average of order $q \in R$):

$$\psi_{av}^\alpha(z^{-1}) = \{p[\psi_0^\alpha(z^{-1})]^q + (1-p)[\psi_1^\alpha(z^{-1})]^q\}^{\frac{1}{q}}, \tag{1.24}$$

where (p, q) are two tuning degrees of freedom, corresponding q to the order of the averaging expression and p to the weighting factor. Therefore, when $q = \{-1, 0, 1\}$, in expression (1.24), we obtain expressions for the {harmonic, geometric, arithmetic} averages.

1.4 Application in Control

The tools of FC can easily be adapted in automatic control systems. Figure 1.3 includes an elemental fractional system in the direct loop with transfer function

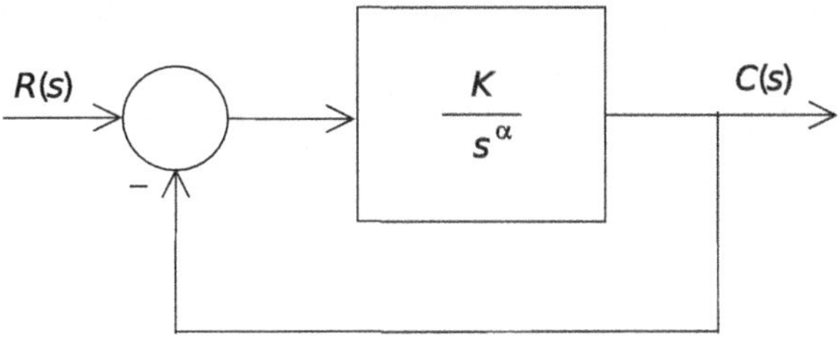

Fig. 1.3 Block diagram for an elemental feedback control system of fractional order α

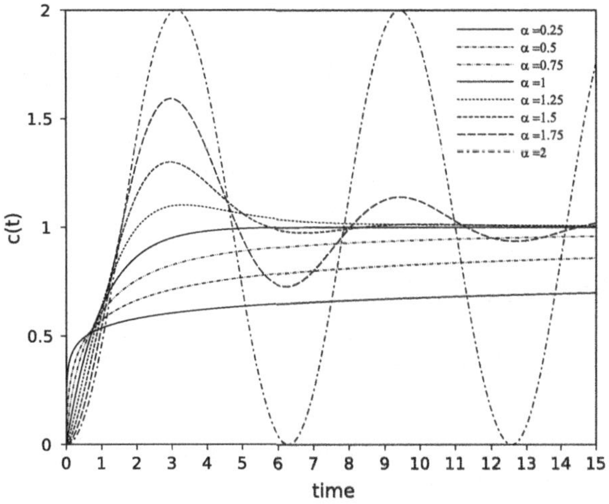

Fig. 1.4 Time response $c(t) = 1 - E\alpha(-t\alpha)$ of the closed-loop system represented in Fig. 1.3 for a unit step reference input and $\alpha = \{0.25, 0.5, 0.75, 1, 1.25, 1.5, 1.75, 2\}$

$G(s) = \frac{K}{s^{\alpha}}$, $1 < \alpha < 2$ and illustrates an important property of fractional controllers. The open-loop Bode diagrams of amplitude and phase have a slope of $-20\,\alpha$ dB/dec and a constant phase of $-\alpha\frac{\pi}{2}$ rad, respectively. Therefore, the closed-loop system has a constant phase margin of $\pi\left(1 - \frac{\alpha}{2}\right)$ rad, independently of the gain K.

Let us now consider $K = 1$, so that $G(s) = \frac{1}{s^{\alpha}}$, and the unit step input $R(s) = \frac{1}{s}$ in the control system represented in Fig. 1.3. The output response is $C(s) = \frac{1}{s(s^{\alpha}+1)}$, or, in the time domain, $c(t) = 1 - E_{\alpha}(-t^{\alpha})$. Figure 1.4 depicts the responses for $\alpha = \{0.25, 0.5, 0.75, 1, 1.25, 1.5, 1.75, 2\}$. We verify that the fractional values 'interpolate' the two integer-order cases for $\alpha = \{1, 2\}$. Furthermore, we note the appearance of a fast initial transient, followed by a very slow convergence toward the steady-state value, which is a typical behavior of many fractional order systems.

Control is a popular application of FC [6, 12, 15, 19]. For example, the generalization of the Proportional, Integral and Derivative (*PID*) algorithm [16], called the fractional *PID*, or $PI^{\lambda}D^{\mu}$, has the transfer function:

$$G_C = K_p + K_I s^{-\lambda} + K_D s^{\mu}, \tag{1.25}$$

where K_P, K_I and K_D are the proportional, integral and differential gains, and λ and μ are the fractional orders of the integral and derivative actions, respectively.

With the $PI^{\lambda}D^{\mu}$ we tune 5 parameters contrary to the standard 3 parameters associated with the integer-order PID. Therefore, in general, the larger number of parameters allows a better system performance.

1.5 Conclusions

This paper presented the fundamental aspects of the FC calculus and two approximation methods for the calculation of fractional derivatives. It was also shown that FC models capture phenomena and properties that classical integer-order simply neglect. The implication of the FC concepts upon the automatic control systems was also explored.

References

1. D. Baleanu, J.T. Machado, A. Luo, *Fractional Dynamics and Control* (Springer, New York, 2011)
2. D. Baleanu, K. Diethelm, E. Scalas, J.J. Trujillo, *Fractional Calculus Models and Numerical Methods* (World Scientific Publishing Company, Amsterdam, 2012)
3. G.E. Carlson, C.A. Halijak, Approximation of fractional capacitors (1/s)(1/n) by a regular Newton process. IEEE Trans. Circuit Theory **10**, 210–213 (1964)
4. R. Hilfer, *Applications of fractional calculus in physics* (World Scientific, Singapore, 2000)
5. A.A. Kilbas, H.M. Srivastava, J.J. Trujillo, *Theory and applications of fractional differential equations. North-Holland Mathematics Studies*, vol. 204 (Elsevier, Amsterdam, 2006)
6. J.T. Machado, Analysis and design of fractional-order digital control systems. Systems Anal. Model. Simul. **27**(2–3), 107–122 (1997)
7. J.T. Machado, Fractional-order derivative approximations in discrete-time control systems. Syst. Anal. Model. Simul. **34**, 419–434 (1999)
8. J.T. Machado, Discrete-time fractional-order controllers. Fract. Calculus Appl. Anal. **4**(1), 47–66 (2001)
9. J.T. Machado, A.M. Galhano, Approximating fractional derivatives in the perspective of system control. Nonlinear Dyn. **56**(4), 401–407 (2009)
10. F. Mainardi, *Fractional Calculus and Waves in Linear Viscoelasticity: An Introduction to Mathematical Models* (Imperial College Press, London, 2010)
11. K.S. Miller, B. Ross, *An Introduction to the Fractional Calculus and Fractional Differential Equations* (Wiley, New York, 1993)
12. C.A. Monje, Y. Chen, B.M. Vinagre, D. Xue, V. Feliu, *Fractional-Order Systems and Controls* (Springer, London, 2010)
13. K.B. Oldham, J. Spanier, *The Fractional Calculus* (Academic Press, New York, 1974)
14. Oustaloup, A.: La Commande CRONE: Commande Robuste d'Ordre Non Entier, Hermes, Paris (1991)
15. I. Petráš: *Fractional-Order Nonlinear Systems: Modeling, Analysis and Simulation* (Springer, Heidelberg, 2011)
16. I. Podlubny, *Fractional Differential Equations* (Academic Press, San Diego, 1999)

17. S.G. Samko, A.A. Kilbas, O.I. Marichev, *Fractional Integrals and Derivatives* (Gordon and Breach Science Publishers, Yverdon, 1993)
18. V.E. Tarasov, *Fractional Dynamics: Applications of Fractional Calculus to Dynamics of Particles* (Fields and Media, Springer, 2010)
19. D. Valerio, J.S. da Costa, *An Introduction to Fractional Control* (IET, Stevenage, 2012)
20. S. Westerlund, *Dead Matter Has Memory* (Causal Consulting, Kalmar, 2002)

Chapter 2
An Approximate Method for Solving a Vibration Equation Involving Fractional Derivatives

Mohsen Razzaghi

Abstract The response of one-degree-of-freedom systems with fractional distributed-order damping is studied. The dynamics of such systems constitutes the problem of the fractional distributed-order oscillator. This paper presents a new numerical method for solving this fractional distributed order oscillator. The method is based on using fractional Taylor vector approximation. The operational matrix of the fractional integration for fractional Taylor vector is given and is utilized to reduce the solution of the fractional distributed order oscillator to a system of algebraic equations. An illustrative example is included to demonstrate the validity and applicability of this technique.

2.1 Introduction

The fractional differential equations (FDEs) have drawn increasing attention and interest due to their important applications in various fields of science and engineering. A history of the development of fractional differential operators can be found in [1–3].

Mathematical modeling has been used in many FDEs. To mention a few, fractional derivatives are used in visco-elastic systems [4], economics [5], continuum and statistical mechanics [6], solid mechanics [7], electrochemistry [8], biology [9] and acoustics [10]. It is known that except for some simple cases, it is either extremely difficult or impossible to analytically solve FDEs. Accordingly, seeking numerical solutions of these equations becomes more and more important [11]. There are various definitions of fractional derivative and integration. The widely used definition of a fractional derivative is the Caputo definition, and for a fractional integration is the Riemann-Liouville definition.

M. Razzaghi (✉)
Department of Mathematics and Statistics, Mississippi State University, Starkville, MS 39762, USA
e-mail: razzaghi@math.msstate.edu

N. Herisanu and V. Marinca (eds.), *Acoustics and Vibration of Mechanical Structures—AVMS-2017*, Springer Proceedings in Physics 198,
https://doi.org/10.1007/978-3-319-69823-6_2

The one degree-of-freedom systems with distributed fractional order dissipation forces have been introduced by Atanackovic and his co-workers [12–14]. They result from the generalization of the multi-term fractional differential viscoelastic model by considering continuous variation of the order of fractional derivative within a closed interval. This model leads to the following initial value problem for the linear fractional distributed-order oscillator (FDO) given in [14] as

$$u^{(2)}(t) + \sigma(t) + \omega^2 u(t) = f(t) \tag{2.1}$$

$$\int_0^1 \phi(p) D^c \sigma dp = \lambda \int_0^1 \psi(p) D^p u dp \tag{2.2}$$

where u and σ(t) are the displacement and the dissipation force, respectively, λ is a constant parameter; φ(p) and ψ(p) are specified functions subjected to certain restrictions resulting from the second law of thermodynamics; ω is the eigenfrequency of the undamped system; f(t) is the external forcing function and Dp is the Caputo type fractional derivative of order p.

The initial conditions

$$u(0) = u_0, \quad u'(0) = u'_0 \tag{2.3}$$

In this paper, we present an alternative numerical method for solving (2.1)–(2.3). In this method, first, the fractional Taylor vector is used to approximate the unknown function. The operational matrix of integration for the fractional Taylor vector is given and is used to reduce the solution of (2.1)–(2.3) to the solution of a system of algebraic equations.

The outline of this paper is as follows: In Sect. 2.2, we provide some basic definitions and properties of fractional calculus. In Sect. 2.3, we present the formulation and the fractional integral operational matrix for the fractional Taylor vector. In Sect. 2.4, we describe the numerical method to solve (2.1)–(2.3). In Sect. 2.5 we report our numerical findings and demonstrate the accuracy of the proposed numerical method by considering a numerical example.

2.2 Preliminaries

2.2.1 *The Fractional Integral and Derivative*

There are various definitions of fractional integration and derivatives. The widely used definition of a fractional integration is the Riemann-Liouville definition and for the fractional derivative is the Caputo definition.

Definition 1 The Riemann-Liouville fractional integral operator of order α is defined as [2]

$$I^{\alpha}y(t) = \tag{2.4}$$

The Riemann-Liouville fractional integral operator have the following properties [2]:

$$\begin{aligned} I^{\alpha}(t) &= \frac{\Gamma(\gamma+1)}{\Gamma(\gamma+\alpha+1)}(t)^{\gamma+\alpha}, \quad && \alpha \geq 0, \quad \gamma > -1 \\ I^{\alpha}I^{\beta}y(t) &= I^{\beta}I^{\alpha}y(t), && \alpha, \beta > 0 \end{aligned} \tag{2.5}$$

Also the fractional integral is a linear operator, that is for constants λ_1 and λ_2, we have

$$I^{\alpha}(\lambda_1 y_1(t) + \lambda_2 y_2(t)) = \lambda_1 I^{\alpha} y_1(t) + \lambda_2 I^{\alpha} y_2(t)$$

Definition 2 The Caputo fractional derivative of order α is defined as [2]:

$$D^{\alpha}y(t) = I^{n-\alpha}\left(\frac{d^n}{dt^n}y(t)\right), \quad n-1 < \alpha \leq n, \quad n \in N \tag{2.6}$$

Some properties:

The fractional integral operator and fractional Caputo's derivative operator do not commute in general, but we have the following property [2]:

$$I^{\alpha}(D^{\alpha}y(t)) = y(t) - \sum_{k=0}^{n-1} y^{(k)}(0)\frac{t^k}{k!} \tag{2.7}$$

2.3 Fractional Taylor Approximation

2.3.1 Fractional Taylor Vector

The fractional Taylor vector is defined as

$$T_{m\gamma}(t) = \left[1, t^{\gamma}, t^{2\gamma}, \ldots, t^{m\gamma}\right]^T \tag{2.8}$$

where m is a positive integer and $\gamma > 0$, is a real number.

2.3.2 Function Approximations

Let $H = L^2[0,1]$, and assume that $T_{m\gamma}(t) \subset H$, $S = span\{1, t^{\gamma}, t^{2\gamma}, \ldots, t^{m\gamma}\}$ and y be an arbitrary element in H. Since S is a finite dimensional vector subspace of H, y has a unique best approximation out of S such as $y_0 \in S$, that is

$$\forall \hat{y} \in S, \quad \|y - y_0\| \leq \|y - \hat{y}\|$$

Since $y_0 \in S$, there exist the unique coefficients $c_0, c_1, c_2, \ldots, c_m$, such that

$$y \simeq y_0 = \sum_{i=0}^{m} c_i t^{i\gamma} = C^T T_{m\gamma}(t) \tag{2.9}$$

where $Tm\gamma(t)$ is given in (2.8) and

$$C^T = [c_0, c_1, c_2, c_m] \tag{2.10}$$

2.3.3 Operational Matrix of Integration for the Fractional Taylor Vector

By using (2.5) and (2.8), we have

$$\begin{aligned} &I^{\alpha}(Tm\gamma(t)) \\ &= \left[\frac{1}{\Gamma(\alpha+1)} t^{\alpha}, \frac{\Gamma(\gamma+1)}{\Gamma(\gamma+\alpha+1)} t^{\gamma+\alpha}, \frac{\Gamma(2\gamma+1)}{\Gamma(2\gamma+\alpha+1)} t^{2\gamma+\alpha}, \ldots, \frac{\Gamma(m\gamma+1)}{\Gamma(m\gamma+\alpha+1)} t^{m\gamma+\alpha}\right]^T \\ &= t^{\alpha} F_{\alpha} T_{m\gamma}(t) \end{aligned} \tag{2.11}$$

where F_α is the operational matrix of integration and is given by

$$F_{\alpha} = diag\left[\frac{1}{\Gamma(\alpha+1)}, \frac{\Gamma(\gamma+1)}{\Gamma(\gamma+\alpha+1)}, \frac{\Gamma(2\gamma+1)}{\Gamma(2\gamma+\alpha+1)}, \ldots, \frac{\Gamma(m\gamma+1)}{\Gamma(m\gamma+\alpha+1)}\right]$$

2.4 Numerical Method

In this section, we use the fractional Taylor vector for solving (2.1) and (2.2) with the initial conditions in (2.3). As suggested in [14], first we convert (2.1) and (2.2) into a single fractional distributed order equation.

Taking the D^p derivative of Eq. (2.1), multiplying it with $\varphi(p)$ and integrating in the interval [0,1] yields

$$\int_0^1 \phi(p)D^{2+p}udp + \int_0^1 \phi(p)D^p\sigma dp + \omega^2 \int_0^1 \phi(p)D^p udp = \int_0^1 \phi(p)D^p f(t)dp \tag{2.12}$$

Using Eq. (2.2) to replace the second integral, we obtain

$$\int_0^1 \phi(p)D^{2+p}udp + \int_0^1 \left[\omega^2\phi(p) + \lambda\psi(p)\right]D^p udp = \int_0^1 \phi(p)D^p f(t)dp \tag{2.13}$$

which is a fractional distributed order equation of the form

$$\int_0^1 \{\phi(p)D^{2+p}u + z(p)D^p u\}dp = f(t) \tag{2.14}$$

where

$$z(p) = \omega^2\phi(p) + \lambda\psi(p), \quad f(t) = \int_0^1 \phi(p)D^p f(t)dp \tag{2.15}$$

By using (2.9) and (2.11), (2.14), can be solved using the numerical solution for fractional distributed-order equations developed in [15, 16].

2.5 Illustrative Example

Consider the FDO given in (2.1) and (2.2) with initial conditions

$$f(0) = f'(0) = 0$$

The forced vibrations of the FDO subjected to the harmonic excitation $f(t) = f_0 \sin(\Omega t)$ is studied. Assuming that $\varphi(p) = a^p$, $\psi(p) = b^p$ with a, b being constants, the solution is obtained with $\Omega = 1.2\omega$, $\omega = 3$, $f_0 = 1$, and $\lambda = 1$. It is noted that for $a = b$ the solution is identical to the elastic with $\omega_{el} = \sqrt{1+\omega^2} = \sqrt{10}$. We solved this problem by using the fractional Taylor vector by selecting $\gamma = 0.5$ and $m = 30$. Figure 2.1 shows the exact and the approximate solution in [0, 30].

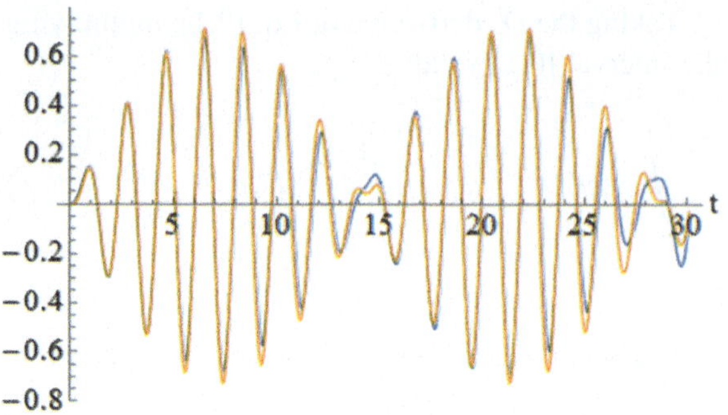

Fig. 2.1 Exact and approximation solution, m = 30 and $\gamma = 0.5$

2.6 Conclusions

In the present work the fractional Taylor vector has been developed for the solution of the equations governing the response of the fractional distributed order oscillator. Here, the system of the two coupled equations governing the response of the FDO oscillator was converted into a single FDO differential equation. Then the operational matrix of integration for the fractional Taylor vector has been utilized to reduce the solution of FDO differential equation to a problem of solving a system of algebraic equations. An illustrative example is examined to assess the performance and effectiveness of the developed approximation technique.

References

1. K.B. Oldham, J. Spanier, *The Fractional Calculus* (Academic Press, New York, 1974)
2. K.S. Miller, B. Ross, *An Introduction to the Fractional Calculus and Fractional Differential Equations* (Wiley, New York, 1993)
3. J.T. Machado, V. Kiryakova, F. Mainardi, Recently history of fractional calculus. Commun. Nonlinear Sci. Numer. Simul. **16**, 1140–1153 (2011)
4. R.L. Bagley, P.J. Torvik, Fractional calculus in the transient analysis of viscoelastically damped structures. AIAA J. **23**(1985), 918–925 (1985)
5. R.T. Baillie, Long memory processes and fractional integration in econometrics. J. Econom. **73**, 5–59 (1996)
6. F. Mainardi, Fractional calculus, Some basic problems in continuum and statistical mechanics, in *Fractals and Fractional Calculus in Continuum Mechanics*, ed. by A. Carpinteri, F. Mainardi (Springer, New York, 1997), pp. 291–348
7. Y.A. Rossikhin, M.V. Shitikova, Applications of fractional calculus to dynamic problems of linear and nonlinear hereditary mechanics of solids. Appl. Mech. Rev. **50**, 15–67 (1997)
8. K.B. Oldham, Fractional differential equations in electrochemistry. Adv. Eng. Softw. **41**, 9–12 (2010)
9. V.S. Erturk, Z.M. Odibat, S. Momani, An approximate solution of a fractional order differential equation model of human T-cell lymphotropic virus I (HTLV-I) infection of CD4 + T-cells. Comput. Math Appl. **62**, 996–1002 (2011)

10. S.A. El-Wakil, E.M. Abulwafa, E.K. El-Shewy, A.A. Mahmoud, Ion-acoustic waves in unmagnetized collisionless weakly relativistic plasma of warm-ion and isothermal-electron using time-fractional KdV equation. Adv. Space Res. **49**, 1721–1727 (2012)
11. M. Khader, N.H. Sweilam, On the approximate solutions for system of fractional integro-differential equations using Chebyshev pseudo-spectral method. Appl. Math. Model. **37**, 9819–9828 (2013)
12. I.M. Atanackovic, A generalized model for the uniaxial isothermal deformation of a viscoelastic body. Acta Mech. **159**, 77–86 (2002)
13. T.M. Atanackovic, M. Budincevic, S. Pilipovic, On a fractional distributed-order oscillator. J. Phys. A, Math. Gen. **38**, 6703–6713 (2005)
14. T.M. Atanackovic, Fractional distributed order oscillator. A numerical solution. J. Serb. Soc. Comput. Mech. **6**, 148–159 (2012)
15. J.T. Katsikadelis, Numerical solution of distributed order fractional differential equations. J. Comput. Phys. **259**, 11–22 (2014)
16. S. Mashayekhi, M. Razzaghi, Numerical solution of distributed order fractional differential equations by hybrid functions. J. Comput. Phys. **315**, 169–181 (2016)

Chapter 3
Acoustic Metamaterials: Theory and Application

Livija Cveticanin and D. Cveticanin

Abstract In this paper the concept of the metamaterial is explained. Using the interaction between the seismic (Rayleigh) wave and the ground-forest system the phenomena of wave stop is clarified. The example of 'seismic metamaterial' in architecture is also shown. The subunit of metamaterial is modeled as a mass-in-mass system. On the mass 1 a small mass with linear elastic spring is connected. The resonance conditions for the two-degrees of freedom system excited with a periodical trigonometric force are obtained. The term of effective mass is introduced. The effect of the negative effective mass on excitation wave spread is shown. The real metamaterial subunit is made of a metal sphere surrounded with rubber on which a thin silicon layer is settled. The change of vibration direction of the core in comparison to the outside layer and wave give the transformation from the positive to negative effective mass. The subunit has only one very narrow frequency band gap. In the paper we considered the subunit with nonlinear elastic connection between basic and added masses. Namely the added mass with non-linear spring acts as a nonlinear vibration absorber. If the excitation has the form of the Jacobi elliptic function and the nonlinear spring property is a cubic function of deflection the solution the frequency of vibration of the system depends on the amplitude of vibration. For certain parameters of the excitation function, the calculated parameters of spring have to be used. It means that the properties of absorber have to be choosing in accordance with excitation parameters. The obtained frequency band gap for frequencies is larger for the nonlinear than for the linear subsystem. The realization of metamaterials by using the subunits incorporated in the basic material or honeycomb structure is shown. The metamaterials are not natural but artificial materials usually composites. The explanation of acoustic metamaterial in noise reduction is considered. The metamaterial produced at the University of Novi Sad, Serbia, is experimentally tested at the Laboratory of

L. Cveticanin (✉)
Obuda University, Nepszinhaz U. 7, Budapest, Hungary
e-mail: cveticanin@uns.ac.rs

D. Cveticanin
Remming, B. Nusica 15, Novi Sad, Serbia
e-mail: dragan.cveticanin@remming.co.rs

N. Herisanu and V. Marinca (eds.), *Acoustics and Vibration of Mechanical Structures—AVMS-2017*, Springer Proceedings in Physics 198,
https://doi.org/10.1007/978-3-319-69823-6_3

Nuclear Sciences in Bucharest, Romania. The experimental rig is shown. The obtained results of energy variation in the metamaterial excited with the white noise are plotted. The results are in a good agreement with theoretically obtained ones.

3.1 Introduction

The term ‘metamaterial’ dates since 1968 when the material with negative electromagnetic properties: negative permittivity and negative magnetic permeability was considered. The result of existence of negative permittivity and permeability is in negative refractive index. These materials are named ‘electromagnetic metamaterials’ and are used to stop the electromagnetic waves.

Motivated by the mathematical analogy between acoustic and electromagnetic waves the ‘acoustic metamaterials’ are introduced. It was asked the material to have negative mass and Young’s modulus. Unfortunately, in the nature there is no material with such negative properties. To obtain the negative mass the artificial material, usually composite, has to be designed. Namely, masses in the system are positive but their combination gives mathematically the negative value of the effective mass. Acoustic metamaterials appear in 2004. Due to negative effective mass the acoustic wave acting on the material stops.

There are some already published papers dealing with the theory of acoustic metamaterials (see [1] and Reference List in it). The aim of this paper is to give the explanation of metamaterials, to give the theory of energy absorption and to show some examples of its design and application. It has to be explained the property of attenuation of waves and existence of stop band in metamaterial for certain mainly low frequencies.

The paper is divided into 5 sections. In Sect. 3.2 we explain the mechanism of metamaterial and give the seismic metamaterial. In Sect. 3.3 the model of the subunit of the acoustic metamaterial is considered. The linear mass-in-mass system is analyzed and the effective mass is introduced. The design of the subunit is presented. In Sect. 3.4, the nonlinear model of the acoustic metamaterial subunit is introduced. The model with cubic nonlinearity is considered. The negative effective mass is calculated and the stop bands of frequency are determined. In Sect. 3.5 the already realized designs of acoustic metamaterials are given. Instead of conclusion, the paper ends with directions of future investigation.

3.2 Explanation of the Mechanism of Metamaterial

One of the most convenient explanation for the property of acoustic metamaterial is due to analogy with seismic metamaterial.

Namely it is known that if on the soft free ground the seismic wave acts the dust is moving freely in the space. In contrary, if on the soft ground the forest of trees is settled the trees bend due to excitation but the ground stay still.

A group of scientists: Colombi et al. [2] was interested in this phenomenon. They planted the pine trees on the average distance of 2.5 m and tested the propagation of the Rayleigh wave along the surface. (A Rayleigh wave is a mechanical wave that travels over the surface of the Earth.)

They put two seismometers S1 and S2 and measure the spectral ratio as the function of the frequency. They found that there are two wide band gaps for the frequencies smaller than 150 Hz. Rayleigh wave propagating in soft sedimentary soil at frequencies lower than 150 Hz, experiences strong attenuation, when interacting with a forest, over two separate large frequency bands. For wavelengths down to 5 m, the resulting band gaps are remarkably large and strongly attenuating when the acoustic impedance of the trees matches the impedance of the soil. Trees which form the forest which act with Rayleigh surface waves and attenuation of the wave for certain frequencies occur are named 'locally resonant metamaterials' in the geophysics context.

During sightseeing in Athen and Rome a certain number of buildings can be find which are similar to the aforementioned seismic metamaterial. Namely, the buildings contain a series of big columns connected with arc. It is known that they are built in BC and since that time a significant number of earthquakes occurred. In spite of that the buildings are not destroyed. The explanation for these phenomena is also given in paper [2]: The columns are vertical resonators settled on the horizontal ground. Since longitudinal resonances of a vertical resonator are inversely proportional to its length, these man-made engineered arrays of resonators attenuate Rayleigh waves at frequency ≤ 10 Hz. Finally, it is suggested that the system has to be designed starting from vertical pillars coupled to the ground with longitudinal resonance ≤ 10 Hz (Fig. 3.3b).

It can be concluded that the term 'metamaterial' is used for those structures where the wave and the material interact and the wave propagation is stopped for certain frequencies. So, we can talk about optic metamaterials, acoustic metamaterials, seismic metamaterials, and so on.

In technics 'metamaterial' is an artificial material, usually a composite, which in interaction with wave gives its attenuation or absorption on certain (mainly law) frequencies.

3.3 Model of the Subunit of the Metamaterial

The basic element of the metamaterial is modeled as a mass-in-mass system (Fig. 3.1). Namely, on the basic mass m_1 a small mass m_2 is connected with an elastic element (spring) with rigidity k_2. Motion of masses is described with displacement coordinates u_1 and u_2. The system has two degrees of freedom.

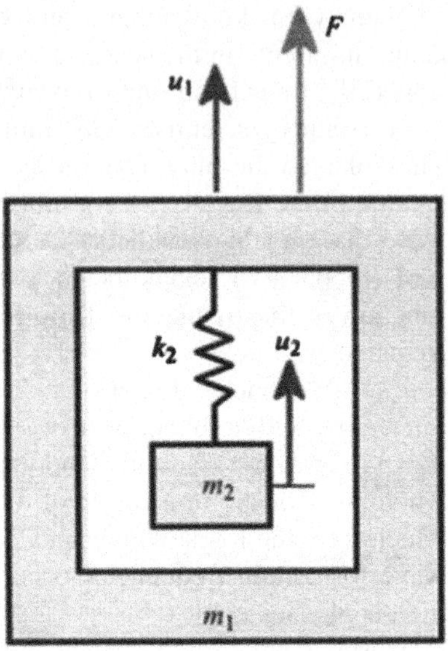

Fig. 3.1 Linear mass-in-mass model

If we excite the system with a periodical force, whose amplitude is F_0 and the frequency ω, mathematical model is

$$m_1\ddot{u}_1 + k_2(u_1 - u_2) = F_0\cos(\omega t), \quad m_1\ddot{u}_2 + k_2(u_2 - u_1) = 0 \tag{3.1}$$

The closed form analytical solution is

$$u_1 = a_1\cos(\omega t), \quad u_2 = a_2\cos(\omega t) \tag{3.2}$$

where the amplitudes of vibration are

$$a_1 = \frac{F_0(k_2 - m_2\omega^2)}{(k_2 - m_1\omega^2)(k_2 - m_2\omega^2) - k_2^2}, \quad a_2 = \frac{F_0 k_2}{(k_2 - m_1\omega^2)(k_2 - m_2\omega^2) - k_2^2} \tag{3.3}$$

Mass 2 with the spring represents the energy absorber if $u_1 = 0$, i.e., if

$$\omega = \omega_2 = \sqrt{k_2/m_2} \tag{3.4}$$

Then (1) transform into

$$-k_2u_2 = F_0\cos(\omega t), \; m_2\ddot{u}_2 = -k_2u_2 \tag{3.5}$$

i.e., the external force is eliminated with the inertia force $m_2\ddot{u}_2$ through the spring k_2

$$m_2 \ddot{u}_2 = F_0 \cos(\omega t) \tag{3.6}$$

Local resonance frequency of the absorber $\omega_l = \sqrt{k_2/m_2}$ transforms the vibration energy to the absorber and stops the main system motion ($u_1 = 0$).

3.3.1 Effective Mass

Very often there is the requirement to identify the mass-in-mass system with the single mass with effective mass m_{eff} whose motion is the same as that of m_1. Assuming that the linear momentum of the effective mass with velocity $\dot{u}_1$ is equal to the sum of linear momentums of the masses m_1 and m_2

$$m_{eff} \frac{du_1}{dt} = m_1 \frac{du_1}{dt} + m_2 \frac{du_2}{dt} \tag{3.7}$$

and using the relations (2) we have

$$m_{eff} = m_1 + m_2 \frac{\omega_2^2}{\omega_2^2 - \omega^2} \tag{3.8}$$

where ω_2 is given with (4). Based on (8), the m_{eff}/m_1 for ω/ω_2 is plotted in Fig. 3.2.

Analyzing the obtained diagram following is concluded:

1. For $\omega = \omega_2$ the 1:1 external resonance between the forcing frequency ω and of the main system exist. The vibration absorber uses the local resonance frequency of the absorber $\omega_l = \sqrt{k_2/m_2}$ which transforms the vibration energy to the absorber and stops the main system motion ($u_1 = 0$).
2. For $\omega < \omega_2$ we have the acoustic mode. The effective mass m_{eff} is positive and the motion u_1 and u_2 are in phase.
3. For $\omega > \omega_2$ we have the optical mode when the effective mass m_{eff} is positive or negative but the displacements u_1 and u_2 are at 180° and out of phase.

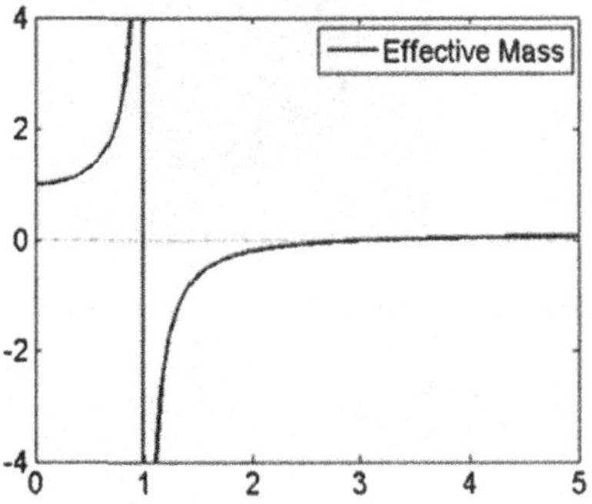

Fig. 3.2 Dimensionless effective mass m_{eff}/m_1 as a function of ω/ω_2 [4]

It is the question how to produce the structure which will respond as if it has 'negative mass' to oscillations at a fixed frequency above resonance.

In [3, 4] it is illustrated how the negative effective mass could occur and what effect it has on wave propagation in the linear one-dimensional lattice system.

3.3.2 Design of Subunit of Metamaterial

A subunit which can behave as to have negative mass contains a spherical lead core usually made of metal, a surrounding rubber layer and a covering thin layer made of silicon or bismuth [5] (see Fig. 3.3).

Let us consider the interaction between the subunit and the wave. For $\omega < \omega_2$ the core particle oscillates in-phase with the wave the dynamic mass density is positive (see Fig. 3.4a).

For $\omega > \omega_2$ the core particle oscillates out of phase with the wave-dynamic mass density can be negative: the density of the coating is sufficiently high. Spherical lead core oscillates out of phase with the rigid, but light, surrounding shell. Harmonic motion on the outside shell is in opposite direction to the acceleration of the 'hidden' lead core. If the body is shaken at that frequency it will feel like it has 'negative mass'.

Experiments made on these subunits give the result shown in Fig. 3.5. It can be seen that the stop band is really much wider than shown in Fig. 3.2.

Our conclusion is that the cause is that the assumed model is very simple and do not take into consideration the elastic property of rubber or silicon. Namely, for these materials the elastic force is a strong nonlinear function of deflection and the assumption of linearity is not enough accurate. It is the reason that we extend the model with nonlinear term.

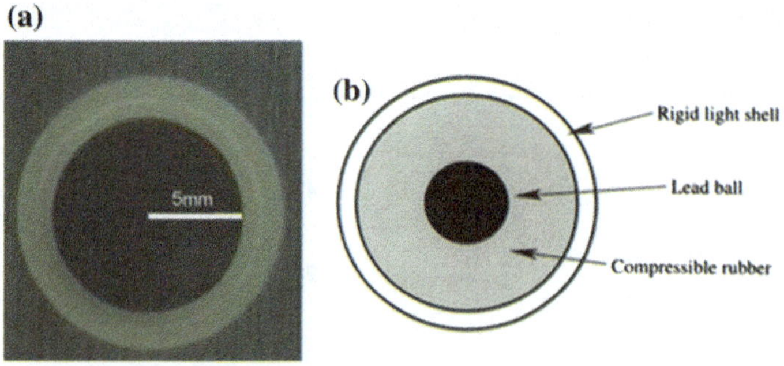

Fig. 3.3 **a** Real subunit for metamaterial, **b** model of subunit [5]

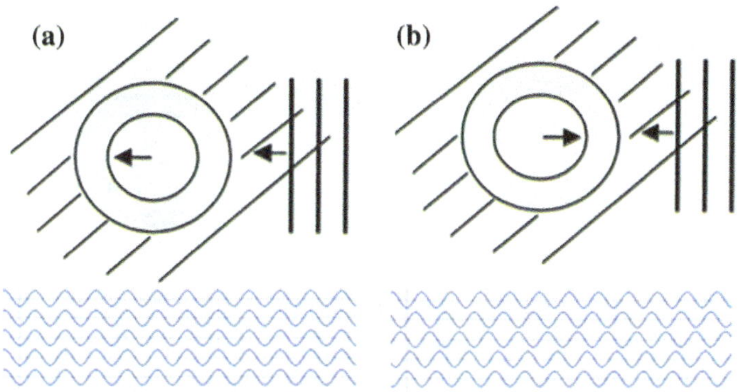

Fig. 3.4 Subunit-wave interaction: **a** in-phase waves, **b** out-of-phase waves

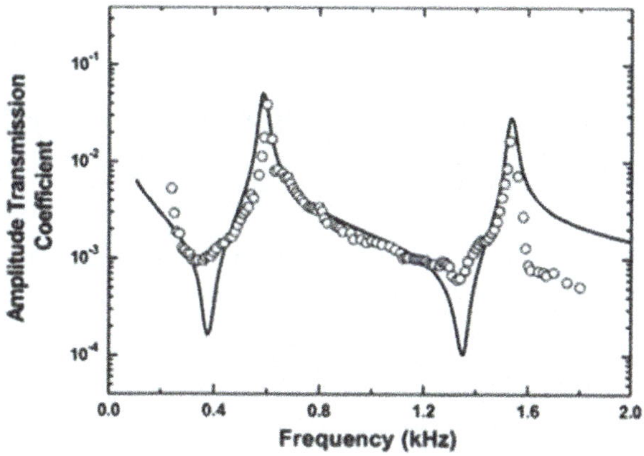

Fig. 3.5 Amplitude-frequency diagram obtained experimentally (circles) and analytically (full line)

3.4 Nonlinear Model of the Metamaterial Subunit

Let us consider that the elastic force is a cubic displacement function. Then, the mathematical model of the nonlinear mass-in-mass system (Fig. 3.6) is

$$\begin{cases} m_1\ddot{u}_1 + k_2(u_1 - u_2) + k_3(u_1 - u_2)^3 = F_0 cn(\omega t, k^2) \\ m_2\ddot{u}_2 + k_2(u_2 - u_1) + k_3(u_2 - u_1)^3 = 0 \end{cases} \tag{3.9}$$

where k_3 is the coefficient of nonlinearity, F0 is the amplitude of the excitation force, *cn* is the cosine Jacobi elliptic function [6] with frequency ω and modulus k.

To solve the system of coupled nonhomogeneous nonlinear differential equations the following transformation is introduced

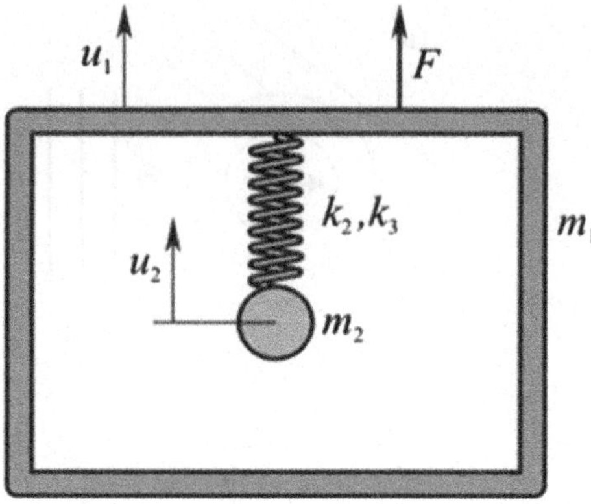

Fig. 3.6 Model of the nonlinear mass-in-mass subunit

$$X = u_1 - u_2, \quad Y = u_2 \tag{3.10}$$

Substituting (10) into (9) we have

$$m_1 m_2 \ddot{X} + (m_1 + m_2) k_2 X + (m_1 + m_2) k_3 X^3 = m_2 F_0 cn(\omega t, k^2), \tag{3.11}$$

$$m_1 \ddot{X} + (m_1 + m_2) \ddot{Y} = F_0 cn(\omega t, k^2). \tag{3.12}$$

Equation (11) is a Duffing differential equation with nonlinearity of third order. The exact closed form solution of (11) is according to [7]

$$X = a_1 cn(\omega t, k^2) \tag{3.13}$$

where

$$a_1 = \frac{m_2 F_0}{(m_1 + m_2) k_2 - m_1 m_2 \omega^2 (1 - 2k^2)} \tag{3.14}$$

The relation (14) is satisfied if

$$k_3 = \frac{2 m_1 \omega^2 k^2 \left[(m_1 + m_2) k_2 - m_1 m_2 \omega^2 (1 - 2k^2) \right]^2}{(m_1 + m_2) m_2 F_0^2} \tag{3.15}$$

Based on (13) the function Y in (12) is obtained. Finally,

$$u_1 = \frac{m_2 a_1 cn}{m_1 + m_2} + \frac{F_0 \int \left(\int cn(\psi, k^2) d\psi \right) d\psi}{(m_1 + m_2) \omega^2} \tag{3.16}$$

$$u_1 = -\frac{m_1 a_1 cn}{m_1 + m_2} + \frac{F_0 \int \left(\int cn(\psi, k^2) d\psi \right) d\psi}{(m_1 + m_2) \omega^2} \tag{3.17}$$

Fig. 3.7 The $u_1 - t$ and $u_2 - t$ diagrams

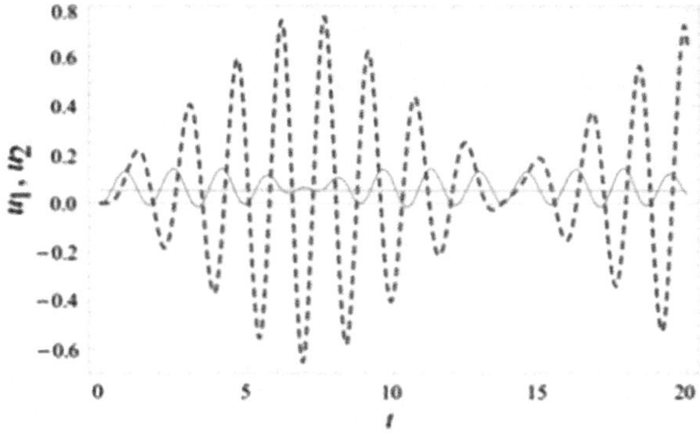

In Fig. 3.7 the $u_1 - t$ and $u_2 - t$ diagrams are plotted. It can be seen that the vibration of the basic mass is with quite small amplitudes, while the absorber uses the energy to vibrate with significant amplitude.

Using the Fourier series expansion of the *cn* Jacobi elliptic function

$$cn(\psi, k^2) = \frac{2\pi}{kK}\sum_{m=0}^{\infty}\frac{q^{m+(1/2)}}{1+q^{2m+1}}\cos\left[(2m+1)\frac{\pi\psi}{2K}\right]$$

where

$$q = \exp(-\pi K'/K) \tag{3.18}$$

K is the total Jacobi integral of the first kind and $K' = K(k') = K\sqrt{1-k^2}$, the displacement functions are

$$\begin{aligned} u_1 = &\frac{m_2 a_1}{m_1+m_2} cn\left(\omega t, k^2\right) \\ &+ \frac{16F_0K}{(m_1+m_2)\omega^2\pi k}\sum_{m=0}^{\infty}\frac{q^{m+(1/2)}}{1+q^{2m+1}}\frac{1}{(2m+1)^2}\sin^2\left[(2m+1)\frac{\pi\omega t}{4K}\right] \end{aligned} \tag{3.19}$$

$$\begin{aligned} u_2 = &-\frac{m_1 a_1}{m_1+m_2} cn\left(\omega t, k^2\right) \\ &+ \frac{16F_0K}{(m_1+m_2)\omega^2\pi k}\sum_{m=0}^{\infty}\frac{q^{m+(1/2)}}{1+q^{2m+1}}\frac{1}{(2m+1)^2}\sin^2\left[(2m+1)\frac{\pi\omega t}{4K}\right] \end{aligned} \tag{3.20}$$

Comparing the approximate solution (19) up to $m = 3$ and the exact solution (16) for u_1 it is obvious that the difference is negligible (Fig. 3.8).

Due to this conclusion in the following analysis the approximate solution will be used.

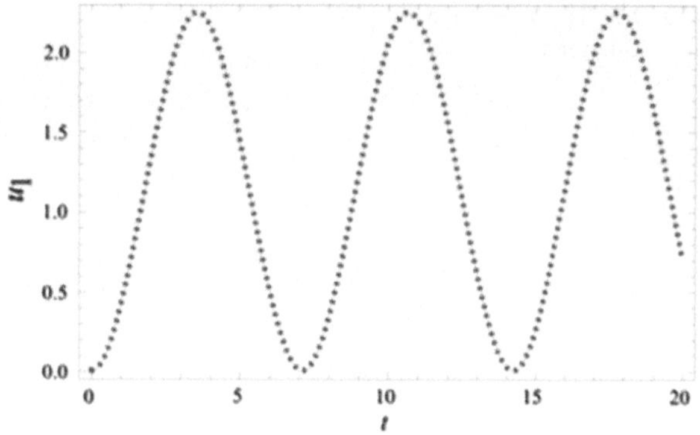

Fig. 3.8 The $u_1 - t$ exact solution (16) (full line) and approximate solution up to $m = 3$ (dotted line)

3.4.1 *Effective Mass*

As it is mentioned for linear systems, the linear momentum K_{eff} of the effective mass with velocity u_1 has to be equal to the sum of linear momentums of the mass 1 and mass 2

$$K_{eff} = m_1\dot{u}_1 + m_2\dot{u}_2 \tag{3.21}$$

gives the following relation

$$K_{eff} = \sum_{m=0}^{\infty} m_{effm} \frac{q^{m+\left(\frac{1}{2}\right)}}{1+q^{2m+1}} \sin\left[(2m+1)\frac{\pi\omega t}{4K}\right] \cdot \left[\left(\frac{4F_0}{k\omega}\right)\left(\frac{1}{2m+1}\right) - m_2 a_1 \left(\frac{\pi^2\omega}{kK^2}\right)(2m+1)\right] \tag{3.22}$$

For series expansion in (22) the corresponding effective masses are

$$m_{effm} = (m_1 + m_2) + \left(\frac{\left(\frac{\pi^2}{4K^2}\right) m_2 (2m+1)^2 \omega^2}{\omega_l^2 - \left[\left(\frac{m_1}{m_1+m_2}\right)(1-2k^2) + \left(\frac{\pi^2}{4K^2}\right)\left(\frac{m_2}{m_1+m_2}\right)(2m+1)^2\right]\omega^2} \right) \tag{3.23}$$

Comparing the relation for effective mass (23) with the previously obtained (8) for linear system, it can be concluded that there is a similarity. For $m = 0$ and $k_3 = 0$ the relations are equal. In Fig. 3.9 the effective masses as frequency functions are plotted.

Analyzing the diagrams it can be seen that the frequency band gap is significantly wider for the nonlinear mass-in-mass unit than it is for linear one (see Fig. 3.5). The condition for effective mass to be zero ($m_{eff} = 0$) is satisfied when

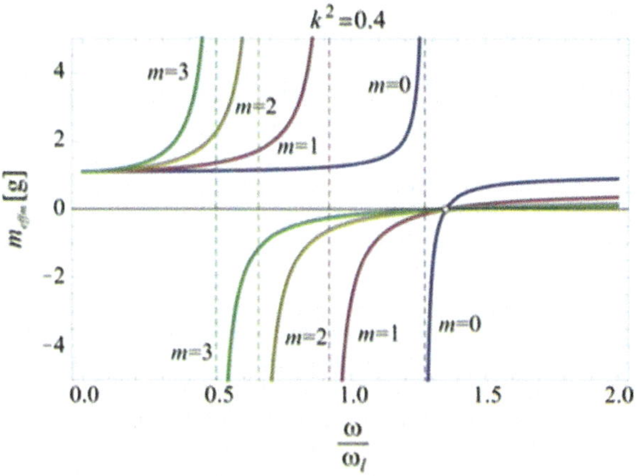

Fig. 3.9 Effective mass-frequency functions

$$\left(\frac{\omega}{\omega_l}\right)_0 = \sqrt{\frac{m_1 + m_2}{m_1(1 - k^2)}} \tag{3.24}$$

For example, for $k^2 = 0.4$ the corresponding value is $\omega/\omega_l = 1.35401$. The resonance regime is

$$1 - \left[\left(\frac{m_1}{m_1 + m_2}\right)(1 - 2k^2) + \left(\frac{\pi^2}{4K^2}\right)\left(\frac{m_2}{m_1 + m_2}\right)(2m + 1)^2\right]\left(\frac{\omega}{\omega_l}\right)_\infty^2 = 0 \tag{3.25}$$

Then, the negative effective mass defined in [8–10] is in the interval $\left(\frac{\omega}{\omega_l}\right)_0$ to $\left(\frac{\omega}{\omega_l}\right)_\infty$. For the negative effective mass ratio, the frequencies which correspond to the negative mass ratio are in the stopping band.

3.5 Future Investigation

Future investigation will be directed toward extending the theoretical knowledge in metamaterials using the damping properties of the subunits and other types of nonlinearity, but also the theoretical consideration of connection of subunits into an unique system (material). Besides, based on the theoretical results, the production of acoustic metamaterials is of interest, but also their experimental investigation and application.

Acknowledgements Investigation is supported by COST CA15125 DENORMS.

References

1. L. Cveticanin, G. Mester, Theory of acoustic metamaterials and metamaterial beams: an overview. Acta Polytech. Hungarica **13**(7), 43–62 (2016)
2. A. Colombi, P. Roux, S. Guenneau, P. Gueguen, R.V. Craster, Forests as a natural seismic metamaterial: rayleigh wave bandgaps induced by local resonances. Sci. Rep. **6**, 19238 (2016). doi:10.1038/srep19238
3. P. Sheng, X.X. Zhang, Z. Liu, C.T. Chan, Locally resonant sonic materials. Phys. B **338**, 201–205 (2003)
4. H.H. Huang, C.T. Sun, G.I. Huang, On the negative effective mass density in acoustic metamaterials. Int. J. Eng. Sci. **47**, 610–617 (2009)
5. G.W. Milton, New metamaterials with macroscopic behavior outside that of continuum elastodynamics. New J. Phys. **9**(359), 1–13 (2007)
6. L. Cveticanin, Analysis technique for the various forms of the duffing equation (Chap. 4), in *The Duffing Equation: Nonlinear Oscillators and their Behaviour*, ed. by I. Kovacic, M. J. Brennan (Wiley, New York, 2011), pp. 81–139
7. L. Cveticanin, M. Zukovic, Negative effective mass in acoustic metamaterial with nonlinear mass-in-mass subsystems. Commun. Nonlinear Sci. Numer. Simul. **51**, 89–104 (2017)
8. E.P. Calius, X. Bremaud, B. Smith, A. Hall, Negative mass sound shielding structures: early results. Phys. Status Solidi B **246**(9), 2089–2097 (2009)
9. R. Zhu, X.N. Liu, G.K. Hu, C.T. Sun, G.L. Huang, A chiral elastic metamaterial beam for broadband vibration suppression. J. Sound Vib. **333**, 2759–2773 (2014)
10. S.H. Lee, C.M. Park, Y.M. Seo, Z.G. Wang, C.K. Kim, Acoustic metamaterial with negative density. Phys. Lett. A **373**, 4464–4469 (2009)

Chapter 4
Analytical Study of the Oblique Impact of an Elastic Sphere with a Rigid Flat

Ozdes Cermik, Hamid Ghaednia and Dan B. Marghitu

Abstract In this study, an analytical model of the elastic impact of a solid rubber sphere with a rigid flat is analyzed. The linear and angular motion of the sphere have been simulated for the oblique (60°) and normal (0°) impact cases. The impact of the sphere with the rigid flat has been represented with a nonlinear contact force. The damping term of a previous normal contact force has been modified with a new expression. The normal contact force as a function of deflection has been studied for different cases.

4.1 Introduction

Impact is an important phenomena which plays an important role in many sports such as baseball, golf, tennis, and soccer since it affects rebound parameters of a ball; speed, angle, and spin etc. A normal and oblique impact of a ball with a flexible or rigid surface have been studied by many researches for decades. Authors mainly focus on dynamic properties of a ball and measure parameters before and after an impact. On the other hand, the behavior of a ball during the impact still attracts researchers. Measuring the motion of the ball during the impact is hard experimentally, since the impact duration is short. The friction force between a ball and a surface is also another challenge because it is difficult to measure.

A spring-damper system has been used by several authors in order to model the impact between a ball and a surface. Dignall and Haake [1] developed an analytical model of the normal impact of a tennis ball on a tennis court using a spring-damper

O. Cermik (✉) · H. Ghaednia · D. B. Marghitu
Department of Mechanical Engineering, Auburn University, 1418 Wiggins Hall, Auburn, AL 36849, USA
e-mail: ozdescermik@auburn.edu

H. Ghaednia
e-mail: hamid.ghaednia@auburn.edu

D. B. Marghitu
e-mail: marghdb@auburn.edu

N. Herisanu and V. Marinca (eds.), *Acoustics and Vibration of Mechanical Structures—AVMS-2017*, Springer Proceedings in Physics 198,
https://doi.org/10.1007/978-3-319-69823-6_4

system. The stiffness and damping coefficients were calculated analytically using the coefficient of restitution and the contact time from the experiments. Haake et al. [2] studied the impact of tennis ball with racket using the model from [1]. It was shown that the stiffness and damping coefficients of the tennis ball increase with higher initial velocity of the ball, [1, 2]. Goodwill and Haake [3] developed a new spring-damper model for the impact between a tennis ball and a head-clamped racket by adding a new linear spring in series. This extra spring is used to simulate the influence of the string-bed. Goodwill and Haake [4] developed a viscoelastic model of normal impact of a tennis ball on a rigid surface. A new damper representing the large deformation on the ball was added in parallel with the classical spring-damper model. Carre et al. [5] studied the impact of a cricket ball on a rigid surface. The spring-damper model was used to predict the force-deflection behavior. Yang et al. [6] developed a nonlinear impact model for a tennis ball and a racket. A nonlinear spring for the string-bed was connected in series with the spring-damper model. Ghaednia et al. [7] studied the oblique impact of a tennis ball with a racket. The impact is presented with a spring damper model. The stiffness and damping coefficients were found experimentally for the normal impact case, and the theory was verified with experiments for the oblique impact.

In this work, the normal and oblique impact of an elastic ball with a rigid flat has been studied theoretically. We have used a spring damper model in order to present the impact. A new expression for the damping term has been proposed in order to overcome the weakness of using a spring damper system for an impact problem.

4.2 Dynamics of the Ball

The elastic ball impacting the rigid surface is shown in Fig. 4.1. The impact angle of the rigid flat with the x_0 horizontal axis is β. R is the radius, and m is the mass of the ball. C is the center of the mass, and E is the contact point between the ball and the rigid flat. A global reference frame of unit vectors [$\mathbf{I}_0$, $\mathbf{J}_0$, $\mathbf{k}_0$] and a local reference frame, with the origin at E, of unit vectors [$\mathbf{I}$, $\mathbf{J}$, $\mathbf{k}$] are considered as shown in Fig. 4.1. The x-axis is tangential, and the y-axis is perpendicular to the rigid flat.

The impact is divided into two phases; compression and restitution. Restitution phase is considered to be fully elastic. Compression phase starts when the ball contacts the surface and at this moment the normal contact force, $\boldsymbol{P}$, and the indentation, δ, are zero. This phase ends at maximum compression, δ_m, and maximum normal force, $\boldsymbol{P}_m$, when the normal component of the velocity of the contact point of the ball is zero. Restitution phase starts at the moment of maximum compression, and the normal contact force decreases from maximum value to zero.

The elasto-plastic normal contact force, $\mathbf{P}$, acts upward alongside the y-axis.

Contact force is defined as:

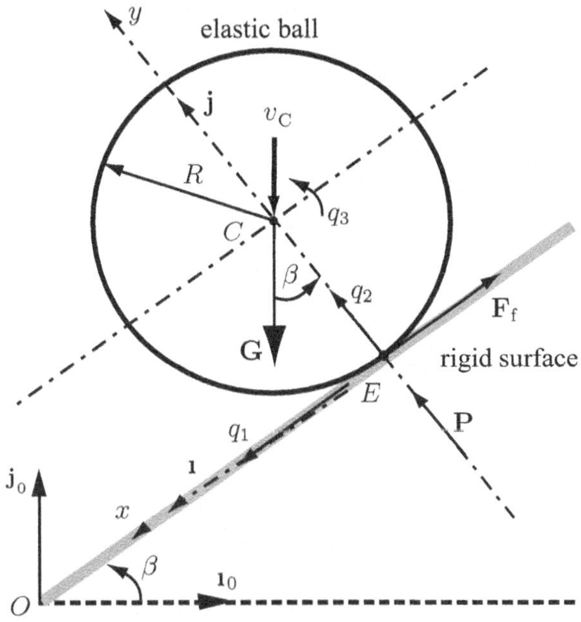

Fig. 4.1 The ball in contact with a rigid surface

$$P = P_{static} + P_{dynamic} \quad P = k\delta - b\dot{\delta} \tag{4.1}$$

where $\delta = q_2(t)$ is the normal elastic displacement during the impact, $\dot{\delta}$ is its rate and k, b are stiffness and damper coefficients, respectively.

The model above is modified with a new expression to overcome discontinuous problem because of the damping term. In reality, both elastic and damping forces should be initially at zero at the beginning of the impact. Equation 4.1 is modified with a new expression for the dynamic part of the contact force and it can be seen below:

$$P = k\delta - b\dot{\delta}(1 - e^{100\delta/R}) \tag{4.2}$$

The total force at the point E is

$$\mathbf{T} = F_f\mathbf{I} + P\mathbf{J} \tag{4.3}$$

where F_f represents the friction force that is opposite to the tangential component of the velocity of the contact point of the ball. The friction force is given by

$$\mathbf{F}_f = \mu P\mathbf{I} \tag{4.4}$$

where μ is the kinetic coefficient of friction. For the case of a rigid ball the equations of motion are

$$m\mathbf{a}_C = \mathbf{G} + \mathbf{T}, \quad \mathrm{I}_C\alpha = \mathbf{r}_{CE} \times \mathbf{T}, \tag{4.5}$$

where G is the weight of the ball, I_C is the mass moment of inertia about C. Diameter, $D = 63$ mm and a mass, $m = 145$ g, kinetic coefficient of friction, $\mu = 0.2$, stiffness coefficient, $k = 60{,}000$ N/m, and damping coefficient, $b = 10.5$ Ns/m have been used for the simulation.

4.3 Results

The impact angle of 0° (normal impact) and 60° are used for the simulations. Initial impact velocity of the center of the ball, $\mathbf{v}_C = -3\mathbf{J}_0$ m/s in the global coordinates is used for both the normal and oblique impacts. Figures 4.2 and 4.3 present the normal and tangential displacement of the center of the ball during the normal and oblique impact, respectively. For the normal impact, the maximum displacement is $\delta_m = -4.31 \times 10^{-3}$ m, and compression phase ends when $t_m = 2.4\times10^{-3}$ s. At this moment, the contact force reaches its highest point, and the restitution phase starts. Tangential displacement stays at zero since there is no force in the tangential direction. For the oblique impact, the maximum normal displacement of the center of the ball is $\delta_m = -2.17 \times 10^{-3}$ m, and compression phase ends when $t_m = 2.44\times10^{-3}$ s. The displacement in the tangential direction increases throughout the impact as seen in Fig. 4.3.

Figure 4.4 shows the normal and tangential velocities of the center of the ball during the normal impact. For the normal impact, initial impact velocity is

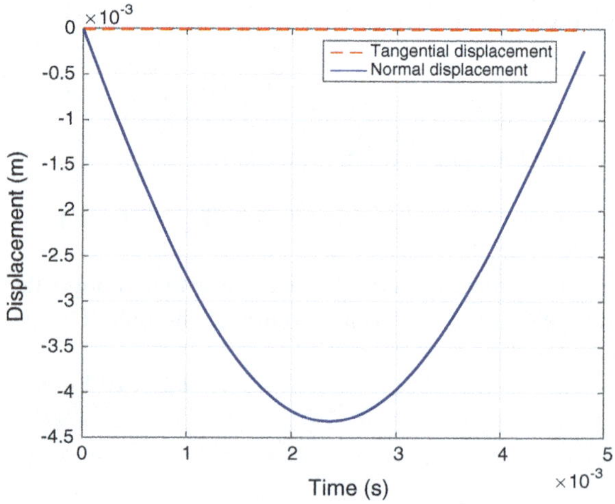

Fig. 4.2 Normal and tangential displacement during the normal impact

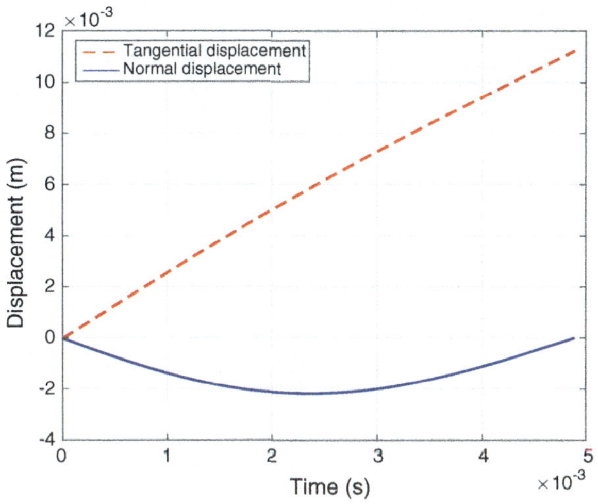

Fig. 4.3 Normal and tangential displacement during the oblique impact at $\beta = 60°$

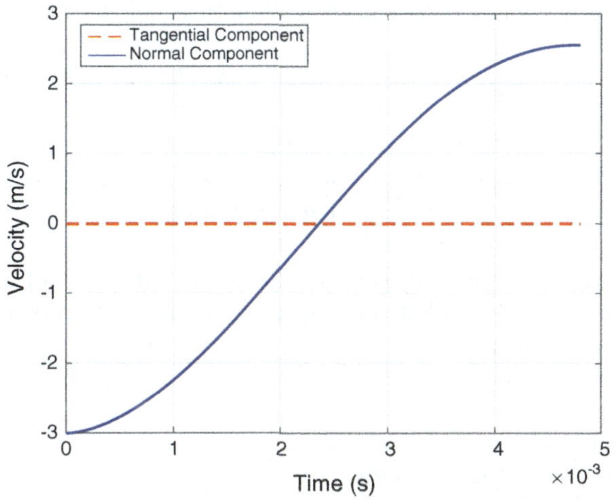

Fig. 4.4 Normal and tangential velocity of the center of the ball during the normal impact

$\mathbf{v}_E(t_0) = -3\mathbf{J}$ m/s, and the impact ends at $t_f = 4.8 \times 10^{-3}$ s. The final velocity at the end of the impact is $\mathbf{v}_E(t_f) = 2.55\mathbf{J}$ m/s. The tangential velocity is zero throughout the impact since no friction affects the ball. Figure 4.5 depicts the normal and tangential velocities of the center of the ball during the oblique impact for 60°. The initial impact velocity for the oblique is $\mathbf{v}_E(t_0) = 2.6\mathbf{I} - 1.5\mathbf{J}$ m/s in local coordinates. The normal component of the velocity becomes zero at $t_m = 2.44 \times 10^{-3}$ s. At this point, maximum displacement and maximum normal contact force occur, compression phase ends, and restitution phase starts. Impact ends at $t_f = 4.9 \times 10^{-3}$ s.

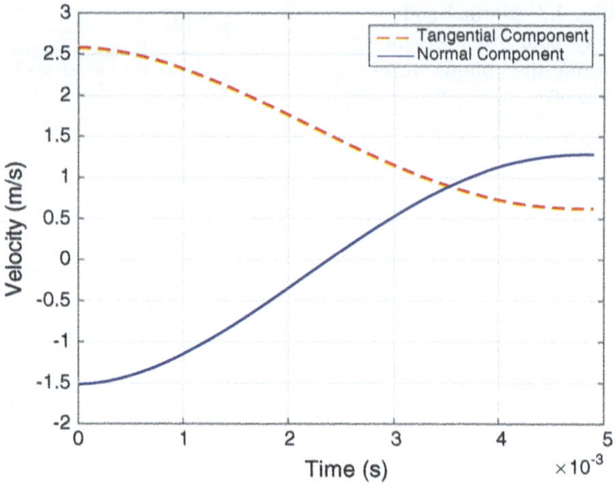

Fig. 4.5 Normal and tangential velocity of the center of the ball during the impact at $\beta = 60^\circ$

The final velocity at the end of the impact is $\mathbf{v}_E(t_f) = 0.64\mathbf{I} + 1.3\mathbf{J}$ m/s. Newtonian, coefficient of restitution is calculated as

$$e = |\mathbf{v}_E(t_f) \cdot \mathbf{J} / \mathbf{v}_E(t_0) \cdot \mathbf{J}|$$

For 60° and 0° impact angles, COR, e, is found 0.86 and 0.85, respectively. Figure 4.6 presents the angular velocity of the ball during the oblique impact for 60°. Angular velocity increases throughout the impact. Rebound angular velocity after the impacts predicted as $\omega_f = 44$ rad/s.

For the normal contact force, P, as a function of displacement, the Eqs. 4.1 and 4.2 have been plotted for both the oblique and normal impact cases. The dashed and solid lines in Figs. 4.7 and 4.8 show the previous model and modified model,

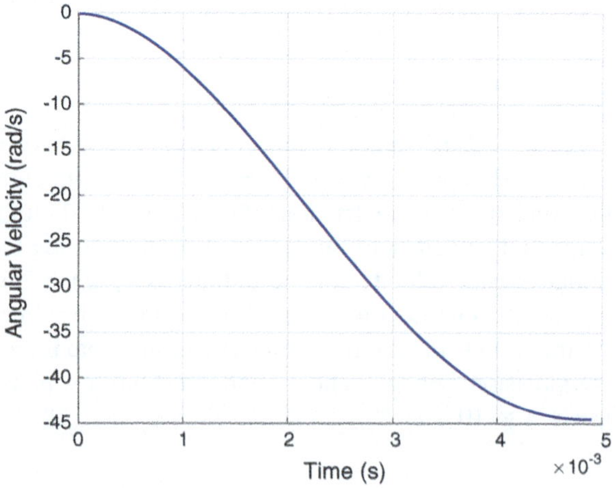

Fig. 4.6 Angular velocity of the ball during the oblique impact at $\beta = 60^\circ$

respectively. Figures 4.7 and 4.8 depict the normal contact force during the normal and oblique impact cases, respectively. Contact force reaches its highest point at the end of compression phase, and at this point, maximum displacement occurs. Restitution phase starts, and contact force starts to decrease from this moment. For the previous model, the contact force is not zero at the beginning of the impact as seen in both Figs. 4.7 and 4.8. This discontinuous comes from the damping term. In reality both elastic and damping forces should be zero at the beginning of the impact. On the other hand, a negative force is present at the end of the impact for the previous model as well. For the modified model with the new expression, the normal contact force starts at zero and decreases back to zero at the end of the impact as seen in both Figs. 4.7 and 4.8.

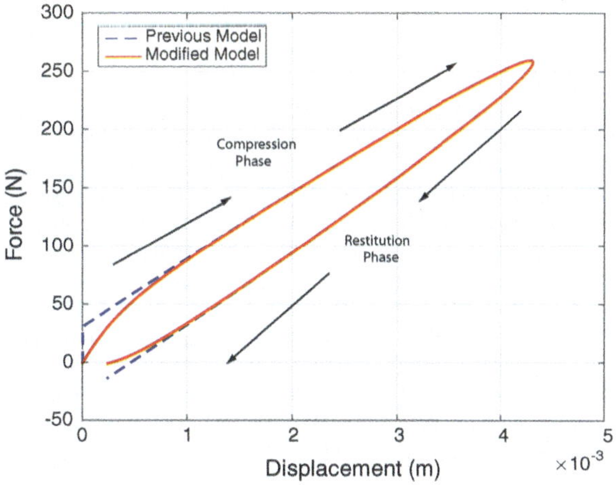

Fig. 4.7 The normal contact force during the normal impact

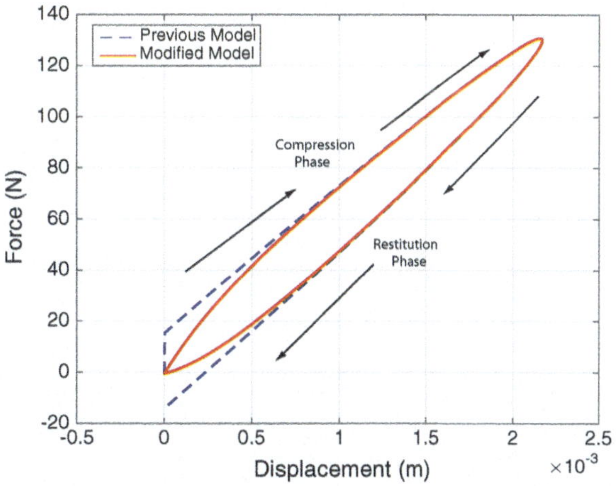

Fig. 4.8 The normal contact force during the oblique impact at $\beta = 60°$

4.4 Conclusions

The normal and oblique impact of an elastic ball with a rigid surface has been studied theoretically. The linear and angular motion of the ball during the impact have been simulated. Previous model consisting linear spring and damper is modified with a new expression for the damping term. With modified model, the normal contact force discontinuity problem is fixed and it starts at zero at the beginning of the impact. A negative contact force at the end of the impact which is seen in the previous model is not present for the modified model.

References

1. R.J. Dignall, S.J. Haake, Analytical modelling of the impact of tennis balls on court surface, in *Tennis Science and Technology Conference*. Blackwell (2000), pp. 155–162
2. S.J. Haake, M.J. Carre, S.R. Goodwill, The dynamic impact characteristics of tennis balls with tennis rackets. J. Sports Sci. **21**(10), 839–850 (2003)
3. S.R. Goodwill, S.J. Haake, Spring damper model of an impact between a tennis ball and racket. Proc. Inst. Mech. Eng. Part C J. Mech. Eng. Sci. **215**(11), 1331–1341 (2001)
4. S.J. Goodwill, S.R. Haake, Modelling of tennis ball impacts on a rigid surface. Proc. Inst. Mech. Eng. Part C J. Mech. Eng. Sci. **218**(10), 1139–1153 (2004)
5. M.J. Carre, D.M. James, S.J. Haake, Impact of a non-homogeneous sphere on a rigid surface. Proc. Inst. Mech. Eng. Part C J. Mech. Eng. Sci. **218**(3), 273–281 (2004)
6. S. Yang, A. Fafitis, A. Wiezel, Nonlinear impact model of a tennis racket and a ball. J. Mech. Sci. Technol. **26**(2), 315–321 (2012)
7. H. Ghaednia, O. Cermik, D.B. Marghitu, Experimental and theoretical study of the oblique impact of a tennis ball with a racket. Proc. Inst. Mech. Eng. Part P J. Sports Eng. Technol. 1754337114567490 (2015)

Chapter 5
Free Oscillations of Euler-Bernoulli Beams on Nonlinear Winkler-Pasternak Foundation

Nicolae Herisanu and Vasile Marinca

Abstract In this paper the free oscillations of a simply supported Euler-Bernoulli beam resting on a nonlinear spring bed with linear and cubic stiffness are studied. The system is discretized by means of the classical Galerkin procedure and its nonlinear dynamic behaviour is analyzed using the Optimal Auxiliary Functions Method (OAFM). Frequency responses are presented in a closed form and their sensitivity with respect to the initial amplitudes are investigated. It is proved that OAFM is a reliable and straightforward approach to solve a set of coupled nonlinear differential equations.

5.1 Introduction

The use of beams of an elastic foundation has recently become widespread in engineering. Several research papers have appeared in literature on this topic. Horibe and Asano proposed a boundary integral equation method for calculating the large deflection of beams on an elastic foundation of the Pasternak type [1]. The governing equation is transformed into a more convenient form such that the iterative scheme can be applied to large deflection problems. Rao [2] studied large amplitude vibration of slender, uniform beams on elastic foundation with edges immovable axially using a direct numerical integration technique. The nonlinear response of a finite beam on nonlinear viscoelastic foundation subjected to a moving load is studied by Ansari et al. [3]. A comprehensive parametric sensitivity analysis is carried out to investigate the effects of different parameters. The response of the nonlinear system is studied by the method of multiple scales. Senalp et al. [4] considered dynamic response of a simply-supported, finite length Euler-Bernoulli

N. Herisanu (✉)
University Politehnica Timisoara, Timişoara, Romania
e-mail: nicolae.herisanu@upt.ro

V. Marinca
Branch of Timisoara, Romanian Academy, Bucharest, Romania
e-mail: vasile.marinca@upt.ro

N. Herisanu and V. Marinca (eds.), *Acoustics and Vibration of Mechanical Structures—AVMS-2017*, Springer Proceedings in Physics 198,
https://doi.org/10.1007/978-3-319-69823-6_5

beam with uniform cross-section resting on a linear and nonlinear viscoelastic foundation acted upon a moving concentrated force. Tsiatas [5] investigated the nonlinear response of beams with variable properties resting on a nonlinear elastic foundation. The variable cross-sectional properties of the beam result in governing differential equations and the solution was achieved using the analogue equation method of Katsikadelis. Bhattiprolu and Davies [6] considered a pinned beam resting on a nonlinear viscoelastic and unilateral foundation. The static and dynamic responses of the beam are studied with Galerkin's method. The equations involve the coordinates of the unknown lift-off-points which are determined as a part of the solution for structural response. Ding et al. [7] studied the convergence of the Galerkin method for the dynamic response of an elastic beam resting on a nonlinear foundation with viscous damping subjected to a moving concentrated load. Younesian et al. [8] applied the Variational Iteration Method for the dynamic response of an elastic beam rested on a nonlinear foundation. Koziol and Hryniewicz [9] presented a new semi-analytical solution of the Timoshenko beam subject to a moving load in case of a nonlinear medium underneath. The solution is obtained using wavelet filters of Coiflet type and Adomian Decomposition Method combined with the Fourier transform. The nonlinear model of microbeam accounts for the slightly curved beam is studied by Sari and Pakdemirli [10] by means of the method of multiple scales. Abdelghany et al. [11] investigated the dynamic response of a non-uniform Euler-Bernoulli simply supported beam subjected to moving load and rested on a nonlinear viscoelastic foundation. Galerkin with Runge-Kutta method are employed in this paper.

The present paper aims to give analytical approximate solutions for free oscillations of the Euler-Bernoulli beam supposed to be isotropic having homogenous mass density cross-section along its length, rested on a nonlinear Winkler-Pasternak foundation. The governing equation is discretized through standard Galerkin procedure and the coupled nonlinear equations are solved by OAFM.

5.2 Nonlinear Euler-Bernoulli Beam Model

The system under investigation is a simply-supported beam resting on a Winkler-Pasternak foundation. The case of a uniform finite beam is governed by the equation [7, 8, 11, 12]

$$EI\frac{\partial^4 \bar{w}}{\partial \bar{x}^4} + \rho A\frac{\partial^2 \bar{w}}{\partial \bar{t}^2} + k_1\bar{w} + k_2\bar{w}^3 - k_3\frac{\partial^2 \bar{w}}{\partial \bar{x}^2} = 0 \quad (5.1)$$

in which E is the Young modulus, I the moment of inertia of the cross-sectional area, A the area of the cross section, ρ the material mass density, k_1, k_2 and k_3 are parts of the Winkler-Pasternak foundation stiffness.

The approximate solution is constructed for a simply supported beam based on the Galerkin approximation:

$$w(x,t) = \sum_{i=1}^{n} X_i(x)T_i(t) \tag{5.2}$$

where n = 2 and $X_i(x) = \sin(i\pi x/L)$, L being the length of the beam.

Using dimensionless quantities one obtains:

$$x = \frac{\bar{x}}{L}, w = \frac{\bar{w}}{L}, t = \bar{t}\omega_i, \quad \alpha = \frac{k_2}{\rho A}, \omega_i^2 = \frac{i^4\pi^4 EI}{\rho AL^4} + \frac{k_1}{\rho A} + \frac{\pi^2 k_3}{\rho AL^2}, \quad i = 1,2 \tag{5.3}$$

By means of Galerkin procedure, one can arrive to the nonlinear equations:

$$\ddot{T}_1 + \omega_1^2 T_1 + \frac{3\alpha}{4}(T_1^3 + 2T_1T_2^2) = 0, \quad \ddot{T}_2 + \omega_2^2 T_2 + \frac{3\alpha}{4}(T_2^3 + 2T_1^2 T_2) = 0 \tag{5.4}$$

with the initial conditions

$$T_1(0) = A_1, \quad \dot{T}_1(0) = 0, \quad T_2(0) = A_2, \quad \dot{T}_2(0) = 0 \tag{5.5}$$

where dot denotes derivative with respect to time. The nonlinear differential equations (5.4) and (5.5) can be solved analytically by means of the OAFM [13].

5.3 OAFM for Free Oscillations of Euler-Bernoulli Beam

If Ω_1 and Ω_2 are the frequencies corresponding to (5.4), then the linear operators are

$$l_1[T_1(t)] = \ddot{T}_1 + \Omega_1^2 T_1, \quad l_2[T_2(t)] = \ddot{T}_2 + \Omega_2^2 T_2 \tag{5.6}$$

Equations (5.4) and (5.5) have the solutions

$$T_1(t) = T_{10}(t) + T_{11}(t), \quad T_2(t) = T_{20}(t) + T_{21}(t) \tag{5.7}$$

where T_{10} and T_{20} are determined from the linear equations

$$\begin{aligned} &\ddot{T}_{10} + \Omega_1^2 T_{10} = 0, \quad T_{10}(0) = A_1, \quad \dot{T}_{10}(0) = 0 \\ &\ddot{T}_{20} + \Omega_2^2 T_{20} = 0, \quad T_{20}(0) = A_2, \quad \dot{T}_{20}(0) = 0 \end{aligned} \tag{5.8}$$

It is clear that

$$T_{10}(t) = A_1 \cos\Omega_1 t, \quad T_{20}(t) = A_2 \cos\Omega_2 t \tag{5.9}$$

The nonlinear operators for (5.4) are respectively

$$N_1(T_1,T_2)=(\omega_1^2-\Omega_1^2)T_1+\tfrac{3\alpha}{4}(T_1^3+2T_1T_2^2)$$
$$N_2(T_1,T_2)=(\omega_2^2-\Omega_2^2)T_2+\tfrac{3\alpha}{4}(2T_1^2T_2+T_2^3) \tag{5.10}$$

Substituting (5.9) into (5.10) we have

$$N_1(T_{10},T_{20})=[(\omega_1^2-\Omega_1^2)A_1+\frac{9}{16}\alpha A_1^3+\frac{3}{4}\alpha A_1A_2^2]\cos\Omega_1 t$$
$$+\frac{3}{16}\alpha A_1^3\cos 3\Omega_1 t+\frac{3}{8}\alpha A_1A_2^2[\cos(2\Omega_2+\Omega_1)t+\cos(2\Omega_2-\Omega_1)t]$$
$$N_2(T_{10},T_{20})=[(\omega_2^2-\Omega_2^2)A_2+\frac{9}{16}\alpha A_2^3+\frac{3}{4}\alpha A_1^2A_2]\cos\Omega_2 t$$
$$+\frac{3}{16}\alpha A_2^3\cos 3\Omega_2 t+\frac{3}{8}\alpha A_1^2A_2[\cos(2\Omega_1+\Omega_2)t+\cos(2\Omega_1-\Omega_2)t] \tag{5.11}$$

The first-order approximate solutions can be found from the linear equations:

$$\ddot{T}_{11}+\Omega_1^2T_{11}=(C_1+2C_2\cos 2\Omega_1 t+2C_3\cos 2\Omega_2 t)\{[A_1(\omega_1^2-\Omega_1^2)$$
$$+\frac{3}{16}\alpha(3A_1^3+4A_1A_2^2)]\cos\Omega_1 t+\frac{3}{16}\alpha A_1^3\cos 3\Omega_1 t\}$$
$$\ddot{T}_{21}+\Omega_2^2T_{21}=(C_4+2C_5\cos 2\Omega_2 t+2C_6\cos 2\Omega_1 t)\{[A_2(\omega_2^2-\Omega_2^2)$$
$$+\frac{3}{16}\alpha(3A_2^3+4A_1A_2^2)]\cos\Omega_2 t+\frac{3}{16}\alpha A_2^3\cos 3\Omega_2 t\} \tag{5.12}$$

with the initial conditions

$$T_{11}(0)=\dot{T}_{11}(0)=0,\quad T_{21}(0)=\dot{T}_{21}(0)=0 \tag{5.13}$$

Avoiding the secular terms, we obtain, respectively

$$\Omega_1^2=\omega_1^2+\frac{3\alpha}{16}(3A_1^2+4A_2^2)+\frac{3\alpha A_1^2}{16}\frac{C_2}{C_1+C_2}$$
$$\Omega_2^2=\omega_2^2+\frac{3\alpha}{16}(3A_2^2+4A_1^2)+\frac{3\alpha A_2^2}{16}\frac{C_5}{C_4+C_5} \tag{5.14}$$

From (5.12) and (5.13), the first-order approximate solutions can be written as

$$
\begin{aligned}
T_{11}(t) = {} & \frac{1}{8\Omega_1^2}\left[A_1(\omega_1^2-\Omega_1^2)+\frac{3\alpha A_1}{16}(3A_1^2+4A_2^2)\right]C_2(\cos\Omega_1 t-\cos 3\Omega_1 t) \\
& + \frac{\alpha A_1^3 C_2}{128\Omega_1^2}(\cos\Omega_1 t-\cos 5\Omega_1 t) \\
& + \frac{\left[A_1(\omega_1^2-\Omega_1^2)+\frac{3\alpha}{16}A_1(3A_1^2+4A_2^2)\right]C_3}{4\Omega_2(\Omega_1+\Omega_2)}[\cos(\Omega_1+2\Omega_2)t-\cos\Omega_1 t] \\
& + \frac{\left[A_1(\omega_1^2-\Omega_1^2)+\frac{3\alpha}{16}A_1(3A_1^2+4A_2^2)\right]C_3}{4\Omega_2(\Omega_2-\Omega_1)}[\cos(\Omega_1-2\Omega_2)t-\cos\Omega_1 t] \\
& + \frac{3\alpha A_1^3 C_3}{4(2\Omega_1^2+3\Omega_1\Omega_2+\Omega_2^2)}[\cos(3\Omega_1+2\Omega_2)t-\cos\Omega_1 t] \\
& + \frac{3\alpha A_1^3 C_3}{4(2\Omega_1^2-3\Omega_1\Omega_2+\Omega_2^2)}[\cos(3\Omega_1-2\Omega_2)t-\cos\Omega_1 t]
\end{aligned}
\tag{5.15}
$$

$$
\begin{aligned}
T_{21}(t) = {} & \frac{1}{8\Omega_2^2}\left[A_2(\omega_2^2-\Omega_2^2)+\frac{3\alpha A_2}{16}(3A_2^2+4A_1^2)\right]C_5(\cos\Omega_2 t-\cos 3\Omega_2 t) \\
& + \frac{\alpha A_2^3 C_5}{128\Omega_2^2}(\cos\Omega_2 t-\cos 5\Omega_2 t) \\
& + \frac{\left[A_2(\omega_2^2-\Omega_2^2)+\frac{3\alpha}{16}A_2(3A_2^2+4A_1^2)\right]C_6}{4\Omega_1(\Omega_1+\Omega_2)}[\cos(\Omega_2+2\Omega_1)t-\cos\Omega_2 t] \\
& + \frac{\left[A_2(\omega_2^2-\Omega_2^2)+\frac{3\alpha}{16}A_2(3A_2^2+4A_1^2)\right]C_6}{4\Omega_1(\Omega_1-\Omega_2)}[\cos(\Omega_2-2\Omega_1)t-\cos\Omega_2 t] \\
& + \frac{3\alpha A_2^3 C_6}{4(2\Omega_2^2+3\Omega_1\Omega_2+\Omega_1^2)}[\cos(3\Omega_2+2\Omega_1)t-\cos\Omega_2 t] \\
& + \frac{3\alpha A_2^3 C_6}{4(2\Omega_2^2-3\Omega_1\Omega_2+\Omega_1^2)}[\cos(3\Omega_2-2\Omega_1)t-\cos\Omega_2 t]
\end{aligned}
\tag{5.16}
$$

From (5.9), (5.15) and (5.16), the first approximations of Eqs. (5.4) are given by Eqs. (5.7).

5.4 Numerical Example

For $\omega_1 = 98.86$, $\omega_2 = 157.8$, $\alpha = 11600$, $A_1 = 0.08$, $A_2 = 0.1$, the optimal values of the convergence-control parameters and then the frequencies could be obtained using the procedures described in [14–16] as: $C_1 = 114.951$, $C_2 = 25.4744$,

$C_3 = 0.211777$, $C_4 = 0.140021$, $C_5 = 0.514075$, $C_6 = -0.0254753$, $\Omega_{1ap} = 99.5218$, $\Omega_{2ap} = 158.237$, which lead to the approximate solutions:

$$\begin{aligned} T1 = {} & 0.08043928\cos\Omega_1 t + 0.000064948\cos 3\Omega_1 t - 0.000119339\cos 5\Omega_1 t \\ & - 2.622\times 10^{-7}\cos(\Omega_1 + 2\Omega_2)t - 1.151\times 10^{-6}\cos(\Omega_1 - 2\Omega_2)t \\ & + 0.0000102434\cos(3\Omega_1 + 2\Omega_2)t - 0.0003937187\cos(3\Omega_1 - 2\Omega_2)t \end{aligned} \tag{5.17}$$

$$\begin{aligned} T2 = {} & 0.100018\cos\Omega_2 t + 4.3869\times 10^{-6}\cos(3\Omega_2)t - 1.8606\times 10^{-6}\cos(5\Omega_2)t \\ & + 4.2439\times 10^{-7}\cos(\Omega_2 + 2\Omega_1)t - 1.863\times 10^{-6}\cos(\Omega_2 - 2\Omega_1)t \\ & - 2.06698\times 10^{-6}\cos(3\Omega_2 + 2\Omega_1)t - 0.0000173991\cos(3\Omega_2 - 2\Omega_1)t \end{aligned} \tag{5.18}$$

To discuss the accuracy of the obtained approximate solutions, in Figs. 5.1 and 5.2 we compare the analytical results (5.17) and (5.18) with numerical results. It is observed that the solutions obtained through the proposed procedure are nearly identical with numerical solutions obtained using a fourth-order Runge-Kutta method.

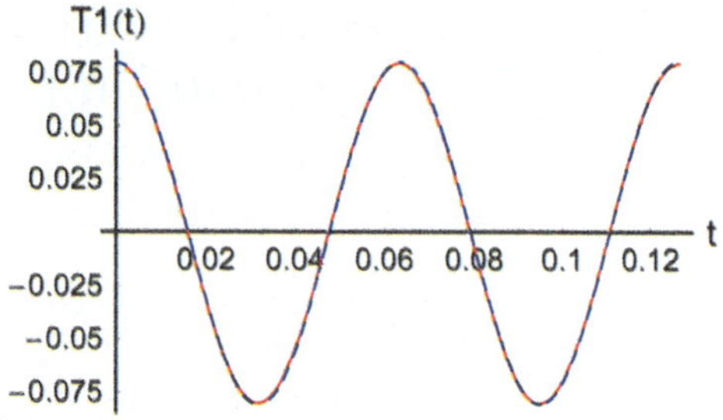

Fig. 5.1 Comparison between the approximate solution (5.17) and numerical results: solid blue lines numerical; dashed red lines approximate

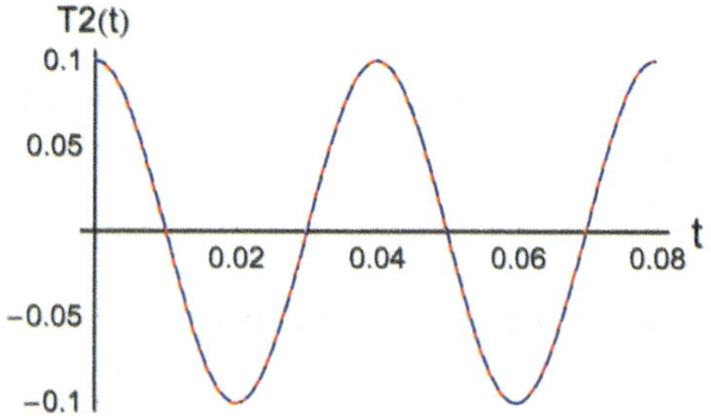

Fig. 5.2 Comparison between the approximate solution (5.18) and numerical results: solid blue lines numerical; dashed red lines approximate

5.5 Conclusions

In this work, analytical solutions are obtained for free oscillations of an elastic beam supported by a nonlinear foundation. By means of OAFM, the dynamic responses were explicitly obtained. Very accurate analytical results were obtained when comparing to numerical ones. The main advantage of OAFM, and also differences from other methods, consists in the involvement of the auxiliary functions which depend on a finite number of convergence-control parameters C_i. These initially unknown parameters are optimally determined using rigorous procedures and lead to a fast convergence of the solution after only one iteration. The application of OAFM does not need the presence of small or large parameters, as is the case of other methods. It was proved that this new method is simple, accurate and very efficient in practice even for strongly nonlinear systems.

References

1. T. Horibe, N. Asano, Large deflection analysis of beams on two-parameter elastic foundation using the boundary integral equation method. JSME Int. J. Series A **44**(2), 231–236 (2001)
2. V. Rao, Large amplitude vibration of slender, uniform beams on elastic foundation. Indian J. Eng. Mater. Sci. **10**, 87–91 (2003)
3. M. Ansari, E. Esmailzadeh, D. Younesian, Internal resonance of finite beams on nonlinear foundations traversed by a moving load, in *ASME Proceedings of IMECE 2008*. Paper no. IMECE2008–68188 (2016), pp. 321–2329
4. A.D. Senalp, A. Arikoglu, I. Ozkol, V.Z. Dogan, Dynamic response of a finite length Euler-Bernoulli beam on linear and nonlinear viscoelastic foundation to a concentrated moving force. J. Mech. Sci. Tech. **24**(10), 19579961 (2010)
5. G.C. Tsiatas, Nonlinear analysis of non-uniform beams on nonlinear elastic foundation. Acta Mech. **209**, 141–152 (2010)
6. U. Bhattiprolu, P. Davies, Response of a beam on nonlinear viscoelastic unilateral foundation and numerical challenges, in *ASME Proceedings of IDETC/CIE*, Paper no. DETC2011-48776 (2011), pp. 813–819
7. H. Ding, L.-Q. Chen, S.-P. Yang, Convergence of Galerkin truncation for dynamic response of finite beams on nonlinear foundation under a moving load. J. Sound Vibr. **331**, 2426–2442 (2012)
8. D. Younesian, B. Saadatnia, H. Askari, Analytical solutions for free oscillations of beams on nonlinear elastic foundation using the variational iteration method. J. Theor. Appl. Mech. **50** (2), 639–692 (2012)
9. P. Koziol, Z. Hryniewicz, Dynamic response of a beam resting on a nonlinear foundation to a moving load: coiflet-based solution. Shock. Vibr. **19**, 995–1007 (2012)
10. G. Sari, M. Pakdemirli, Vibration of a slightly curved microbeam resting on an elastic foundation with nonideal boundary conditions. Math. Probl. Eng. ID736148 (2013)
11. S.M. Abdelghany, K.M. Ewis, A.A. Mahmoud, M.M. Nassar, Dynamic response of non uniform beam subjected to moving load and resting on a nonlinear viscoelastic foundation. J. Basic Appl. Sci. **4**, 192–199 (2015)
12. F. Pellicano, F. Mastrodi, Nonlinear dynamics of a beam on elastic foundation. Nonlinear Dyn. **14**(4), 335–355 (1997)

13. N. Herisanu, V. Marinca, Approximate analytical solutions to Jerk equations, in *Springer Proceedings in Mathematics and Statistics*, vol. 182 (2016), pp. 169–176
14. V. Marinca, N. Herisanu, *Nonlinear Dynamical Systems in Engineering. Some approximate Approaches* (Springer, Berlin, 2011)
15. V. Marinca, N. Herisanu, *The Optimal Homotopy Asymptotic Method. Engineering Applications* (Springer, Berlin, 2015)
16. N. Herisanu, V. Marinca, Gh Madescu, An analytical approach to non-linear dynamical model of a permanent magnet synchronous generator. Wind Energy **18**, 1657–1670 (2015)

Chapter 6
Vibration of Nonlinear Nonlocal Elastic Column with Initial Imperfection

Vasile Marinca and Nicolae Herisanu

Abstract This paper is focused on obtaining explicit analytical approximate solutions via a simple but rigorous new procedure, namely the Optimal Auxiliary Functions Method (OAFM) for the post buckling behavior of an initially imperfect nonlocal elastic column. The column is simply supported at one end and subjected to an axially force at the other movable end. We assume that Euler-Bernoulli hypothesis holds and Eringen's nonlocal elasticity is adopted. The accuracy of our procedure is proved by numerical simulations developed in order to validate approximate analytical results.

6.1 Introduction

The nonlocal elasticity theory pioneered by Eringen [1] is a modification of classical elasticity theory and has been widely accepted and attracted a growing attention in the last years. This theory is often applied to analyze the vibration behavior of nanostructures, and has been used in many areas including light and high toughness fibers, statistical mechanics, ocean engineering, micro and nano electromechanical systems, and so on. For a detailed literature readers are referred to Gosh and Ray [2] and Askari et al. [3]. The study of Challamel [4] is focused on the geometrically exact elastic stability of two kinematically constrained flexible columns, modeled by the Euler-Bernoulli beam theory. The validity of parallel and translational beam assumptions is discussed. Using nonlocal elasticity theory, Hashemi et al. [5] considered surface effects including surface elasticity, surface stress and surface density on the free vibration analysis of Euler-Bernoulli and Timoshenko nanobeams. The governing equations are obtained and solved for

---

V. Marinca
Branch of Timisoara, Romanian Academy, Bucharest, Romania
e-mail: vasile.marinca@upt.ro

N. Herisanu (✉)
University Politehnica Timisoara, Timişoara, Romania
e-mail: nicolae.herisanu@upt.ro

N. Herisanu and V. Marinca (eds.), *Acoustics and Vibration of Mechanical Structures—AVMS-2017*, Springer Proceedings in Physics 198,
https://doi.org/10.1007/978-3-319-69823-6_6

silicon and aluminum nanobeams with three different boundary conditions. Xu et al. [6] presented a semianalytical treatment for calculating large elastic deformation of an initially imperfect nonlocal elastic column which is considered to be prismatic and inextensible. The constitutive equation corresponds to a differential type of Eringen's nonlocal elasticity theory and Euler-Bernoulli assumption is used. The size effect and the initial imperfection are also discussed. Togun [7] analyzed nonlinear vibration of a nanobeam resting on an elastic foundation. The nonlinear equation of motion includes stretching of the neutral axis (that introduces cubic nonlinearity into the equation), forcing and damping. The buckling and post-buckling behavior of a nonlinear discrete repetitive system is studied by Challamel et al. [8]. Ebrahimi and Nasirzadeh [9] used the Timoshenko beam theory for free vibration of nonlocal beam via Hamilton's principle. The solution is obtained by means of differential transformation method. A discrete column model on the basis of the central finite difference formulation and the equivalent Hencky bar-chain model are presented by Wang et al. [10] for the elastic buckling of Eringen's column with allowance for self-weight. More sophisticated recent models aimed on fitting the nano-/micro-structural behavior are presented in [11–14], where the continuous structural members are considered as those with infinite degree-of-freedom, offering reliable results.

In this paper, the nonlocal elastic column is considered and the thickness to length ratio is assumed to be small such that the effect of transverse shear deformation is neglected. The nonlinear differential equation is solved using OAFM.

6.2 Equation of Motion

In what follows, we consider the vibration of a nonlocal elastic column with a slight geometrical curvature as an imperfection (Fig. 6.1). This inextensible, simply supported column of uniform cross-section A and length L is subjected to a conservative force P at its right movable end.

The Cartesian coordinate system is chosen such that the abscissa x coincides with the line connecting the two hinged ends, the origin is located at its left end. If θ is the angle of inclination of the arc length reckoned along the column, then the general governing equation can be written in the form [6]:

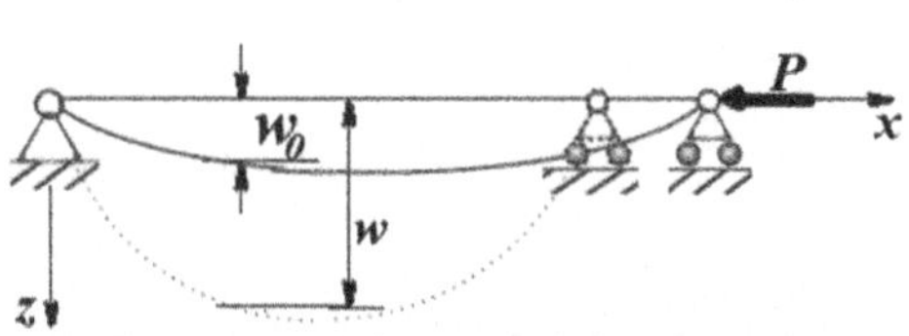

Fig. 6.1 Initially imperfect nonlocal elastic column—geometry and coordinate system

$$EI\frac{d^2\theta}{ds^2}+P\sin\theta=P(e_0a)^2\cos\theta\left[\frac{d^2\theta}{ds^2}-\tan\theta\left(\frac{ds}{d\theta}\right)^2\right]+EI\frac{d^2\theta_0}{ds^2} \tag{6.1}$$

where EI is the flexural rigidity, e_0a is the parameter that allows for the size effect, s is the arc length, and θ_0 is the initial imperfection. If a_0 is the column midspan initial size, then [6]

$$\theta^*(s)=\sin^{-1}\left[a_0\frac{d\sin\pi s/L}{ds}\right] \tag{6.2}$$

Using the dimensionless parameters

$$x=\frac{s}{L},\quad \mu=\frac{e_0a}{L},\quad d=\sqrt{\frac{PL^2}{EI}},\quad \beta=\frac{a_0\pi}{L} \tag{6.3}$$

and the following result obtained from (6.2):

$$\frac{d^2\theta^*}{ds^2}=\pi^2\beta(\beta^2-1)\left(1+\frac{9}{4}\beta^2\right)\cos\pi x+\frac{3}{4}\pi^2\beta^3(\beta^2-1)\cos 3\pi x \tag{6.4}$$

Equation (6.1) can be rewritten as

$$\begin{aligned}&\theta''+\lambda^2\sin\theta-\mu^2\lambda^2(\theta''\cos\theta-\theta'^2\sin\theta)\\&-\pi^2\beta(\beta^2-1)\left(1+\frac{9}{4}\beta^2\right)\cos\pi x-\frac{3}{4}\beta^3\pi^2(\beta^2-1)\cos 3\pi x=0\end{aligned} \tag{6.5}$$

where prime denotes derivative with respect to x. If α is the angle of the rotation at the left end of the column, then the initial conditions for (6.5) are

$$\theta(0)=\alpha,\quad \theta'(0)=0 \tag{6.6}$$

For solving the above nonlinear differential equation, we will use OAFM [15].

6.3 The Optimal Auxiliary Functions Method—OAFM

The nonlinear differential equations with variable coefficients (6.5) and the initial conditions (6.6) are of the general form

$$l(\theta(x))+n(\theta(x))=0,\quad x\in D,\quad b\left(\theta(x),\frac{d\theta}{dx}\right)=0 \tag{6.7}$$

where l and n are linear and nonlinear operator, respectively.

The approximate analytical solution of (6.7) can be expressed as

$$\tilde{\theta}(x) = \theta_0(x) + \theta_1(x, C_i), \quad i = 1, 2, \ldots, n \tag{6.8}$$

where the initial approximation $\theta_0(x)$ can be obtained from the linear equation

$$l[\theta_0(x)] = 0, \quad b\left(\theta_0(x), \frac{d\theta_0}{dx}\right) = 0 \tag{6.9}$$

and the first approximation $\theta_1(x,C_i)$ can be obtained from the equation

$$l[\theta_1(x, C_i)] + A_1[\theta_0(x, C_k)]P(n(\theta_0(x)) + A_2[\theta_0(x, C_j)] = 0, b\left(\theta_1(x, C_i)\frac{d\theta_1}{dx}\right) = 0 \tag{6.10}$$

where A_1 and A_2 are arbitrary auxiliary functions depending on the initial approximation and several unknown parameters C_k and C_j. $P(n[\theta_0(x)])$ means a part of the nonlinear operator. The auxiliary functions A_1 and A_2 are not unique, and are of the same form like $\theta_0(x)$. The parameters C_i can be optimally identified via different methods [16–18] and after that, the first-order approximate solution is well-defined. It should be also emphasized that our procedure is independent of the presence of small or large parameters in the governing equation or in boundary/ initial conditions.

6.4 Application of OAFM to Vibration of Nonlinear Nonlocal Elastic Column with Initial Imperfection

In the following we apply our procedure to obtain analytic approximate solution of (6.5) and (6.6). For this purpose we choose the linear operator of the form

$$l[\theta(x)] = \theta'' + \omega^2\theta \tag{6.11}$$

The initial approximation can be obtained from (6.9):

$$\theta_0'' + \omega^2\theta_0 = 0, \quad \theta_0(0) = \alpha, \quad \theta'(0) = 0 \tag{6.12}$$

where ω is the unknown frequency of the system. Equation (6.12) has the solution:

$$\theta_0(x) = \alpha \cos \omega x \tag{6.13}$$

The nonlinear operator corresponding to (6.5) is defined by

$$\begin{aligned} n[\theta(x)] &= -\omega^2\theta + \lambda^2 \sin\theta - \mu^2\lambda^2(\theta'' \cos\theta - \theta'^2 \sin\theta) \\ &- \pi^2\beta(\beta^2 - 1)\left(1 + \frac{9}{4}\beta^2\right) \cos \pi x - \frac{3}{4}\beta^3\pi^2(\beta^2 - 1) \cos 3\pi x \end{aligned} \tag{6.14}$$

By substituting (6.13) into (6.14) it holds that

$$n[\theta_0(x)] = M_1 \cos\omega x + M_2 \cos 3\omega x + M_3 \cos 5\omega x + N_1 \cos \pi x + N_2 \cos \pi x \tag{6.15}$$

where

$$\begin{aligned} M_1 &= \lambda^2\left(\alpha - \frac{\alpha^3}{8} + \frac{\alpha^5}{192}\right) - \omega^2\alpha + \mu^2\lambda^2\omega^2\left(\alpha - \frac{\alpha^3}{8} + \frac{\alpha^5}{192}\right), \\ M_2 &= \lambda^2\left(\frac{\alpha^5}{384} - \frac{\alpha^3}{24}\right) + \mu^2\lambda^2\omega^2\left(-\frac{3}{8}\alpha^3 + \frac{3}{128}\alpha^5\right), N_2 = -\frac{3}{4}\pi^2\beta^3(\beta^2 - 1), \\ M_3 &= \frac{\lambda^2\alpha^5}{1920} + \mu^2\lambda^2\omega^2\frac{5\alpha^5}{384}, N_1 = -\pi^2\beta(\beta^3 - 1)\left(1 + \frac{9}{4}\beta^2\right) \end{aligned} \tag{6.16}$$

Within the expressions (6.16) we used the expansions

$$\begin{aligned} \sin\theta &= \left(\alpha - \frac{\alpha^3}{8} + \frac{\alpha^5}{192}\right) \cos\omega x + \left(-\frac{\alpha^3}{24} + \frac{\alpha^5}{384}\right) \cos 3\omega x + \frac{\alpha^5}{1920} \cos 5\omega x \\ \cos\theta &= 1 - \frac{\alpha^2}{4} + \frac{\alpha^4}{64} + \left(-\frac{\alpha^2}{4} + \frac{\alpha^4}{48}\right) \cos 2\omega x + \frac{\alpha^4}{192} \cos 4\omega x \end{aligned} \tag{6.17}$$

The (6.10) can be written as

$$\begin{aligned} &\theta_1'' + \omega^2\theta_1 + A_1(\alpha \cos\omega x, C_k)P(n[\theta_0(x)]) + A_2(\alpha \cos\omega x, C_j) = 0, \\ &\theta_1(0) = \theta_1'(0) = 0 \end{aligned} \tag{6.18}$$

where the optimal auxiliary functions A_1 and A_2 can be chosen in the form

$$\begin{aligned} &A_1(\alpha \cos\omega x, C_k) = -(C_1 + 2C_2 \cos 2\omega x + 2C_3 \cos 4\omega x) \\ &A_2(\alpha \cos\omega x, C_j) = 0 \\ &P(n[\theta_0(x)]) = M_1 \cos\omega x + M_2 \cos 3\omega x \end{aligned} \tag{6.19}$$

Considering (6.19), (6.18) can be rewritten as

$$\begin{aligned}&\theta_1'' + \omega^2\theta_1 = [M_1C_1 + (M_1 + M_2)C_2 + M_2C_3]\cos\omega x + (M_2C_1 + M_1C_2 \\ &\quad + M_1C_3)\cos 3\omega x + (M_2C_2 + M_1C_3)\cos 5\omega x + M_2C_3\cos 7\omega x \\ &\theta_1(0) = \theta_1'(0) = 0\end{aligned} \tag{6.20}$$

Avoiding secular terms in (6.20), we obtain the condition

$$\omega^2 = \frac{\lambda^2\left[\left(\frac{\alpha^3}{24} - \frac{\alpha^5}{384}\right)(C_2 + C_3) - \left(\alpha - \frac{\alpha^3}{8} + \frac{\alpha^5}{192}\right)(C_1 + C_2)\right]}{\mu^2\lambda^2\left[\left(\alpha - \frac{\alpha^3}{8} + \frac{\alpha^5}{192}\right)(C_1 + C_2) - \left(\frac{3\alpha^3}{8} + \frac{3\alpha^5}{128}\right)(C_2 + C_3)\right] - \alpha(C_1 + C_2)} \tag{6.21}$$

The solution of (6.20) is given by

$$\begin{aligned}\theta_1(x, C_1, C_2, C_3) &= \frac{M_2C_1 + M_1(C_2 + C_3)}{8\omega^2}(\cos\omega x - \cos 3\omega x) \\ &+ \frac{M_2C_2 + M_1C_3}{24\omega^2}(\cos\omega x - \cos 5\omega x) + \frac{M_2C_3}{48\omega^2}(\cos\omega x - \cos 7\omega x)\end{aligned} \tag{6.22}$$

From (6.13), (6.22) and (6.8), the first-order approximate solution becomes

$$\tilde{\theta}(x, C_1, C_2, C_3) = \alpha\cos\omega x + \theta_1(x, C_1, C_2, C_3) \tag{6.23}$$

6.5 Numerical Example

Considering $\alpha = \pi/6$, $\lambda = 1$, $\mu = 0.1$ and $\beta = 0.01$, by minimizing the residual of the initial equation, the optimal values of the convergence-control parameters are obtained as $C_1 = 3.45524$, $C_2 = -2.12519$, $C_3 = 2.69221$, and the approximate frequency is $\omega = 0.985034$. In this case, the first-order approximate solution given by (6.23) becomes (Fig. 6.2)

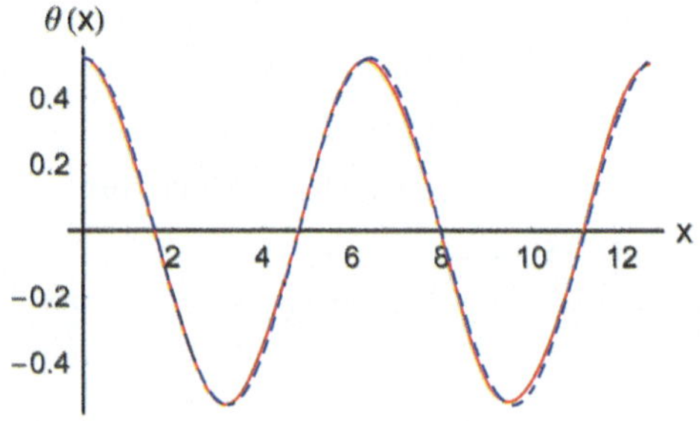

Fig. 6.2 Comparison between the approximate solution (6.24) and numerical integration results: dashed line approximate solution, solid line numerical solution

$$\tilde{\theta}(x) = \frac{1}{6}\pi\cos\omega x - 0.00264622(\cos\omega x - \cos 3\omega x) + 0.000898379(\cos\omega x - \cos 5\omega x) - 0.0003694909(\cos\omega x - \cos 7\omega x) \quad (6.24)$$

6.6 Conclusions

In the present work we proposed a new and reliable procedure—OAFM. This approach accelerates the rapid convergence of the approximate analytical solution of nonlinear equation with variable coefficients. Our procedure is valid even if the nonlinear equation does not contain any small or large parameters. It is clear that the construction of the first iteration is different from any other approaches. We refer to the optimal auxiliary functions A_1 and A_2, which depend on several initially unknown parameters C_i. The convergence-control parameters ensure a fast convergence of the approximate analytical solution using only one iteration.

It should be emphasized that the main advantage of the OAFM is the possibility to optimally control and adjust the convergence of the solutions through the auxiliary functions. The approximate analytical solution is found in an excellent agreement with the numerical integration results, which proves the validity of this procedure.

References

1. A.C. Eringen, *Nonlocal continuum fields theory* (Springer, NY, 2002)
2. S. Gosh, D. Roy, Numeric-analytic form of the Adomian decomposition method for two-point boundary value problems in nonlinear mechanics. J. Eng. Mech. **133**(10), 1124–1133 (2007)
3. H. Askari, E. Esmailzadeh, D. Zhang, Nonlinear vibration analysis of nonlocal nanowires. Composites B **67**, 60–71 (2014)
4. N. Challamel, On geometrically exact post-buckling of composite columns with interlayer slip —the partially composite elastic. Int. J. Nonlin. Mech. **47**, 7–17 (2007)
5. S.H. Hashemi, M. Fakher, R. Nazemnezhad, Surface effects on free vibration analyses of nanotubes using nonlocal elasticity: a comparison between Euler-Bernoulli and Timoshenko. J. Solid Mech. **5**(3), 290–304 (2013)
6. S.P. Xu, M.R. Xu, C.M. Wang, Stability analysis of nonlocal elastic columns with initial imperfection. Math. Prob. Eng. Article ID 341232 (2013)
7. N. Togun, Nonlocal beam theory for nonlinear vibration of a nanobeam resting on elastic foundation. Bound. Value Probl. (2016). doi:10.1186/s13661-016-0561-3
8. N. Challamel, A. Kocsis, C.M. Wang, Higher-order gradient elasticity models applied to geometrically nonlinear discrete systems. Theor. Appl. Mech. **42**(4), 223–248 (2015)
9. F. Ebrahimi, P. Nasirzadeh, A nonlocal Timoshenko beam theory for the vibration analysis of thick nanobeams using differential transform method. J. Theor. Appl. Mech. **53**(4), 1041–1052 (2015)

10. C.M. Wang, H. Zhang, N. Challamel, Y. Xiang, Buckling of nonlocal columns with allowance for selfweight. J. Eng. Mech. (2016). doi:10.1061/(ASCE)EM.1943-7889.0001088
11. A.V. Krysko, J. Awrejcewicz, O.A. Saltykova, S.S. Vetsel, V.A. Krysko, Nonlinear dynamics and contact interactions of the structures composed of beam-beam and beam-closed cylindrical shell members. Chaos Solitons Fractals **91**, 622–638 (2016)
12. J. Awrejcewicz, A.V. Krysko, S.P. Pavlov, M.V. Zhigalov, V.A. Krysko, Stability of the size-dependent and functionally graded curvilinear Timoshenko beams. J. Comput. Nonlinear Dynam **12**(4), 041018 (2017)
13. J. Awrejcewicz, A.V. Krysko, S.P. Pavlov, M.V. Zhigalov, V.A. Krysko, Chaotic dynamics of size dependent Timoshenko beams with functionally graded properties along their thickness. Mech. Syst. Signal Process. **93**, 415–430 (2017)
14. A.V. Krysko, J. Awrejcewicz, S.P. Pavlov, M.V. Zhigalov, V.A. Krysko, Chaotic dynamics of the size-dependent non-linear micro-beam model. Commun. Nonlin. Sci. Numer. Simul. **50**, 16–28 (2017)
15. N. Herisanu, V. Marinca, Approximate analytical solutions to Jerk equations, in *Springer Proceedings of Mathematics and Statistics*, vol. 182, (2016), pp. 169–176
16. V. Marinca, N. Herisanu, *Nonlinear Dynamical Systems in Engineering. Some approximate Approaches* (Springer, Berlin, 2011)
17. V. Marinca, N. Herisanu, *The Optimal Homotopy Asymptotic Method. Engineering Applications* (Springer, Berlin, 2015)
18. N. Herisanu, V. Marinca, Gh Madescu, An analytical approach to non-linear dynamical model of a permanent magnet synchronous generator. Wind Energy **18**, 1657–1670 (2015)

Chapter 7
Flap-Wise Vibrations of Axially Functionally Graded and Centrifugally Stiffened Beams with Multiple Masses Having Rotary Inertia

Kemal Mazanoğlu and Tolga Karakuzu

Abstract This paper is on flap-wise vibrations of centrifugally stiffened axially functionally graded beams with multiple lumped masses including rotary inertia effects. Material properties are defined along beams' length using the function appropriate to the power law distribution. Energy expressions of Euler–Bernoulli beam model are stated for flexural vibrations of beam with additional multiple masses rotating around a hub. Energy formulations are discretised to apply Rayleigh–Ritz solution method which uses admissible polynomial mode shape function. Analyses are conducted by changing number, size and inertias of additional masses, taper ratios of beam, rotating speed and radius of hub and the power of distribution function of beam material. Non-dimensional parameters are used to reflect results that are validated by those given in current literature. Unique results are also given for the homogeneous and functionally graded beam rotating with multiple masses. Effects of varying parameters and advantages/shortcomings of present solution are reflected and discussed.

7.1 Introduction

Beam elements rotating together with a hub may have additional masses in some applications such as pendulum type systems, robot arms, crank arms basically modelled with additional balancing mass, and etc. The beam in rotation should have proper modal characteristics arranged using FG (functionally graded) material and additional masses, if other parameters are strictly limited.

K. Mazanoğlu (✉)
Engineering Faculty, Department of Mechanical Engineering,
Usak University, 64200 Usak, Turkey
e-mail: kemal.mazanoglu@usak.edu.tr

T. Karakuzu
Institute of Natural and Applied Sciences, Usak University, 64200 Usak, Turkey
e-mail: tolgakarakuzu01@gmail.com

N. Herisanu and V. Marinca (eds.), *Acoustics and Vibration of Mechanical Structures—AVMS-2017*, Springer Proceedings in Physics 198,
https://doi.org/10.1007/978-3-319-69823-6_7

Early sources in literature include multi degree of freedom or continuous models using fundamental solution methods for the vibration of uniform beam carrying masses without inertia effects. However, some of recent works proposing improved analytical solutions are quoted here. Low [1] presents continuous solution with a modified Dunkerley formula to obtain natural frequencies of beams carrying two concentrated masses. Same author also solved the problem using the Rayleigh method [2]. A closed form exact solution is conducted by Maiz et al. [3] for vibration of the beam carrying concentrated masses with rotary inertia effects.

In literature, the flexural vibration analysis of centrifugally stiffened beams has been frequently presented. Although the most of these works use the finite element based analyses [4, 5], dynamic stiffness matrix method [6] and Rayleigh–Ritz method [7] are also put to use. Bhat [8] uses orthogonal polynomials in the Rayleigh–Ritz method to inspect natural frequencies of rotating uniform cantilever beam with a tip mass. Similar problem is investigated by Khulief [9] who employs the finite element model.

Vibration of non-rotating axially functionally graded (AFG) beam has just been inspected in several papers and one of them uses Fredholm integral equation [10]. Besides, vibration of rotating AFG beams is presented by Zarrinzadeh et al. [11] and Rajasekaran [12] who apply finite element based approach and differential quadrature method respectively. Ritz method is also performed using simple [13] and Chebyshev polynomials [14].

To the best of author's knowledge, this is the first work on flap-wise vibration analysis of rotating AFG tapered beams carrying multiple concentrated masses at arbitrary locations. Rotary inertia effects are also taken into consideration for additional masses located on the Euler–Bernoulli beam. Energy expressions are stated and Rayleigh–Ritz solution procedure is applied.

7.2 Theoretical Formulations

Cross-sectional properties of the tapered beam linearly change along the x axis as shown in Fig. 7.1.

Variations of height, h, and width, b, of the tapered beam are expressed as $h(x) = h_0(1 - c_h(x/L))$, $b(x) = b_0(1 - c_b(x/L))$. Here h_0 and b_0 symbolize the dimensions at the root of beam. c_h and c_b are given as taper ratios in terms of the height and width respectively. Cross-sectional area, $A(x)$ and area moment of inertia, $I(x)$, are written as $A(x) = h(x)b(x)$, $I(x) = b(x)h^3(x/12)$. AFG beam's material parameter, p_m, which may be modulus of elasticity E, or density ρ, varies along the beam. Variation of material parameter is formulated as $p_m(x) = (p_{m1} - p_{m2})(x/L)^n + p_{m2}$, where, p_{m1} and p_{m2} are the parameters of two different materials combined based upon material non-homogeneity factor, n. Hub rotating speed is taken constant for the model of centrifugally stiffened bending beam with additional concentrated mass. In addition, bending-torsion and

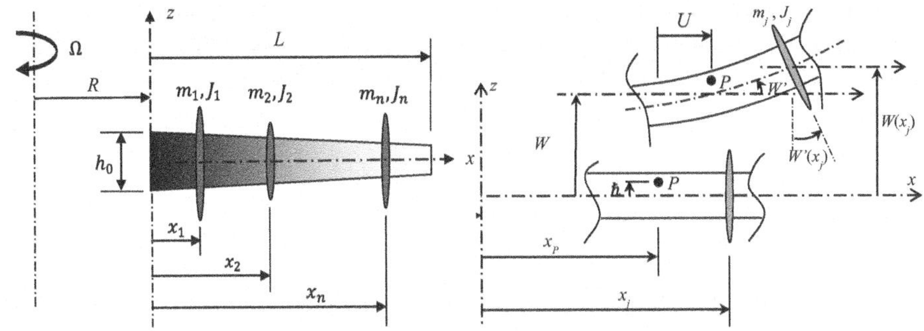

Fig. 7.1 Model of rotating AFG beam with multiple additional masses

bending-bending coupling effects are assumed negligible for the beam selected. While the beam carries lumped masses and rotates around a hub with a constant speed, Ω, centrifugal force, $F(x)$, which leads to increase of the strain energy, is explained for each beam part separated by additional masses, M_j, as follows:

$$F(x) = \int_x^L \rho(x)A(x)\Omega^2(R+x)\,dx + \sum_{j=j}^{N} M_j\Omega^2(R+x_j), \quad x_{j-1} < x \le x_j \tag{7.1a}$$

$$F(x) = \int_x^L \rho(x)A(x)\Omega^2(R+x)\,dx, \quad x_N < x \le L \tag{7.1b}$$

where R and N symbolize the radius of hub and number of lumped masses respectively. x_j denotes mass positions except for x_0, which is reference position at the root of beam. As given and explained in previous works [6, 13], strain energy arising from the flap-wise bending is expressed as follows:

$$S = \frac{1}{2}\int_0^L \left\{E(x)I(x)\left(W''(x)\right)^2 + F(x)\left(W'(x)\right)^2\right\}dx + C_1 \tag{7.2}$$

where W' and W'' stand for the first and second derivatives of flap-wise displacement, W. As is known, constant term, C_1, does not affect the results.

On the other hand, kinetic energy of rotating beam vibrating in flap-wise direction is expressed by defining motion of the particle, p on the beam. By neglecting the rotation around neutral axis corresponding to the Euler–Bernoulli beam theory, velocity vector $\boldsymbol{v}$ of particle in flap-wise vibration is constituted as follows [13]:

$$v = \left(\dot{U} - \eta\,\Omega\right)\boldsymbol{i} + \left(R + x_p + U\right)\Omega\boldsymbol{j} + \dot{W}\boldsymbol{k} \tag{7.3}$$

where $\boldsymbol{i}, \boldsymbol{j}, \boldsymbol{k}$ denote unit vectors in coordinate axes x, y, z respectively. Besides, η is the distance from neutral x-z plane. Constant rotating speed leads constant axial displacement, U, and thus vanished axial velocity, $\dot{U}$. Integrating kinetic energy of particle gives kinetic energy of the rotating beam without additional mass.

$$T_b = \frac{1}{2}\int_0^L \omega^2 \rho(x)A(x)(W(x))^2\,dx + C_2 \tag{7.4}$$

where ω denotes natural frequencies of flap-wise vibrations. Ineffective constant term C_2 is eliminated. However, the system has additional kinetic energy T_M due to inertia effects of masses M_j and mass moment of inertias J_j.

$$T_M = \frac{1}{2}\sum_{j=1}^{N} M_j\omega^2 (W(x_j))^2 + \frac{1}{2}\sum_{j=1}^{N} J_j\omega^2 (W'(x_j))^2 \tag{7.5}$$

Thus, total kinetic energy is expressed as $T = T_b + T_M$. Results are obtained by assuming flexural displacement function, W, given as follows:

$$W(x) = \sum_{i=1}^{m} \kappa_i \lambda_i(x) \tag{7.6}$$

where κ denotes the coefficient of terms that indexed by "i" with total number, m. $\lambda(x)$ is general term that forms admissible polynomial functions that are given as $(x/L)^2(1-x/L)^{i+1}$ for fixed-fixed, $(x/L)^1(1-x/L)^i$ for pinned-pinned, and $(x/L)^2(1-x/L)^{i-1}$ for fixed-free boundaries.

Based upon the conservation of energy law, the energy difference is made minimum by differentiating it with respect to the coefficients of shape functions.

$$\frac{\partial}{\partial \kappa_i}\sum (S - T) = 0, \quad i = 1, 2, \ldots, m \tag{7.7}$$

Energy expressions are written in discretised form and energy difference is partially differentiated with respect to coefficients κ_{i1} and κ_{i2}.

$$S = \frac{1}{2}\int_0^L E(x)I(x)\sum_{i1=1}^{m} \kappa_{i1}\lambda''_{i1}\sum_{i2=1}^{m} \kappa_{i2}\lambda''_{i2}\,dx + \frac{1}{2}\int_0^L F(x)\sum_{i1=1}^{m} \kappa_{i1}\lambda'_{i1}\sum_{i2=1}^{m} \kappa_{i2}\lambda'_{i2}\,dx \tag{7.8a}$$

$$T = \frac{1}{2}\omega^2\int_0^L \rho(x)A(x)\sum_{i1=1}^{m} \kappa_{i1}\lambda_{i1}\sum_{i2=1}^{m} \kappa_{i2}\lambda_{i2}\,dx + \frac{1}{2}\omega^2\sum_{j=1}^{N} M_j (W(x_j))^2 + \frac{1}{2}\omega^2\sum_{j=1}^{N} J_j (W'(x_j))^2 \tag{7.8b}$$

$$d_{i1,i2} = \frac{\partial^2}{\partial \kappa_{i1} \partial \kappa_{i2}} \sum (S - T) \quad i1 = 1, 2, \ldots, m \quad i2 = 1, 2, \ldots, m \tag{7.9}$$

$d_{i1,i2}$ terms are located into a matrix, D, and the determinant of matrix is equated to zero ($\det(D) = 0$) to obtain natural frequencies.

7.3 Results and Discussion

Results are given by the non-dimensional hub radius δ, hub rotating speed α, natural frequency μ, mass β, and inertia γ, parameters that are expressed as follow: $\delta = R/L$, $\alpha = \Omega L^2((\rho_0 A_0)/(E_0 I_0))^{0.5}$, $\mu = \omega L^2((\rho_0 A_0)/(E_0 I_0))^{0.5}$, $\beta = M/(\rho_0 A_0 L)$, $\gamma = r_g/L(J/(ML^2))^{0.5}$. The subscript "0" represent the properties at $x = 0$. Material properties of the beams made of Aluminium and Zirconia (ZrO_2) are given as: $E_{al} = 70\,\text{GPa}$, $\rho_{al} = 2702\,\text{kg/m}^3$, $E_{zr} = 200\,\text{GPa}$, $\rho_{zr} = 5700\,\text{kg/m}^3$. Success of the Rayleigh–Ritz method applied using admissible functions with 12 polynomial terms are discussed in following case studies.

Natural frequency parameters obtained for non-rotating tapered cantilever beam with additional tip mass are shown in Table 7.1. Point mass assumption is made and thus rotary inertia effects are neglected. Results agree well with the non-dimensional natural frequencies given by Khulief [9]. Frequency parameters decrease with ascending taper ratio and augmenting tip mass.

Table 7.2 shows non-dimensional natural frequencies of non-rotating cantilever beam with two additional masses having rotary inertias. Obtained frequencies

Table 7.1 Effects of taper ratio and additional tip mass on natural frequency parameters of non-rotating cantilever beam ($\beta_1 = \beta_2 = \beta,\ c_h = c_b = c,\ n = 0$)

β	c	Method	Natural frequency parameters				
			μ_1	μ_2	μ_3	μ_4	μ_5
0	0.1	Present	3.67370	21.5502	59.1886	115.397	190.357
		Khulief [9]	3.67371	21.5506	59.1960	115.451	190.593
	0.5	Present	4.62515	19.5476	48.5788	91.8127	149.390
		Khulief [9]	4.62517	19.5480	48.5855	91.8575	149.580
1	0.1	Present	1.50353	15.6426	48.4801	100.019	170.296
		Khulief [9]	1.50354	15.6428	48.4843	100.056	170.471
	0.5	Present	1.18689	13.0798	38.3995	78.0368	132.013
		Khulief [9]	1.18690	13.0804	38.4050	78.0746	132.183

sufficiently approach to those given by Maiz et al. [3]. Results show that the approximation level of method slightly decreases with increasing mass and inertia values. Secondly, greater masses and rotary inertias cause smaller frequency parameters. Thirdly, mass position is more influential on lower vibration frequencies.

The method is also tested for tapered AFG beams rotating around a hub having radius. Results shows good agreement with non-dimensional frequencies given by Rajasekaran [12] as illustrated in Table 7.3. It is obvious that increase of hub radius

Table 7.2 Effects of additional masses, and inertias on natural frequency parameters of non-rotating cantilever beam ($\gamma_1 = \gamma_2 = \gamma$, $\beta_1 = \beta_2 = \beta$, $c = 0$, $n = 0$)

γ	β	x/L	Method	Natural frequency parameters		
				$\mu_1^{0.5}$	$\mu_2^{0.5}$	$\mu_3^{0.5}$
0.05	0.5	0.25, 0.75	Present	1.5976	4.2473	6.3000
			Maiz et al. [3]	1.5976	4.2466	6.2961
		0.25, 0.5	Present	1.7677	3.8764	6.4275
			Maiz et al. [3]	1.7677	3.8759	6.4207
	1	0.25, 0.75	Present	1.4500	3.9422	5.7873
			Maiz et al. [3]	1.4499	3.9408	5.7797
		0.25, 0.5	Present	1.6829	3.5817	5.7508
			Maiz et al. [3]	1.6828	3.5809	5.7418
0.1	0.5	0.25, 0.75	Present	1.5903	4.0535	6.0909
			Maiz et al. [3]	1.5903	4.0495	6.0715
		0.25, 0.5	Present	1.7579	3.8397	6.1714
			Maiz et al. [3]	1.7579	3.8384	6.1329
	1	0.25, 0.75	Present	1.4411	3.6943	5.4196
			Maiz et al. [3]	1.4411	3.6874	5.3853
		0.25, 0.5	Present	1.6677	3.5283	5.5156
			Maiz et al. [3]	1.6676	3.5261	5.4687

Table 7.3 Effects of hub rotating speed and hub radius on fundamental frequency parameter of tapered AFG beam ($c_h = 0.5$, $c_b = 0$, $n = 2$)

α	δ	Method	μ_1 for several boundary conditions		
			Fixed-free	Fixed-fixed	Pinned-pinned
1	0	Present	4.7452	15.4096	6.8560
		Rajasekaran [12]	4.7453	15.4096	6.8560
	1	Present	4.9280	15.5323	7.1018
		Rajasekaran [12]	4.9281	15.5324	7.1018
10	0	Present	11.8954	21.5971	14.8931
		Rajasekaran [12]	11.8955	21.5971	14.8930
	1	Present	17.6147	28.5078	22.6625
		Rajasekaran [12]	17.6148	28.5077	22.6624

Table 7.4 Effects of non-homogeneity parameter on natural frequency parameters of rotating AFG beam ($\delta = 0,\ \alpha = 5,\ c_h = c_b = 0$)

Boundary conditions	n	Method	Natural frequency parameters			
			μ_1	μ_2	μ_3	μ_4
Fixed-free	0	Present	6.4495	25.4461	65.2050	124.5664
		Rajasekaran [12]	6.4494	25.4461	65.2050	124.5664
	1	Present	6.9716	25.6519	63.3085	119.5305
		Rajasekaran [12]	6.9717	25.6522	63.3094	119.5321
Pinned-pinned	0	Present	13.0953	43.3513	92.8561	162.0019
		Rajasekaran [12]	13.0953	43.3513	92.8561	162.0019
	1	Present	12.1032	40.6536	87.5306	153.0311
		Rajasekaran [12]	12.1033	40.6541	87.5317	153.0331

and rotating speed lead to augmentation of fundamental frequency parameters. It is also seen that speed and radius parameters are more effective on the frequency parameters when the beams have boundaries leading low rigidity.

Table 7.4 tabulates the variation of non-dimensional frequencies reflecting the effects of non-homogeneity parameter, n. It is seen that obtained results closely match with the results of Rajasekaran [12]. Besides, the frequency parameters of non-homogeneous beams are lower than those of homogeneous beams except for the first and second frequency parameters of rotating cantilever beam.

Table 7.5 Effects of non-homogeneity, additional masses, rotating speed and boundary on natural frequency parameters ($\delta = 0,\ \gamma = 0.01,\ x_1/L = 0.25,\ x_2/L = 0.75$)

Boundary conditions	α	β	n	Natural frequency parameters		
				$\mu_1^{0.5}$	$\mu_2^{0.5}$	$\mu_3^{0.5}$
Fixed-fixed	1	0.1	0	4.5772	7.2001	10.2418
			1	4.3647	6.7929	9.7807
		1	0	3.7515	5.1917	8.7500
			1	3.4079	4.7490	8.3875
	5	0.1	0	4.8135	7.4017	10.4197
			1	4.5776	6.9773	9.9501
		1	0	4.1095	5.5177	9.1996
			1	3.7457	5.1462	8.9467
Fixed-free	1	0.1	0	1.8493	4.6251	7.3288
			1	1.9825	4.6852	7.1423
		1	0	1.5415	4.0647	5.9906
			1	1.5512	4.0299	5.9124
	5	0.1	0	2.5095	4.9979	7.5515
			1	2.5773	5.0230	7.3391
		1	0	2.3975	4.6756	6.3343
			1	2.4075	4.6642	6.2876

Novel results are given in Table 7.5 for rotating AFG beams with multiple masses having rotary inertia. Effects of non-homogeneity are explicitly seen as decreases of non-dimensional frequencies for the beams with fixed-fixed boundaries. However, the non-homogeneity is not very influential on frequencies of cantilever beams. Besides, it is shown again that additional masses have decreasing effects on the frequency parameters in contrast to the increasing effects of rotating speed.

7.4 Conclusions

In this paper, flap-wise vibration analyses are conducted for centrifugally stiffened axially functionally graded beams with additional masses including rotary inertia effects. Comparison of results shows that the Rayleigh–Ritz approximation can be applied for the problem considered. In addition, it is obtained that frequency parameters decrease with ascending taper ratio and augmenting mass and inertia values. Mass position has less influence on higher vibration frequencies. Hub radius and rotating speed increase the frequency parameters and they are more effective for the beams with low rigidity. Non-homogeneity has obvious effects on frequency parameters of the beam with boundaries leading to high rigidity.

Acknowledgements The research topic in this work is supported by "The Scientific and Technological Research Council of Turkey (TUBITAK)" within the framework of national project as "1002—Short Term R&D Funding Program" (Project no: 215M756).

References

1. K.H. Low, A modified Dunkerley formula for eigenfrequencies of beams carrying concentrated masses. Int. J. Mech. Sci. **42**, 1287–1305 (2000)
2. K.H. Low, Natural frequencies of a beam-mass system in transverse vibration: Rayleigh estimation versus eigenanalysis solutions. Int. J. of Mech. Sci. **45**, 981–993 (2003)
3. S. Maiz, D.V. Bambill, C.A. Rossit, P.A.A. Laura, Transverse vibration of Bernoulli–Euler beams carrying point masses and taking into account their rotatory inertia: exact solution. J. Sound Vib. **303**, 895–908 (2007)
4. J. Chung, H.H. Yoo, Dynamic analysis of a rotating cantilever beam by using the finite element method. J. Sound Vib. **249**, 147–164 (2002)
5. J.B. Yang, L.J. Jiang, D.C.H. Chen, Dynamic modelling and control of a rotating Euler-Bernoulli beam. J. Sound Vib. **274**, 863–875 (2004)
6. J.R. Banerjee, Free vibration of centrifugally stiffened uniform and tapered beams using the dynamic stiffness method. J. Sound Vib. **233**(5), 857–875 (2000)
7. H.H. Yoo, S.H. Shin, Vibration analysis of rotating cantilever beams. J. Sound Vib. **212**(5), 807–828 (1998)
8. R.B. Bhat, Transverse vibrations of a rotating uniform cantilever with tip mass as predicted by using beam characteristics orthogonal polynomials in the Rayleigh–Ritz method. J. Sound Vib. **105**, 199–210 (1986)

9. Y.A. Khulief, Vibration frequencies of a rotating tapered beam with end mass. J. Sound Vib. **134**, 87–97 (1989)
10. Y. Huang, X.-F. Li, A new approach for free vibration of axially functionally graded beams with non-uniform cross-section. J. Sound Vib. **329**, 2291–2303 (2010)
11. H. Zarrinzadeh, R. Attarnejad, A. Shahba, Free vibration of rotating axially functionally graded tapered beams. Proc. Inst. Mech. Eng. Part G: J. Aerosp. Eng. **226**, 363–379 (2012)
12. S. Rajasekaran, Differential transformation and differential quadrature methods for centrifugally stiffened axially functionally graded tapered beams. Int. J. Mech. Sci. **74**, 15–31 (2013)
13. K. Mazanoglu, S. Guler, Flap-wise and chord-wise vibrations of axially functionally graded tapered beams rotating around a hub. Mech. Syst. Sign. Process. **89**, 97–107 (2017)
14. J. Fang, D. Zhou, Free vibration analysis of rotating axially functionally graded-tapered beams using Chebyshev–Ritz method. Mater. Res. Innov. **19**, 1255–1262 (2015)

Chapter 8
Study of Vibrations of a Robotic Arm, Using the Lagrange Equations with Respect to a Non-inertial Reference Frame

Andrei Craifaleanu and Ion Stroe

Abstract The paper studies the free vibrations of a robotic arm, located on a rotating platform. The robotic arm is modeled as a system with a finite number of degrees of freedom, consisting of elastically coupled rigid bodies. The differential equations of the free vibrations are obtained using the Lagrange equations formalism, in a generalized form, valid with respect to a non-inertial reference frame. The influence of the kinematic parameters of the platform motion upon the eigenfrequencies of the robotic arm is analyzed, considering two configurations of the system.

8.1 Introduction

Traditionally, the Lagrange equations formalism is used to study the dynamics of mechanical systems with respect to inertial reference frames, conventionally considered as fixed. However, in various applications concerning devices located on airspace structures or other vehicles in motion, a generalized form of the formalism, valid with respect to movable, non-inertial frames, can prove beneficial. Such an approach was adopted in [1] to deduce the equations of the relative motion of a material point and in [2] to study the dynamics of a discrete system of material points in motion with respect to a translating or a rotating frame. Reference [3] presents a general expression of the kinetic energy of a system in relative motion. A modified form of the formalism, valid with respect to a central reference frame, i.e. to a particular non-inertial frame, is given in [4] for a system of two particles. The authors of the present paper used in [5] a different version of the same approach, based on an equivalent kinetic energy, as well as on some simplifying assumptions, in order to study the relative motion of a robotic arm. They later

A. Craifaleanu (✉) · I. Stroe
University "Politehnica" of Bucharest, Bucharest, Romania
e-mail: andrei.craifaleanu@upb.ro

I. Stroe
e-mail: ion.stroe@gmail.com

N. Herisanu and V. Marinca (eds.), *Acoustics and Vibration of Mechanical Structures—AVMS-2017*, Springer Proceedings in Physics 198,
https://doi.org/10.1007/978-3-319-69823-6_8

proposed a general form of the formalism, valid for arbitrary relative motions and non-inertial reference frames [6].

The paper studies the free vibrations of a robotic arm located on a rotating platform, by using another version of the generalized Lagrange formalism, equivalent with the previously proposed one, but more convenient for systems of rigid bodies, as it introduces original calculation formulae of the generalized Coriolis force.

8.2 Lagrange Equations with Respect to a Non-inertial Frame

A system of material points is considered, in motion with respect to movable reference frame, generally non-inertial. The motion of the movable frame relative to a fixed one is defined by the velocity $\bar{v}_O$ and the acceleration $\bar{a}_O$ of the origin, by the angular velocity, $\bar{\omega}_0$, and by the angular acceleration, $\bar{\varepsilon}_0$. The position of a point P_i of the system, with respect to the movable frame, is defined by the position vector $\bar{r}_i$, while its relative motion is governed by the equation

$$m_i\bar{a}_{ir} = \bar{F}_i + \bar{F}_{it} + \bar{F}_{ic}, \tag{8.1}$$

where $\bar{F}_i$ is the applied force, $\bar{F}_{it}$ the force of transport and $\bar{F}_{ic}$ the Coriolis force:

$$\bar{F}_{it} = -m_i[\bar{a}_O + \bar{\varepsilon}_0 \times \bar{r}_i + \bar{\omega}_0 \times (\bar{\omega}_0 \times \bar{r}_i)]_{ir}, \quad \bar{F}_{ic} = -2m_i\bar{\omega}_0 \times \bar{v}_{ir}, \tag{8.2}$$

By expressing the relative position of the material points with the generalized coordinates, q_k $(k = 1, 2, \ldots, h)$ and processing (8.1) in the standard way [2], the Lagrange equations with respect to a non-inertial frame are obtained

$$\frac{\mathrm{d}}{\mathrm{d}t}\left(\frac{\partial E_r}{\partial \dot{q}_k}\right) - \frac{\partial E_r}{\partial q_k} = Q_k + Q_{kt} + Q_{kc} \quad (k = 1, 2, \ldots, h), \tag{8.3}$$

where the relative kinetic energy, the generalized applied force, the generalized transport force and the generalized Coriolis force, respectively, were introduced:

$$E_r = \sum_{i=1}^{n} \frac{1}{2} m_i v_{ir}^2, \tag{8.4}$$

$$Q_k = \sum_{i=1}^{n} \bar{F}_i \cdot \frac{\partial \bar{r}_i}{\partial q_k} = \frac{\delta_k L}{\delta q_k}, \tag{8.5}$$

$$Q_{kt} = -\sum_{i=1}^{n} m_i \bar{a}_O \cdot \frac{\partial \bar{r}_i}{\partial q_k} - \sum_{i=1}^{n} m_i (\bar{\varepsilon}_0 \times \bar{r}_i) \cdot \frac{\partial \bar{r}_i}{\partial q_k} - \sum_{i=1}^{n} m_i [\bar{\omega}_0 \times (\bar{\omega}_0 \times \bar{r}_i)] \cdot \frac{\partial \bar{r}_i}{\partial q_k}, \tag{8.6}$$

$$
\begin{aligned}
Q_{kc} &= -2\sum_{i=1}^{n} m_i(\bar{\omega}_0 \times \bar{v}_{ir}) \cdot \frac{\partial \bar{r}_i}{\partial q_k} = -2\bar{\omega}_0 \cdot \sum_{i=1}^{n} m_i \bar{v}_{ir} \times \frac{\partial \bar{r}_i}{\partial q_k} \\
&= -2\bar{\omega}_0 \cdot \sum_{i=1}^{n} m_i \bar{v}_{ir} \times \frac{\partial \bar{v}_{ir}}{\partial \dot{q}_k}.
\end{aligned} \tag{8.7}
$$

8.3 Calculation of the Generalized Force of Transport

The three terms in expression (8.6) of the generalized transport force take the forms

$$
Q_{kt}^{a} = -\sum_{i=1}^{n} m_i \bar{a}_O \cdot \frac{\partial \bar{r}_i}{\partial q_k} = -\bar{a}_O \cdot \sum_{i=1}^{n} m_i \frac{\partial \bar{r}_i}{\partial q_k} = -\bar{a}_O \cdot \frac{\partial}{\partial \dot{q}_k} \sum_{i=1}^{n} m_i \bar{v}_{ir} = -\bar{a}_O \cdot \frac{\partial \bar{H}_r}{\partial \dot{q}_k}, \tag{8.8}
$$

$$
Q_{kt}^{\varepsilon} = -\sum_{i=1}^{n} m_i(\bar{\varepsilon}_0 \times \bar{r}_i) \cdot \frac{\partial \bar{r}_i}{\partial q_k} = -\bar{\varepsilon}_0 \cdot \sum_{i=1}^{n} m_i \left(\bar{r}_i \times \frac{\partial \bar{v}_{ir}}{\partial \dot{q}_k}\right) = -\bar{\varepsilon}_0 \cdot \frac{\partial \bar{K}_{Or}}{\partial \dot{q}_k}, \tag{8.9}
$$

$$
Q_{kt}^{\omega} = -\sum_{i=1}^{n} m_i[\bar{\omega}_0 \times (\bar{\omega}_0 \times \bar{r}_i)] \cdot \frac{\partial \bar{r}_i}{\partial q_k} = -\sum_{i=1}^{n} m_i \left(\frac{\partial \bar{r}_i}{\partial q_k} \times \bar{\omega}_0\right) \cdot (\bar{\omega}_0 \times \bar{r}_i) = \frac{\partial E_c}{\partial q_k}, \tag{8.10}
$$

where the relative linear momentum, the relative angular momentum and the kinetic energy of the circular velocities, respectively, were introduced:

$$
\bar{H}_r = \sum_{i=1}^{n} m_i \bar{v}_{ir}, \quad \bar{K}_{Or} = \sum_{i=1}^{n} \bar{r}_i \times m_i \bar{v}_{ir}, \quad E_c = \frac{1}{2} \sum_{i=1}^{n} m_i (\bar{\omega}_0 \times \bar{r}_i)^2. \tag{8.11}
$$

Terms (8.8) and (8.10) were also found in [2], but term (8.9) was included together with the generalized Coriolis force in a more complicated expression.

8.4 The Generalized Coriolis Force in the Case of a Rigid Body

For a rigid body in relative translation, $\bar{v}_{ir} = \bar{v}_r$ and formula (8.7) becomes

$$
Q_{kc} = -2m\bar{\omega}_0 \cdot \left(\bar{v}_r \times \frac{\partial \bar{v}_r}{\partial \dot{q}_k}\right) = -2\bar{\omega}_0 \cdot \left(\bar{H}_r \times \frac{\partial \bar{v}_r}{\partial \dot{q}_k}\right) = -2\bar{\omega}_0 \cdot \left(\bar{v}_r \times \frac{\partial \bar{H}_r}{\partial \dot{q}_k}\right). \tag{8.12}
$$

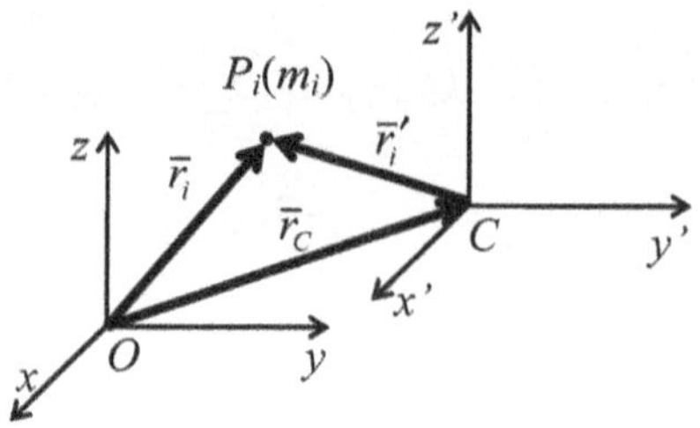

Fig. 8.1 Central reference frame in relative translation

For a rigid body in relative motion with a fixed point in the origin of the movable frame, $\bar{v}_{ir} = \bar{\omega}_r \times \bar{r}_i$ and expression (8.7) becomes, after certain calculations,

$$Q_{kc} = -2\bar{\omega}_0 \cdot \sum_{i=1}^{n} m_i(\bar{r}_i \otimes \bar{r}_i) \cdot \left(\bar{\omega}_r \times \frac{\partial\bar{\omega}_r}{\partial\dot{q}_k}\right) = -2\bar{\omega}_0 \cdot \bar{\bar{P}}_O \cdot \left(\bar{\omega}_r \times \frac{\partial\bar{\omega}_r}{\partial\dot{q}_k}\right), \quad (8.13)$$

where the tensor of the moments of inertia with respect to planes *Oyz*, *Ozx* and *Oxy*, respectively, and of the products of inertia was introduced:

$$\bar{\bar{P}}_O = \sum_{i=1}^{n} m_i\bar{r}_i \otimes \bar{r}_i, \quad \left[\bar{\bar{P}}_O\right] = \begin{bmatrix} J_{Oyz} & J_{xy} & J_{xz} \\ J_{yx} & J_{Ozx} & J_{yz} \\ J_{zx} & J_{zy} & J_{Oxy} \end{bmatrix}. \quad (8.14)$$

For an arbitrary system of material points in relative motion, the central reference frame $Cx'y'z'$ is considered, in relative translation with respect to the movable reference frame *Oxyz* (Fig. 8.1).

By denoting with $\bar{r}_i'$ and $\bar{v}_{ir}'$, respectively, the position vector and the velocity of the material point P_i with respect to reference frame $Cx'y'z'$, after certain calculations it follows that

$$Q_{kc} = -2\bar{\omega}_0 \cdot m\bar{v}_{Cr} \times \frac{\partial\bar{v}_{Cr}}{\partial\dot{q}_k} - 2\bar{\omega}_0 \cdot \sum_{i=1}^{n} m_i\bar{v}_{ir}' \times \frac{\partial\bar{r}_i'}{\partial q_k}, \quad (8.15)$$

where $\bar{v}_{Cr}$ is the velocity of the mass center with respect to reference frame *Oxyz*.

The last term in expression (8.15) represents the generalized Coriolis force in the relative motion about the center of mass.

It results that for a rigid body in arbitrary relative motion,

$$Q_{kc} = -2\bar{\omega}_0 \cdot m\bar{v}_{Cr} \times \frac{\partial\bar{v}_{Cr}}{\partial\dot{q}_k} - 2\bar{\omega}_0 \cdot \bar{\bar{P}}_C' \cdot \left(\bar{\omega}_r \times \frac{\partial\bar{\omega}_r}{\partial\dot{q}_k}\right), \quad (8.16)$$

where $\bar{\bar{P}}_C'$ is the tensor (8.14) expressed with respect to the central frame $Cx'y'z'$.

8.5 The Kinetic Energy of Circular Velocities in the Case of a Rigid Body

For a rigid body, the kinetic energy of the circular velocities takes the form

$$E_c = \frac{1}{2} m(\bar{\omega}_0 \times \bar{r}_C)^2 + \frac{1}{2} \bar{\omega}_0 \cdot \bar{\bar{J}}'_C \cdot \bar{\omega}_0, \tag{8.17}$$

where $\bar{\bar{J}}'_C$ is the inertia tensor with respect to a central reference frame.

8.6 Robotic Arm

A robotic arm is considered, consisting of homogeneous bar OA elastically hinged in O on a rotating platform and in A on another homogeneous bar AB (Fig. 8.2). The length of the two bars are l_1 and l_2 and their masses m_1 and m_2, respectively. The elastic hinges are modeled by the torsion springs of rigidities k_1 and k_2.

The bars move in the plane Oxy, while the platform rotates with angular velocity Ω, first about Oz axis (Fig. 8.2a) and next about Ox axis (Fig. 8.2b).

In the reference position, with non-twisted springs, the bars make the angles α_1 and α_2, respectively, with Ox direction.

When the system vibrates, these angles vary with φ_1 and φ_2.

If the platform is at rest ($\Omega = 0$), the differential equations of the free vibrations of the robotic arm are obtained using the traditional form of Lagrange equations, resulting the differential equations

$$\begin{cases} \left(\frac{m_1}{3} + m_2\right) l_1^2 \ddot{\varphi}_1 + \frac{1}{2} m_2 l_1 l_2 \ddot{\varphi}_2 \cos(\alpha_2 - \alpha_1 + \varphi_2 - \varphi_1) \\ \quad - \frac{1}{2} m_2 l_1 l_2 \dot{\varphi}_2^2 \sin(\alpha_2 - \alpha_1 + \varphi_2 - \varphi_1) + (k_1 + k_2)\varphi_1 - k_2 \varphi_2 = 0 \\ \frac{1}{3} m_2 l_1 l_2 \ddot{\varphi}_1 \cos(\alpha_2 - \alpha_1 + \varphi_2 - \varphi_1) \\ \quad + \frac{1}{2} m_2 l_2^2 \ddot{\varphi}_2 + m_2 l_1 l_2 \dot{\varphi}_1^2 \sin(\alpha_2 - \alpha_1 + \varphi_2 - \varphi_1) - k_2 \varphi_1 + k_2 \varphi_2 = 0. \end{cases} \tag{8.18}$$

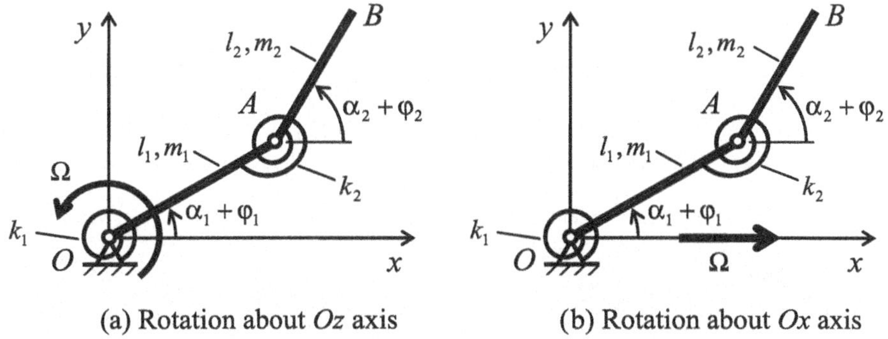

Fig. 8.2 Robotic arm on a rotating platform

If the platform rotates, the Lagrange equations with respect to a non-inertial reference frame (8.3) can be used.

If the rotation takes place about Oz axis, the correction terms

$$\begin{cases} Q_{1t} = \frac{1}{2} m_2 l_1 l_2 \Omega^2 \sin(\alpha + \phi_2 - \phi_1) \\ Q_{2t} = -\frac{1}{2} m_2 l_1 l_2 \Omega^2 \sin(\alpha + \phi_2 - \phi_1), \end{cases} \quad \begin{cases} Q_{1c} = m_2 l_1 l_2 \Omega \dot{\phi}_2 \sin(\alpha + \phi_2 - \phi_1) \\ Q_{2c} = -m_2 l_1 l_2 \Omega \phi_1 \sin(\alpha + \phi_2 - \phi_1). \end{cases} \tag{8.19}$$

shall be added to the right side of system (8.18).

If the rotation takes place about Ox axis, the correction terms are:

$$\begin{cases} Q_{1t} = \frac{1}{2}\left(\frac{m_1}{3} + m_2\right) l_1^2 \Omega^2 \sin 2(\alpha_1 + \varphi_1) + \frac{1}{2} m_2 l_1 l_2 \Omega^2 \sin(\alpha_2 + \varphi_2) \cos(\alpha_1 + \varphi_1) \\ Q_{2t} = \frac{1}{6} m_2 l_2^2 \Omega^2 \sin 2(\alpha_2 + \varphi_2) + \frac{1}{2} m_2 l_1 l_2 \Omega^2 \sin(\alpha_1 + \varphi_1) \cos(\alpha_2 + \varphi_2), \end{cases} \tag{8.20}$$

$$Q_{1c} = 0, \quad Q_{2c} = 0. \tag{8.21}$$

8.7 Numerical Application

An application was considered, with the following numerical values: $l_1 = l_2 = 1$ m, $m_1 = m_2 = 1$ kg, $k_1 = k_2 = 2000$ Nm.

The variations of the circular eigenfrequencies, obtained from the linearized system of differential equations, with respect to the angular velocity of the platform, are plotted in Fig. 8.3, for the two analyzed cases.

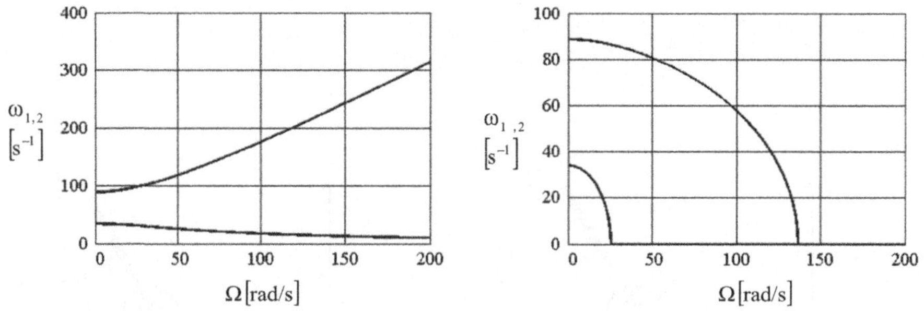

(a) Rotation about Oz axis (b) Rotation about Ox axis

Fig. 8.3 Variation of circular eigenfrequencies with respect to the platform angular velocity

8.8 Conclusions

The paper presented a generalized form of the Lagrange equations, valid with respect to a non-inertial reference frame. Calculation formulae were given for the generalized transport force, determined for a discrete system or material points, but valid without any modification also for rigid bodies.

Original calculation formulae of the generalized Coriolis force were deduced for a rigid body: (8.12) for relative translation, (8.13) for relative motion with a fixed point in the origin of the movable frame and (8.16) for general relative motion.

The formalism was used to study the free vibrations of a robotic arm located on a rotating platform.

The effect of the platform rotation about an axis perpendicular to the plane of the arm consists in reducing the fundamental eigenfrequency and increasing the second one. The rotation about an axis in the plane of the arm determines the reduction of both eigenfrequencies, up to the disappearance of the vibration.

References

1. L.D. Landau, E.M. Lifshitz, in *Mechanics* (translated from Russian), 3rd edn. Volume 1 of *Course of Theoretical Physics* (Butterworth, Heinemann, 2000)
2. C. Gignoux, B. Silvestre-Brac, *Solved Problems in Lagrangian and Hamiltonian Mechanics* (Springer, Dordrecht, Heidelberg, London, New York, 2009), pp. 23, 48
3. A.I. Lurie, *Analytical Mechanics* (Springer, New York, 2002)
4. A. Brizard, *Introduction to Lagrangian and Hamiltonian Mechanics* (World Scientific Publishing Company, Singapore, 2008)
5. I. Stroe, A. Craifaleanu, Calculus of a compass robotic arm using Lagrange equations in non-inertial reference frames, in *Proceedings of the International Conference of Aerospace Sciences "AEROSPATIAL 2012"* pp. 137–141, Bucharest, 11–12 Oct 2012
6. I. Stroe, A. Craifaleanu, in *Generalization of the Lagrange equations formalism, for motions with respect to non-inertial reference frames*. Applied Mechanics and Materials, vol. 656 (Trans Tech Publications, Switzerland, 2014), pp. 171–180

8.5 Conclusions

Part II
Environmental and Occupational Noise

Chapter 9
Selection of Measurement Strategy for the Assessment of Long-Term Environmental Noise Indicators Using Multi-criteria Optimization

Darko Mihajlov, Momir Praščević and Nicolae Herisanu

Abstract This paper presents a methodology for determining the values of environmental noise indicators based on measurement time intervals that are considerably shorter than one year, whereby the measurement results correspond to the actual state of acoustic load at the measured location on a yearly basis, with satisfactory accuracy and precision. According to the set requirements, the optimal time interval for noise indicator measurement was selected through multi-criteria optimization by means of the PROMETHEE method.

9.1 Introduction

According to Directive 2002/49/EC relating to the assessment and management of environmental noise [1], the compound noise indicator and the night-time noise indicator represent the fundamental environmental noise indicators. Their values, required for the calculation of strategic noise maps, can be obtained through measurements and calculations based on a predefined procedure applicable for all EU members [2]. For the purpose of acoustic spatial planning and noise protection zoning (quiet zones), additional noise indicators are used—day-time indicator and evening-time indicator—whose values can also be obtained through prescribed measurements and calculations. Verification of the calculated noise indicator values, i.e. their calibration, is performed through direct measurement of noise indicators for the observation period of one year.

D. Mihajlov (✉) · M. Praščević
Faculty of Occupational Safety, University of Niš, Niš, Serbia
e-mail: darko.mihajlov@znrfak.ni.ac.rs

M. Praščević
e-mail: momir.prascevic@znrfak.ni.ac.rs

N. Herisanu
Faculty of Mechanical Engineering, University Politehnica Timisoara, Timisoara, Romania
e-mail: nicolae.herisanu@upt.ro

N. Herisanu and V. Marinca (eds.), *Acoustics and Vibration of Mechanical Structures—AVMS-2017*, Springer Proceedings in Physics 198,
https://doi.org/10.1007/978-3-319-69823-6_9

The presented approach to determining the degree of environmental noise pollution involves two measurement strategies for assessing annual noise indicators at a given location [3].

The first strategy is long-term measurement of equivalent noise level during a sufficiently long period that would encompass all variations in the operation of the noise source, as well as all meteorological variations.

The second strategy is short-term measurement of equivalent noise level conducted under a specific operating mode of the noise source and under specific meteorological conditions.

Long-term measurement of equivalent noise level can be conducted using two measurement strategies—permanent noise monitoring or semi-permanent noise monitoring.

Permanent monitoring is the constant measurement of equivalent noise level at a given, fixed, measurement point, 24 h a day and 365 days a year, conducted by means of permanently installed noise monitoring stations. Permanent monitoring provides the actual insight into noise trends at the examined location and precise data for the calibration of strategic noise maps.

Semi-permanent monitoring, lasting continuously from several days to several weeks or months, is used for assessing annual noise indicators, economically tracking environmental noise trends, determining whether noise level limits are adhered to, informing the public and presenting the current issues, extending the knowledge about the relationship between population noise exposure and dose and about the related effects, and calibrating strategic noise maps. Semi-permanent monitoring requires noise monitoring stations that are easily moved from one location to another.

The relevant standards do not define the adequate measurement time intervals for semi-permanent noise monitoring or the number of measurements of equivalent noise level at the same location during a year that would yield representative sets of data for annual noise indicator assessment. The authors were thus compelled to establish the duration of semi-permanent measurements on the basis of experimental studies, while taking into account the various operating modes of noise sources, e.g. the structure and volume of traffic for road traffic investigation, in order to obtain a representative measurement time interval in case slight corrections were subsequently required. If the conditions of noise propagation or emission vary significantly during different seasons, e.g. due to the use of winter tires on vehicles or due to snowy roads in the winter, in order to obtain precise data, the researcher chooses a dynamic plan to assess when the measurements will be required during different seasons and selects measurement time intervals so that they encompass as many different meteorological conditions as possible.

9.2 Methodology

The initial assumption in this paper is that, based on the results of long-term permanent noise monitoring [4, 5], it is possible to determine the optimal duration of the measurement interval, which will then serve as the basis for assessing long-term annual noise indicators at the observed locations [5], burdened prevalently by road traffic noise. The assessment would be accurate to 1.5 dB with a precision of 1.0 dB.

The required accuracy implies that no value of any noise indicators measured by one of the strategies of semi-permanent monitoring should vary in the absolute amount by more than 1.5 dB from the real (actual) value of the noise indicator obtained through long-term permanent monitoring.

The required precision of 1.0 dB implies that the standard deviation value of the measurement results of any noise indicators measured using one of the strategies of semi-permanent monitoring should not exceed 1.0 dB.

The defined values of 1.5 dB for accuracy and 1.0 dB for precision, as quality parameters for the adopted solutions to the problem considered in the paper, originate from the requirements of the Directive 2002/49/EC that the resolution for the design of strategic noise maps should be 5 dB.

Solving the problem using Multi-criteria optimization (MCO) involves four basic stages [6]:

1. Problem formulation;
2. Determination of relative importance (weight) of the criteria;
3. Selection of an adequate MCO method for the solution of a specified problem;
4. Investigation of the stability of the solution.

Stage 1

In accordance with the defined research aim—finding the minimum measurement time interval that will meet the set criteria—the following alternatives have been formulated:

- **Alternative a_1**—Measurement strategy with a measurement time interval of one week (from Monday at 00:00 to Sunday at 24:00);
- **Alternative a_2**—Measurement strategy with a measurement time interval of one month (from 00:00 on the first day of the month to 24:00 on the last day of the month);
- **Alternative a_3**—Measurement strategy with a measurement time interval of six months;
- **Alternative a_4**—Measurement strategy with a measurement time interval of one year (from January 1 at 00:00 to December 31 at 24:00).

The preference towards certain alternatives according to the presented order ($a_1 \rightarrow a_2 \rightarrow a_3 \rightarrow a_4$) stems from the research aim: to use the shortest possible measurement time interval for a monitoring station to obtain satisfactory results for annual noise indicator values at a given measurement point, whose accurate value can be obtained only by measurement over a period of one year. The core of the problem is the fact that in the case of long-term permanent monitoring of noise over one year (alternative a_4) the monitoring station is located in the same place for the entire period, which significantly reduces its usability.

In order to select the optimal measurement strategy using MCO, a set of nine criteria is considered in the paper.

The selected criteria are divided into three categories, whereby criteria $f_1 \div f_4$ pertain to standard deviation of noise indicator measurement results, as the measure of average deviation of all data from the mean value within individual variant solutions, represented by different measurement time intervals.

The second category of criteria comprises criteria $f_5 \div f_8$, which represent the maximum deviation of noise indicator values for specific measurement time intervals from their annual value and their six-month value.

The third category of criteria comprises only the criterion f_9, which defines the degree of measuring equipment utilization. As opposed to all the other criteria, the criterion f_9 is represented by qualitative attributes due to its nature.

The manner of quantification of qualitative attributes is given in Table 9.1, whereby the Saaty scale was used to quantify the relationships between criterion pairs [7].

Stage 2

Since the defined criteria are not of equal importance, it is necessary to define the importance factors of specific criteria by using the appropriate weight coefficients or weights.

Relative criterion weights were determined using the *Delphi* method [8], through consideration of the criterion weights proposed by ten selected experts in this field (excluding the person who processed the data).

Stage 3

The optimal measurement strategy for a multi-criteria-defined problem can be selected using the PROMETHEE (Preference Ranking Organization Method for Enrichment of Evaluations) [9], as an acceptable method in terms of possible choices of how to determine the inter-comparison of alternatives in relation to the nature and values of specific criteria. This involves a selection of a suitable type of preference function and a determination of the values of indifference and preference

Table 9.1 Quantification of qualitative attributes according to criterion goal for the defined alternatives

	Criterion f_9	
	Criterion goal: *max*	
	Qualitative evaluation	Quantitative evaluation
a_1	Very high	9
a_2	Medium	4
a_3	Low	2
a_4	Poor	1

parameters, which, together with the weight coefficients of the criteria, have a key role in producing the final solution to the problem.

Using the PROMETHEE method, the measurement strategy is selected in five steps:

1. Definition of the type of general criterion and preference and indifference parameters for each individual criterion;
2. Determination of preference for each pair of alternatives according to each criterion seriatim;
3. Determination of preference indices for each pair of alternatives and creation of a preference index table;
4. Calculation of the values of input and output flow for each alternative and partial ranking of compared alternatives (PROMETHEE 1);
5. Calculation of the values of clean flows for all alternatives and complete ranking of alternatives (PROMETHEE 2).

9.3 Conclusions

The values of weight coefficients of set criteria, as the dominant factors for the given case of decision making, are the result of detailed analysis of the problem at hand, analysis of individual importance, and their comparison by a team of experts in the field, which significantly eliminated the researcher's subjectivity in the decision-making process.

The research results greatly depend on the selection of the type/form of preference function, since it indicates smaller, greater, or equal importance of one alternative over another according to a given criterion. The selected preference functions are the result of detailed analysis of criteria quantities, their values, and potential sensitivity of the function itself to slightest changes in the criteria values.

The third factor significantly influencing the outcome of the optimization is the definition of the values of indifference parameters q and preference parameters p, which, in the case of attempting to generalize their values for all measurement points with same or similar characteristics in terms of noise, represents perhaps the most demanding portion of the multi-criteria analysis performed in this research.

The only factors of measurement uncertainty for the experimental results are the uncertainty due to the measurement chain and the uncertainty due to non-stationary traffic load, which are represented by a standard deviation of the weekly values of noise indicators at specific measurement points.

The accuracy and precision of noise indicator values at the observed measurement points, obtained through one-week measurement interval, correspond to the initial hypothesis: up to 1.5 dB deviation of mean values of specific noise indicators obtained during semi-permanent monitoring from the actual values of those noise indicators obtained during permanent monitoring; and up to 1 dB standard deviation of the measured levels of specific noise indicators obtained during semi-permanent monitoring.

Acknowledgements This research is part of the project "Development of methodology and means for noise protection from urban areas" (No. TR-037020) and "Improvement of the monitoring system and the assessment of a long-term population exposure to pollutant substances in the environment using neural networks" (No. III-43014. The authors gratefully acknowledge the financial support of the Serbian Ministry for Education, Science and Technological Development for this work.

References

1. Directive 2002/49/EC of the European Parliament and the Council relating to the assessment and management of environmental noise. Official J. Eur. Commun. L 189 (2002)
2. S. Kephalopoulos, M. Paviotti, F.A. Lédée, Common noise assessment methods in Europe (CNOSSOS-EU). Reference Report by the Joint Research Centre of the European Commission. (Publications Office of the European Union, Luxembourg, 2012). doi:10.2788/31776
3. M. Praščević, D. Mihajlov, Noise indicators determination based on long-term measurements. Facta univ., Series: Working Living Environ. Prot. **11**(1), 1–11 (2014)
4. D. Mihajlov, Praščević M, Permanent and semi-permanent road traffic noise monitoring in the city of Nis (Serbia). J. Low Freq. Noise, Vibr. Act. Control. **34**(3), 251–268 (2015). doi:10.1260/0263-0923.34.3.251
5. M. Praščević, D. Mihajlov, D. Cvetković, Permanent and semi-permanent noise monitoring—first results in the city of Niš. in Proceedings of 24th International Conference on "Noise and Vibration" (Niš, 2014), pp. 33–40. ISBN: 978-86-6093-062-2
6. I. Nikolić, S. Borović, *Višekriterijumska optimizacija [Multi-criteria Optimization]* (Medija centar Odbrana, Beograd, 1996)
7. T.L. Saaty, *Analytic hierarchy process* (McGrawHill, New York, 1980)
8. C. Hsu, B. Sandford, The Delphi technique: making sense of consensus. Pract. Assess. Res. & Eval. **12**(10), 1–8 (2007)
9. Visual PROMETHEE 1.4 Manual. www.promethee-gaia.net

Chapter 10
On Acoustic Comfort in Urban Transport on Rails

Vasile Bacria, Eugen Ghita and Nicolae Herisanu

Abstract An important part of urban transport is performed using rail vehicles that, consequently, represent a major source of urban noise. This noise affects both people inside and outside the vehicle. In the paper we present the results obtained in investigating the acoustic comfort from rail vehicles in urban transportation system. In this respect, there are investigated the noise sources, the generated acoustic field, the noise level, the degree of discomfort, and some measures to increase the level of acoustic comfort are established.

10.1 Introduction

An important part of the transportation system in urban environment is based on rail transportation means, including tramways and subways. The rail vehicles during motion have several sources of noise and vibration that affect passengers and inhabitants from neighbourhood, producing acoustic discomfort. It is worthy to mention that extensive research on railway noise has been performed in the past, especially within the EU funded projects IMAGINE and HARMONOISE.

In this paper, we presented an investigation of the noise generated by rail transportation means by identifying and investigating the noise sources, the generated acoustic field, the levels and characteristic frequency spectra, and the degree of acoustic discomfort, establishing measures for ensuring acoustic comfort.

V. Bacria · E. Ghita · N. Herisanu (✉)
University Politehnica Timisoara, Timisoara, Romania
e-mail: nicolae.herisanu@upt.ro

V. Bacria
e-mail: vasile.bacria@upt.ro

E. Ghita
e-mail: eugen.ghita@upt.ro

N. Herisanu and V. Marinca (eds.), *Acoustics and Vibration of Mechanical Structures—AVMS-2017*, Springer Proceedings in Physics 198,
https://doi.org/10.1007/978-3-319-69823-6_10

10.2 Noise Sources to the Rail Transportation Means

Following the functioning of some parts of a tramway during its motion, two categories of noise and vibration sources were identified: exterior sources and interior sources. In this respect, the most important exterior noise and vibration sources are: wheels running on rails, electric motors drive, transmission gears, suspension systems, the air compressor, ventilation system, braking action, and air resistance.

The exterior noise is transmitted inside the tram either as airborne noise through the ventilation channels, various holes and leaks, and as structure-born noise, transmitted through the structure of the tram which vibrates.

As interior noise sources one can be identified the opening and closing of the doors, passengers' speech and motions, etc.

The noise and vibration generated by trams depends on their technical state, speed of motion, nature and quality of railway. Possible non-uniformities of the railway generate movements of rotation, roll and pitch movements, which causes both damages of some parts of the tram and a higher noise level.

The trams employed in Timisoara urban transport system under investigation are characterized by total levels of acoustic pressure ranging between 79 and 93 dB, evaluation being performed by measurements taken at tram passing with an average speed of 40–50 km/h at 7–10 m distance from the microphone [1]. Besides, within the frequency spectra predominate low and medium frequencies.

Analyzing the noise produced by rolling of wheels on the railway, it could be emphasized that all elements which are in contact during the motion have an important contribution, which is increased by an increasing speed.

The influence of the rolling way is dependent on the quality of its surface, irregularities of running surfaces, imperfect joints of the rails, roughness and unevenness, as well as the wavy wear of the rail. All these elements generate a mechanical noise. In this case, the main source of noise is the wheel-rail contact. Due to the wavy wear of the rail, there is produced noise having frequencies ranging between 70 and 1000 Hz.

It is also observed that once with increasing the speed of motion, the highest components from the noise spectra are moved toward the domain of high frequencies.

Concerning the noise generated by the electric drive motor in functioning, it is observed an overlap of aerodynamic noise, electromagnetic noise, and mechanical noise.

10.3 Investigation of the Acoustic Field

The noise generated by the sources mentioned above propagates inside and outside of tram, producing a discomfort for passengers as well as for inhabitants from the neighborhood.

The study of the acoustic field from inside of the tram could be performed considering the differential equation of acoustic waves written in Cartesian coordinates [1, 2, 3]:

$$\frac{\partial^2 \phi}{\partial x^2} + \frac{\partial^2 \phi}{\partial y^2} + \frac{\partial^2 \phi}{\partial z^2} = \frac{1}{c^2} \frac{\partial^2 \phi}{\partial z^2} \tag{10.1}$$

where $\phi = \phi(x, y, z, t)$ is the speed potential in a point of coordinates x, y, z, from inside of the tram considered as a closed space at a moment of time t.

Using the expressions

$$p = -\rho_0 \frac{d\phi}{dt}, \quad v_x = \frac{d\phi}{dx}, \quad v_y = \frac{d\phi}{dy}, \quad v_z = \frac{d\phi}{dz} \tag{10.2}$$

One can determinate the acoustic pressure and the components of the speed of a particle. Frequencies associated with the closed space from inside of the tram can be determined as [1]:

$$f = \frac{\omega}{2\pi} = \frac{c}{2} \sqrt{\left(\frac{n_x}{l_x}\right)^2 + \left(\frac{n_y}{l_y}\right)^2 + \left(\frac{n_z}{l_z}\right)^2} \tag{10.3}$$

Taking into account the multitude of values that n_x, n_y and n_z may have, it results that there exist an infinite number of vibration modes, their frequencies depending on the shape and size of the tram.

After the integration of (10.1), one can obtain, by means of (10.2), the acoustic pressure in a specific point. These considerations are useful for investigation of the acoustic field from the inside of the tram when the bounding surfaces are perfectly reflective from acoustical point of view, and at the impact with the incidence waves, the acoustical energy is not dissipating. However, in real conditions, the stationary acoustic waves formed in the closed space are damped, which implies the presence of an additional damping term $e^{-\delta t}$ in the expression of the acoustic pressure. Also the phase angles are significant in this situation since the bounding surfaces are not perfectly rigid. Therefore the study is very complex and it is worthy to be accompanied by experimental investigations.

10.4 Experimental Investigations

As mentioned above, inside the trams forms a very complex acoustic field, as a result of propagation of waves from different sources and multiple reflections [4]. Taking into account this complexity, an experimental approach has been resorted to. In this respect, measurements have been performed for the noise generated by trams from Timisoara public transportation system. These measurements were performed inside the tram during motion. Besides the equivalent noise levels, it was measured the maximum and minimum noise levels, the noise at starting, during the motion and as well as at stopping in the station. It was also measured the outdoor noise in the tram station at arrival and departure of the tram, as well as the noise generated during motion at 2 m distance from residential buildings. Measurements were performed according to [5, 6].

The measurements have covered the entire network of trams from Timisoara city, being investigated old Romanian trams and imported trams.

The equipment employed for measurements consisted on a Bruel & Kjaer noise analyzer 2250 which has been mounted inside the trams on the place of a passenger, at 1.20 m high and 0.6 m distance from the walls of trams. For outdoor measurements, the microphone has been mounted at 2.5 m distance from the tram and 1.3 m high from the ground. The main results are presented in Fig. 10.1.

All the results including all measured parameters were centralized in a database designed for managing the noise produced on the entire network of trams from Timisoara. Tables 10.1 and 10.2 contain a brief summary of the results. Table 10.1 presents the values obtained for the measurements performed inside the trams, while Table 10.2 presents the results obtained by outdoor measurements at arrival and departure of old Romanian trams from the station.

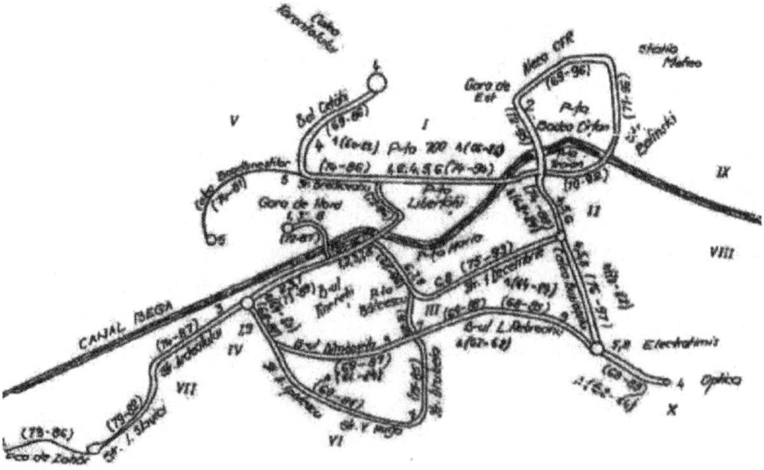

Fig. 10.1 Results of measurements on the Timisoara tram network

Table 10.1 Values of the noise inside the trams

Transportation means	L [dB(A)]		
	At starting	During motion	In the station
Old Romanian trams	85–102	68–96	60–85
Imported trams	60–88	60–88	51–70

Table 10.2 Values of the noise outside the tram

Position of tram	L [dB(A)]						
	Line 1	Line 2	Line 4	Line 6	Line 7	Line 8	Line 9
At arrival	73–81	70–74	70–78	65–73	70–74	72–79	68–78
At departure	75–85	73–82	68–72	72–75	64–69	70–74	65–74

It was observed that for the imported trams, the obtained values were 4–8 dB lower than those obtained for old Romanian trams.

Also, the noise level measured in the station during the stopping of the tram was 60–74 dB for old Romanian trams and 8–12 dB lower for the imported trams. In the same time, the noise level measured outdoor during the passing of old Romanian trans trams, was 75–80 dB at the pavement border, and 74–78 dB at 2 m distance from the walls of the residential buildings, while these levels were 5–9 dB lower for imported trams.

The investigation of the noise generated by trams included also a recording of time-history, a spectral analysis and a statistical distribution, inside and outside the tram. In Figs. 10.2, 10.3 and 10.4 is presented a set of these diagrams for a specific tram.

Following the obtained results, it can be concluded that the noise generated by trams produce acoustic discomfort for both passengers and inhabitants from residential buildings from neighborhood. Consequently, it is necessary to implement some measures intended to obtain acoustic comfort.

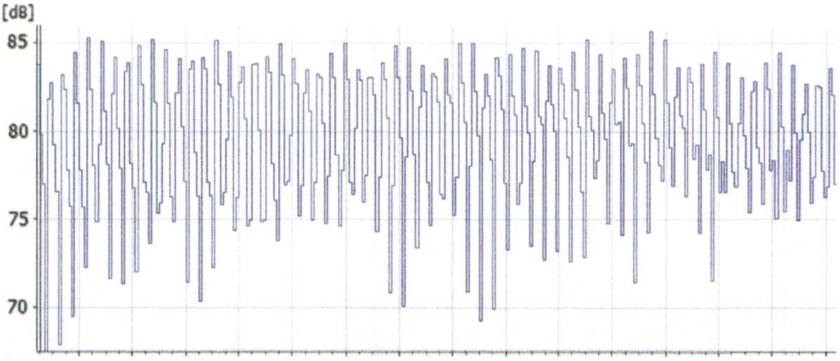

Fig. 10.2 Time-history of the noise inside an old Romanian tram during motion

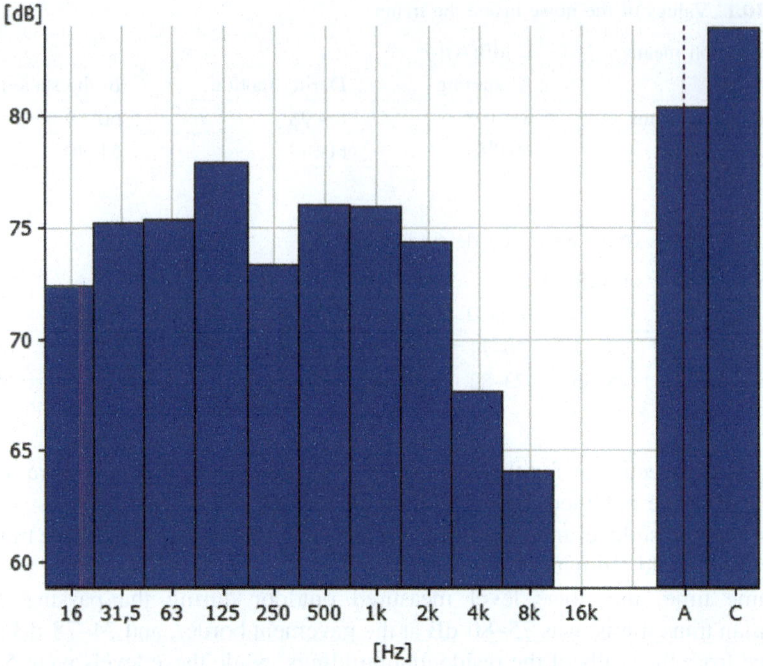

Fig. 10.3 Spectral distribution of the noise inside an old Romanian tram during motion

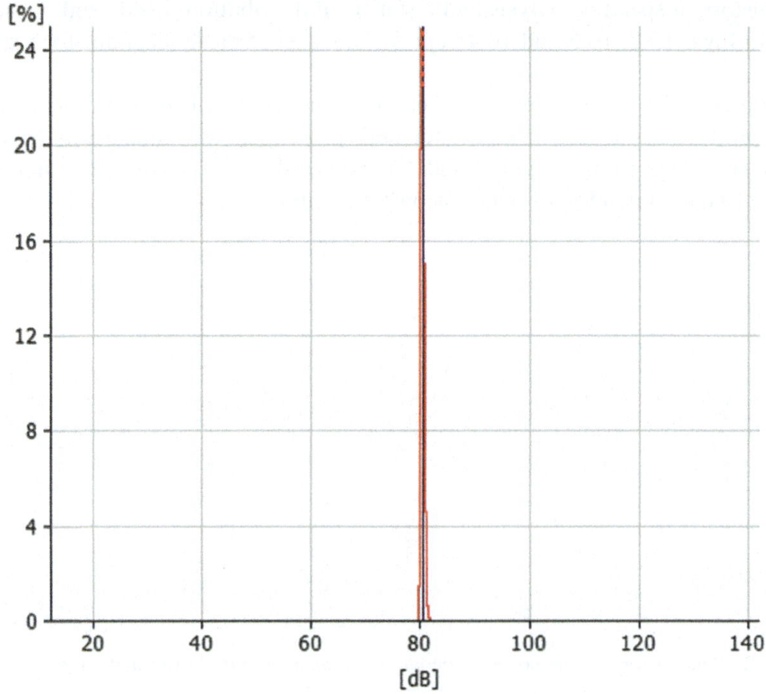

Fig. 10.4 Statistical distribution of the noise inside the tram during motion

10.5 Measures for Realizing the Acoustic Comfort

In case of trams, the solutions intended to realize the acoustic comfort and to mitigate the generated noise and vibration belong to two categories: measures which refers to the vehicle and measures related to the rolling way and the environment.

In order to realize a silent tram, efficient technical solutions should be applied against noise and vibration by improving the rolling wheels, the suspension system, the wheel drive system, the breaking system, and the structure of the tram.

A supplementary comfort could be ensured if the period of fundamental vibration of the structure is greater than the forced vibration generated by the irregularities of the rolling way, which is possible by using dual suspension bogies. Meandering movements, roll and pitch movements, should be avoided. Using conical gears with curved teeth, using rubber bands instead of metal and covering the bogies with absorbing layers could be very useful in improving the acoustic comfort.

In order to improve the acoustic comfort, inside the tram the walls should be covered by absorbent materials and all small holes must be sealed. Concerning the outdoor noise, acoustic arrangements specific to urban areas should be applied.

10.6 Conclusions

During the motion of transportation means on rails, it is important to realize the acoustic comfort of the passengers. In order to establish the conditions of acoustic comfort, it is necessary to develop experimental investigations. Following the identification of these conditions of comfort, some measures to improve the acoustic comfort could be recommended for improving the situation. These measures should be implemented and their efficiency should be verified by new measurements.

In this paper, an application is presented to the case of the tram network of Timisoara city. As an interesting conclusion of this study, it is found that once with increasing the speed, the highest components from the noise spectra are moved from low and middle frequencies toward the domain of high frequencies.

References

1. M. Grumazescu, A. Stan, N. Wegener, V. Marinescu, *Combating the noise and vibration in technics, Ed. Tehnica*, Bucharest, 1964
2. E.I. Iudin, *Isolation against noise, Ed. Tehnica*. (Bucharest, 1968)
3. M. Gafitanu, V. Merticaru, L. Bilborosch, *Noise and vibration, Ed. Junimea*. (Iasi, 1980)

4. M. Toader, L. Brindeu, V. Bacria, N. Herisanu, C. Stefan, Noise and vibration produces by trams, ICCA Conference, 2nd edn. (Timisoara, 2002), pp. 455–460
5. STAS 10009-88, Urban acoustics. Admissible limits of noise levels
6. ISO 1996-2:2007, Acoustics—description, measurement and assessment of environmental noise—Part 2: determination of environmental noise levels

Chapter 11
Application of Artificial Neural Network to Prediction of Traffic Noise Levels in the City of Niš, Serbia

Jelena Tomić, Nebojša Bogojević and Zlatan Šoškić

Abstract Road traffic noise is one of the main sources of environmental pollution in urban areas. Since the early 1950s many mathematical models for traffic noise prediction have been developed. Available models are usually based on regression analysis of experimental data. This paper presents the application of multilayer feedforward neural network in prediction of equivalent continuous level of road traffic noise in urban areas of the city of Niš, Serbia. Predictions of developed neural network are compared to data collected by traffic noise monitoring, as well as to predictions of commonly used traffic noise models. Obtained results show that the application of artificial neural network may improve accuracy of traffic noise prediction.

11.1 Introduction

Road traffic represents dominant source of noise pollution in urban environments. According to the latest data obtained from the strategic noise mapping for the EU agglomerations and major roads, close to 68 million people are exposed to daytime road traffic noise levels above the excess exposure threshold, fixed by the EU at 55 dB(A) [1]. In the past decades many studies [2–4] have shown that long term noise exposure significantly affects human health and productivity, causing stress, sleep disturbance, cardiovascular problems, etc. Moreover, the environmental noise pollution has a great impact on real estate prices [5].

J. Tomić (✉)
School of Electrical Engineering, University of Belgrade, Belgrade, Serbia
e-mail: tomic.j@mfkv.kg.ac.rs

J. Tomić · N. Bogojević · Z. Šoškić
Faculty of Mechanical and Civil Engineering in Kraljevo, University of Kragujevac, Kragujevac, Serbia
e-mail: bogojevic.n@mfkv.rs

Z. Šoškić
e-mail: soskic.z@mfkv.kg.ac.rs

N. Herisanu and V. Marinca (eds.), *Acoustics and Vibration of Mechanical Structures—AVMS-2017*, Springer Proceedings in Physics 198,
https://doi.org/10.1007/978-3-319-69823-6_11

In order to control road traffic noise, it is necessary to have a suitable calculation method for noise levels prediction. The calculation method can be of fundamental importance in the process of urban planning and designing, as well as for the traffic noise reduction through processes of traffic management. Software packages for noise prediction are numerous, but their price is usually high and their usage is highly complex. For that reason were developed mathematical models for traffic noise prediction based on the experimental results of noise level measurements. Since early 1950s, many authors have offered a large number of mathematical models, and most of the available models are based on linear regression analysis. On the other hand, several publications proposed use of "soft computing" methods, such as artificial neural networks (ANN) [6–8] or genetic algorithms [9, 10], for prediction of the traffic noise levels. The critical review of some of the most used models are given in [11], and one of the conclusions of this study was that each model is strongly influenced by certain characteristics of the traffic and the environment, such as pavement type, driving skills, road and vehicles maintenance, etc.

This paper presents a novel model for traffic noise prediction which reflects the peculiarities of the traffic flow at the territory of the city of Niš, Serbia. The model uses a neural network in order to establish a complex functional relationship between traffic flow parameters and A-weighted equivalent noise level (LAeq). The obtained results of traffic noise prediction are compared to experimental results of noise level measurements, as well as to predictions of several most commonly used traffic noise models.

11.2 Measurements

In order to develop a novel model for prediction of traffic noise levels in urban areas of the city of Niš, measurements of A-weighted equivalent noise levels were performed with Brüel & Kjær 2260 and 2250 sound level meters in 18 streets of this city. All measurements were carried out in dry weather conditions, without snow coverage, and with wind speeds lower than 5 m/s. The measurements were performed at distances 7–15 m from the axis of the road, and at height 1.5 m above the ground. All the measurement sites were at flat ground and above asphalt or concrete surface. Measurement positions were distant from the intersections and traffic-control lights, so it was assumed that all vehicles moved at a steady speed of 50 km/h, which is the speed limit for the territory of Niš. A total of 270 measurements were carried out, each with duration of 15 min. In addition, during each of the measurements, the traffic density data, consisting of the number of light motor vehicles N_c, medium trucks N_{mt}, heavy trucks N_{ht}, buses N_b and motorcycles N_m, were collected.

11.3 Neural Network Model

An artificial neural network (ANN) is a computational tool inspired by biological neural systems. It is a massive parallel distributed processor with the ability to learn and generalize. Therefore, ANNs are usually used to model complex functional relationships between input and output data, or to find patterns in data. A neural network consists of interconnected group of artificial neurons (nodes), organized into multiple layers: one input layer, one or more hidden layers, and one output layer. These nodes are connected to each other via synapses (also known as connections or links). The knowledge of the neural network is contained in the values of the synaptic weights.

Model of an ANN node is shown in Fig. 11.1. Node represents a processing element of artificial neural network. Node's input signals x_i, $i = 1, \ldots, m$, are processed by its integration and activation functions. The inputs and the weights are combined and summed by node's integration function:

$$f = \sum_{i=1}^{m} w_i x_i + \theta \tag{11.1}$$

where w_i denotes synaptic weight, while θ represents bias of the node. The obtained sum is passed through node's activation function. Commonly, the used activation functions are:

1. Step function: $a(f) = \begin{cases} 1, & f \geq 0 \\ 0, & f < 0 \end{cases}$
2. Hard limiter: $a(f) = \mathrm{sgn}(f) = \begin{cases} 1, & f \geq 0 \\ -1, & f < 0 \end{cases}$
3. Ramp function: $a(f) = \begin{cases} 1, & f > 1 \\ f, & 0 \leq f \leq 1 \\ 0, & f < 0 \end{cases}$
4. Unipolar sigmoid function: $a(f) = \frac{1}{1 + \exp(-\lambda f)}$

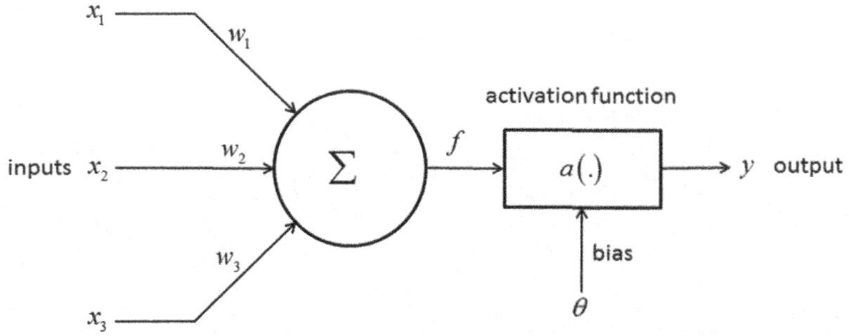

Fig. 11.1 ANN node model

5. Bipolar sigmoid function: $a(f) = \frac{2}{1+\exp(-\lambda f)} - 1$
6. Linear function: $a(f) = f$

The relationships between ANN input and output parameters are established in a network training process which involves tuning the values of the synaptic weights between the network nodes. Backpropagation algorithm is a commonly used method for ANN training process. This algorithm involves two phases:

1. propagation of the input data forward through the network and comparison of obtained network output to desired output;
2. calculation of the prediction error and its propagation backward through the network for the purpose of synaptic weights update.

A feedforward neural network is developed for the purpose of road traffic noise prediction for the territory of the city of Niš. Developed neural network estimates equivalent A-weighted sound level as its only output, based on the input data set which consists of the number of light motor vehicles, medium trucks, heavy trucks, buses and motorcycles during a 15-min period, and distance from the road central line. Since ANN with only one hidden layer is capable of approximating any arbitrary function [12], developed neural network has one hidden layer, which includes 7 nodes. Bipolar sigmoid function and linear function are used as transfer functions of the hidden layer and the output layer, respectively. The neural network topology is shown in Fig. 11.2.

During the neural network training process backpropagation algorithm updates weight and bias values according to Levenberg–Marquardt optimization. A total of 188 randomly selected data sets were used for network training, while 82 data sets were used for testing and validation of the developed neural network. Validation set is also used in order to prevent overfitting (i.e. poor generalization performance) by applying the concept of early stopping. Prediction error obtained for validation set is monitored during network training process, and when the error increases for a specified number of successive iterations, the training is stopped, and ANN weights and biases are updated to values which gave the minimum of the validation error.

Variations of the mean squared error (MSE) during the neural network training process are shown in Fig. 11.3. Minimum MSE for validation data set is achieved in the 20th epoch.

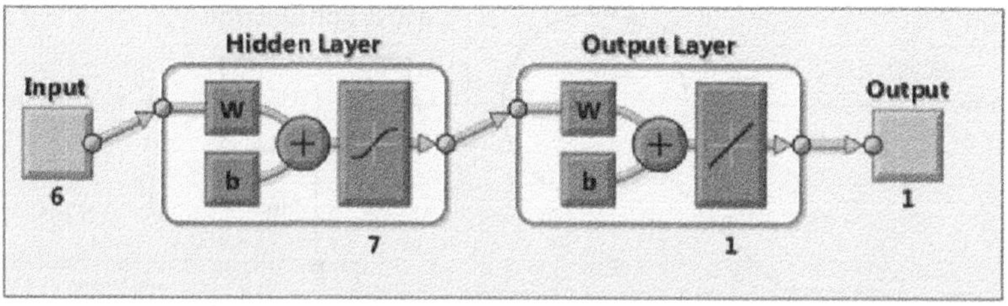

Fig. 11.2 ANN topology

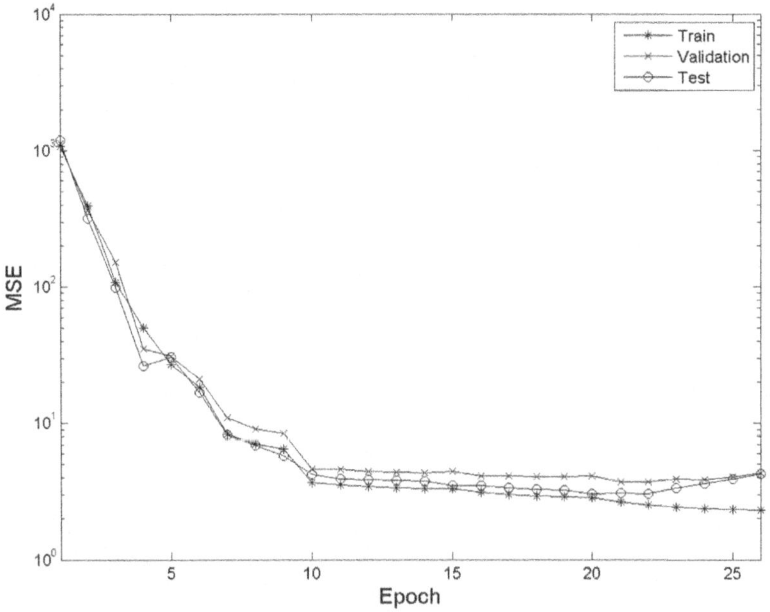

Fig. 11.3 Mean squared error during neural network training

Figure 11.4 shows the linear regression between the network outputs (predicted LAeq) and corresponding targets (measured LAeq).

11.4 Results

The results of traffic noise prediction of the developed ANN are compared to experimental results of noise level measurements, as well as to the results of traffic noise prediction obtained by the following models for traffic noise prediction: German (RLS 90) [13], Nordic Prediction Model [14], Italian model (denoted with CNR in the following text) [15], French Prediction Model [16]. Statistical analyses of differences between the measured and calculated noise levels were carried out for all 270 measurement data sets. The mean values of the absolute differences between noise levels ($\Delta\overline{L}$) and standard deviations of the differences (σ) were calculated. Further, the total number of predictions m with error larger than 3 dB was determined for each of the applied models.

The results of statistical analyses obtained for training (index "train") and testing ("test") data sets are separately presented in the Table 11.1. Figure 11.5 shows a comparative chart of measured and calculated noise levels. The results are shown for randomly selected 30 of 82 measurements used for the neural network testing.

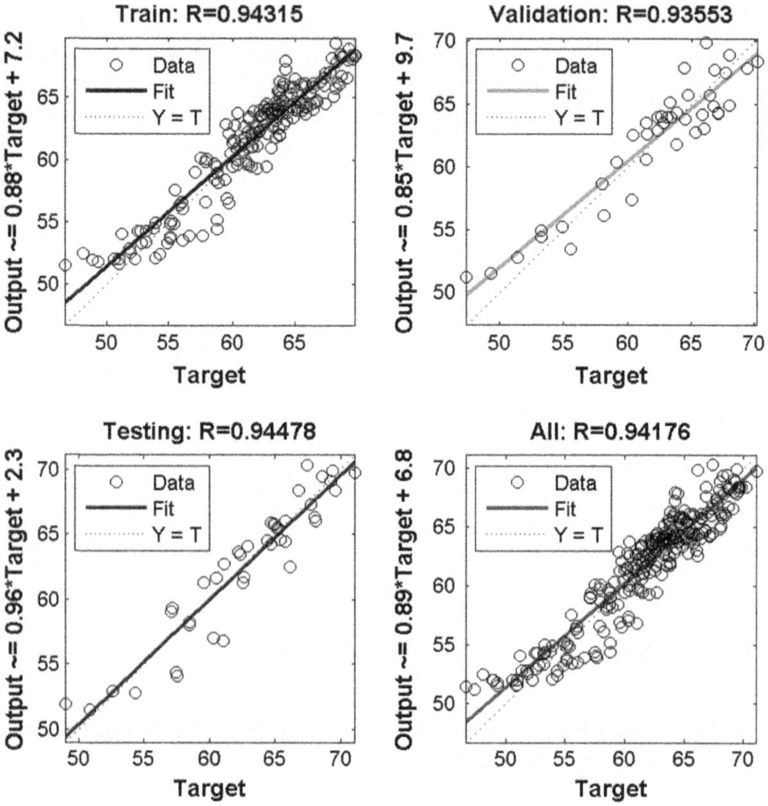

Fig. 11.4 Linear regression between the measured values ("target") and predictions of the ANN ("output"); "Y = N" line indicates an ideal ANN with outputs (Y) equal to targets (T)

Table 11.1 Comparison of different models for noise prediction

Model	$\Delta\overline{L}$			σ			m		
	Exp	Train	Test	Exp	Train	Test	Exp	Train	Test
French	2.64	2.66	2.59	1.43	1.40	1.50	116	78	38
CNR	4.44	4.49	4.33	1.96	1.90	2.09	210	149	61
Nordic	2.35	2.35	2.33	1.35	1.35	1.35	93	60	33
RLS 90	2.90	2.89	2.94	1.69	1.60	1.90	128	88	40
ANN	1.38	1.31	1.53	0.97	0.95	1.00	22	13	9

The statistical analysis shows that the application of developed neural network for traffic noise prediction presented in this paper leads to predictions of noise levels in the urban areas of the city of Niš that are more accurate than predictions of the other models of traffic noise. It is important to notice that the number of predictions with an error larger than 3 dB is significantly reduced.

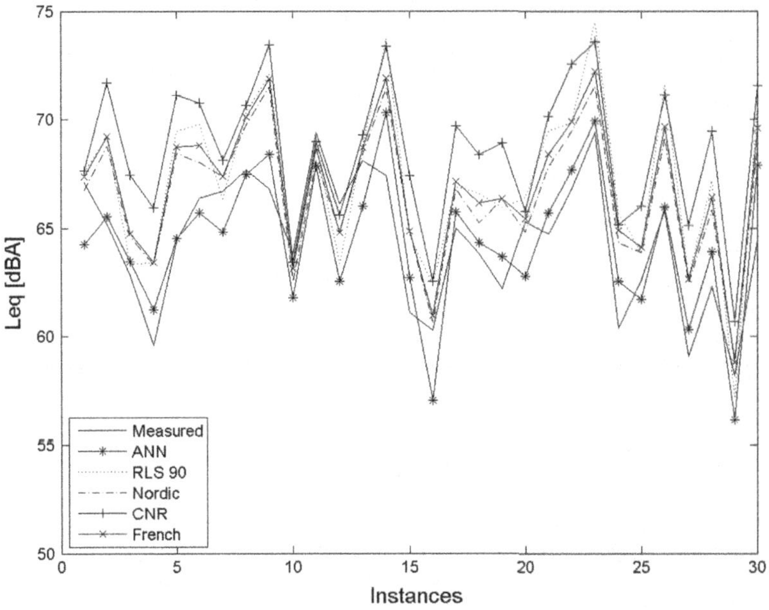

Fig. 11.5 Comparative chart of noise levels

The reason for the superior performance of the presented neural network is its ability to find a complex functional relationship between the input and the output data. However, the developed ANN is tuned to the specific conditions in the city of Niš.

11.5 Conclusions

Since the application of the extensively used mathematical models for traffic noise prediction does not provide satisfying results for the territory of the city of Niš, in order to develop more precise model, a total of 270 noise level measurements were performed in 18 streets of this city. The multilayer feedforward neural network for traffic noise prediction is developed and trained with experimental results of traffic noise monitoring.

The noise levels estimated by developed ANN were compared to experimental results, as well as to results of widely used mathematical models. The statistical analysis shows that the application of ANN enables not only better prediction of traffic noise levels, but also significantly lower number of predictions with error larger than 3 dB.

Developed neural network for traffic noise prediction may be used in the process of urban planning, designing of new traffic lanes, widening or moving the existing lanes, as well as for the noise prediction in the case of changes in traffic regime, or for the noise level reduction through the processes of traffic management.

Acknowledgements The authors wish to express their gratitude to Ministry for education, science and technology of Republic of Serbia for support through the research grant TR37020.

References

1. P. De Vos, G. Licitra, Noise maps in the European Union: an overview, in Noise Mapping in EU: Models and Procedures, ed. by G. Licitra, (CRC Press, USA, 2013), pp. 285–310
2. E. Ohrstrom, R. Rylander, Sleep disturbance by road traffic noise—a laboratory study on number of noise events. J. Sound Vib. **143**(1), 93–101 (1990)
3. A. Fyhri, R. Klaeboe, Road traffic noise, sensitivity, annoyance and self-reported health—a structural equation model exercise. Environ. Int. **35**(1), 91–97 (2009)
4. S. Pirreera, E.D. Valck, R. Cluydts, Nocturnal road traffic noise: a review on its assessment and consequences on sleep and health. Environ. Int. **36**(5), 492–498 (2010)
5. M.A. Theebe, Planes, trains, and automobiles: the impact of traffic noise on house prices. J. Real Estate Finance Econ. **28**(2), 209–234 (2004)
6. G. Cammarata, S. Cavalieri, A. Fichera, A neural network architecture for noise prediction. Neural Netw. **8**(6), 963–973 (1995)
7. N. Gearo, A. Torija, A. Ramos-Ridao, I. Requena, D.P. Ruiz, M. Zamorano, A neural network based model for urban noise prediction. J. Acoust. Soc. Am. **128**, 1738–1746 (2010)
8. S. Givargis, H. Karimi, A basic neural traffic noise prediction model for Tehran's roads. J. Environ. Manage. **91**, 2529–2534 (2010)
9. O. Gundogdu, M. Gokdag, F. Yuksel, A traffic noise prediction method based on vehicle composition using genetic algorithms. Appl. Acoust. **66**(7), 799–809 (2005)
10. S. Rahmani, S. Mousavi, M.J. Kamali, Modeling of road-traffic noise with the use of genetic algorithm. Appl. Soft Comput. **11**(1), 1008–1013 (2011)
11. J. Quartieri, N.E. Mastorakis, G. Iannone, C. Guarnaccia, S. D'Ambrosio, A. Troisi, T.L.L. Lenza, A review of traffic noise predictive models, in Proceedings of the 5th WSEAS International Conference on "Applied and Theoretical Mechanics" (MECHANICS'09), (Puerto De La Cruz, Canary Islands, Spain, December 14–16, 2009), pp. 72–80. ISBN: 978-960-474-140-3/ISSN: 1790-2769
12. B. Irie, S. Miyake, Capabilities of three-layered perceptrons, in Proceedings of the IEEE International Conference on Neural Networks (1988), pp. 1–641
13. Richtlinien für den Lärmschutz an Strassen (RLS 90) (Germany, 1990)
14. Nordic prediction method for road traffic noise (Statens Planverk 96, Nordic Countries, 1996)
15. G.B. Canelli, K. Gluck, S. Santoboni, A mathematical model for evaluation and prediction of the mean energy level of traffic noise in Italian towns. Acustica **53**, 31–36 (1983)
16. F. Besnard, J.F. Hamet, J. Lelong, E. Le Duc, V. Guizard, N. Furst, S. Doisy, G. Dutilleux, Road noise prediction. 1—Calculating sound emissions from road traffic (2009)

Chapter 12
Noise and Aggressiveness in the Workplace

Miodrag Milenović, Snežana Živković and Milan Veljković

Abstract In this paper, we look at the relationship between noise and aggressiveness at a production line at the Industry of Mechanical Engineering Niš. The study was carried out on a sample of 60 manual workers who were exposed to a noise level between 70 and 90 dB; the control group involved 60 administration workers, where noise levels did not exceed 55 dB. Noise levels were measured with a fonometer (Benetech GM 1351), which registers noise levels between 30 and 130 dB. Aggressiveness was measured using SIGMA questionnaire (which indicates the level of activity of the self-regulatory reaction to attack), and is an integral part of the Cybernetic Model of Personality Dimensions. Results showed there was a difference between the study and control groups (M exp = 61.30; M con = 51.44), which was statistically significant (t-test = 5.478, df = 98, $p < 0.01$); it is also noted that length of employment did not affect levels of aggressiveness. Results of this study support earlier finds that high noise levels not only affect employees' psycho-physiology, but they correlate to their increased readiness to react aggressively to everyday life situations.

12.1 Introduction

Noise is defined as unpleasant, undesired sound in an environment where people live and work, which can have an adverse impact on one's health. The impact on health of this pollutant has been long neglected, even though there are numerous ways to measure and calculate noise levels. Noise has a wide spectre of adverse

M. Milenović (✉)
Department of Psychology, Faculty of Philosophy, Niš, Serbia
e-mail: mbmilenovic@gmail.com

S. Živković · M. Veljković
Faculty of Occupational Safety, Niš, Serbia
e-mail: snezana.zivkovic@znrfak.ni.ac.rs

M. Veljković
e-mail: milan.veljkovic@znrfak.ni.ac.rs

N. Herisanu and V. Marinca (eds.), *Acoustics and Vibration of Mechanical Structures—AVMS-2017*, Springer Proceedings in Physics 198,
https://doi.org/10.1007/978-3-319-69823-6_12

health, social and economic impacts. Evolution has "programmed" humans to be aware of sounds which can warn of potential danger. Human body would react to noise in an appropriate way, e.g. fight or flee, resulting in nervous, endocrine and cardio changes that can have long-lasting consequences. Although man can get used to noise exposure, level of habituation differs between individuals and is rarely complete.

It is estimated that 2–8% of European population lives in areas where the level of daily noise exceeds 60 dB, depending on the country. We talk about general environment noise, noise at work and noise during social and recreational activities. Noise level ranges from everyday situations, such as leave dropping, which is very low at 10–20 dB; vacuum cleaner at 55–65 dB; proximity to main roads and freeways at 70–80 dB; to rock concerts with 100–110 dB [1].

12.2 Noise at Work

Workplace noise is the most researched type of noise, although the focus is moving to noise during social activities, such as bars and disco clubs, as well as general environmental noise. Despite existing standards for noise management and hearing protection, workplace noise reduction and public health activities, hearing loss caused by noise exposure is still a huge problem in developed countries.

In the USA, hearing loss caused by noise is the most common professional disability, with roughly 22 million workers being exposed to harmful noise levels [2]. According to epidemiological studies carried out in offices with noise levels higher than 55 dB, workers may experience uneasiness. Several studies showed that 35–40% of office workers feel edgy when noise reaches 55–60 dB [3]. In the UK, The Control of Noise at Work Regulations sets 80 dB as a limit, where ear protection must be available, and 85 dB where ear protection is compulsory. In the USA the limit is slightly higher at 90 dB [4].

It is a generally accepted view that noise level below 70 dB does not cause hearing damage, regardless of the length of exposure. It is also accepted that an 8-h exposure to noise of 85 dB is potentially harmful. At level higher than 85 dB, damage is in direct proportion to noise pressure and length of exposure [5].

As a result, workplaces where noise exceeds allowed levels have to assess potential harm caused by noise. The first indicator of potential excessive noise is when it is not possible to communicate at a distance of 1–2 m. If a subjective assessment cannot rule out a harmful impact of noise on hearing, then a total noise level needs to be measured, regardless of its spectral content. It is measured with a sound level meter.

The extent of damage caused by noise will also depend on its acoustic spectrum, i.e. whether noise is made of low frequency tones or more harmful middle and high frequency. For that reason, noise should be measured by octave. In measuring by octave, a sound level meter fitted with octave band filters, which have the ability the ability to split the audible spectrum into smaller bands, thus identifying the frequency content of the noise. We are talking about octaves with middle frequency (geometric median of extreme frequencies). If the noise level in any of the octaves exceeds allowed levels, this noise is harmful during a full-time shift [6].

12.3 Impact of Noise

As already mentioned, noise has a wide spectrum of adverse impact. It can generally be divided into auditory and non-auditory. Auditory impacts include hearing impairment caused by an auditory trauma and tinnitus, experienced as noises or ringing in the ears or head (when no such external physical noise is present) [7]. Non-auditory, on the other hand, includes a wide range of psycho-physiological stressful effects. The most common ones affect cardiovascular [8] and endocrine systems [9], sleeping [10], cognitive functions [11], motivation [12], and psychopathological disorders in particular [13]. Here, it is necessary to say that high noise levels are linked to subsyndromal conditions, such as psychiatric symptoms and anxieties, rather than specific syndromes, e.g. depressions or phobias [14].

12.4 Methods and Instruments

In this study, our aim was to establish whether there is a link between noise levels and increased aggressiveness in industrial workers. Aggressiveness is interpreted as a reactive modality (rather than instrumental modality characteristic of a psychopathic personality), where an increased activity of amygdala is present on a neural level, thus generating emotions which can be expressed as an aggressive behavior [15].

Consequently, aggressiveness was measured using SIGMA questionnaire from the Cybernetic Model of Personality Dimensions—KON6 [16]. SIGMA (a regulator of reaction to attack) is located in the limbic system; it modulates primary tonic arousal, based on the program for destructive reactions, formed during the ontogenetic development. This SIGMA model allows not only a direct activation of this program, when referring to the so called primary aggressiveness; it also allows a secondary activation, based on signals sent by the centre that regulates the reaction to attack, in situations when there is a secondary, or reactive, aggressiveness. Due to its importance for a necessary aggressive reaction, this model assumes a significant link between a regulator for reaction to attack and regulator of a person's general activity. Distortions of this regulator are manifested in various modulated aggressive reactions of a wide spectrum. Here are some examples of questionnaire statements: "I get enraged easily but it doesn't last"; "A person who does not enjoy life does not deserve to live". "I have to get what I want, no matter what", etc. In the questionnaire, the respondent answers a question by selecting one of out of five statements on a Likert scale, ranging from "fully agree" to "fully disagree".

The study intensity of noise level was measured with a fonometer (Benetech GM 1351), which registers a range between 30 and 130 dB. Noise was measured during various periods, i.e. at 10 min intervals during the second, fourth, and sixth working hour every day for the duration of 3 working weeks. The study was carried out at the Industry of Mechanical Engineering Niš, on 60 manufacturing and 60 administrative workers respectively.

12.5 Results and Discussion

During measuring intervals, the fonometer detected noise intensity in the range between 70 and 92 dB, or the average of 81 dB, in manufacturing plants, whilst their colleagues in administration did not experience a level higher than 55 dB. Results showed different SIGMA arithmetic medians for these groups: M exp = 61.30 and M con = 51.44 respectively. These differences are also statistically significant, t-test = 5.478 for df = 98, for an important level of significance of $p < 0.01$.

If we look at all available studies of physiological reactions to noise exposure, it is evident that they accept that noise affects the release of stress hormones, mainly cortisol, which plays a crucial role in the hypothalamus-pituitary gland-adrenaline activity axis, which is of the regulators of psychophysiological homeopath. Distortion of this homeopath due to noise has an impact on a whole spectrum of maladaptive body reactions [17, 18], such as queasiness, headache, mood changes [19]; reduced concentration, fatigue, irritability, psychomotor problems [3], etc.

The emphasis of our research was on the excessive impulsion amygdala receives from thalamus, the first subcortical instance reached by electrical impulse from the periphery of the body, i.e. the senses, including hearing. This impulsion triggers hypersensitivity and a set of negative emotions, rage in particular, which some studies was identified in as many as 50% of cases [20]. Hypersensitivity and negative emotions become a basis for reactive type of aggressiveness, especially in workplace, where there few opportunities for venting the rage [21], which is then most often manifested as negativity, edginess or low morale.

Outside work, reactive aggressiveness caused by noise "colours" a wide spectrum of emotions and behaviours, such as anxiety, argumentative behaviour, reduced frustration tolerance or reduced altruism [22, 23]. Depending on one's personality type and strategies for dealing with stress, a person can develop psychosomatic reactions and illness [24].

12.6 Conclusions

Noise as a stressor at work can have various negative effects. One of them is an increase of reactive aggressiveness, which was the topic of our study. We showed that a tendency for reactive aggressiveness increases with noise intensity, at least in cases of the 70–90 dB interval. These results support a view that in manufacturing plants with high noise levels workers must legally use adequate protection to avoid their negative impact, which is often not registered on a conscious level.

In a future study, it would be desirable to explore other problems related to noise and aggressiveness, such as personal experience of noise as a stress factor in relation to physical characteristics and noise exposure itself. Another topic to would be to study the link between reactive aggressiveness and frequency, which is another important noise parameter. Yet another topic could be to explore how the degree to which a person is able to control the source of noise is related to their level of sensitivity to noise; also, to which extent this personality trait is a result of physiological or environmental factors, and, in the first place, learning. A research in these and other questions related to non-auditory impact of noise would assist with lessening undesired consequences and would improve working conditions in manufacturing plants.

References

1. W. Passchier-Vermeer, W.F. Passchier, Noise exposure and public health. Environ. Health Persp. Suppl. **108**(1), 123–131 (2000)
2. M. Basner, W. Babisch, A. Davis, M. Brink, C. Clark, S. Janssen, S. Stansfeld, Auditory and non-auditory effect of noise on health. Lancet **383**, 1325–1332 (2014)
3. A.P. Smith, D.E. Broadbent, Non-auditory effects of noise at work: a review of the literature. HSE Contract Research Report No. 30, HMSO, London, 1992
4. H. Miadema, Noise and health: how does noise affect us? Proceedings of Inter-noise, Hague **1**, 3–20 (2001)
5. L. Goines, L. Hagler, Noise pollution: a modern plague. South. Med. J. **100**(3), 287–294 (2007)
6. D. Beritić-Stahuljak, E. Ţuškin, F. Valić, J. Mustajbegovič, Buka, in *Medicina rada* (Medicinska naklada, Zagreb, 1999)
7. V. Đukić, L. Janošević, N. Arsović, *Otorinolaringologija sa maksilofacijalnom hirurgijom* (Zavod za udžbenike i nastavna sredstva, Beograd, 2014)
8. T. Lang, C. Fouriaud, M.C. Jacquinet, Length of occupational noise exposure and blood pressure. Int. Arch. Occup. Environ. Health **63**, 369–372 (1992)
9. A. Cavatorta, M. Falzoi, A. Romanelli, Adrenal response in the pathogenesis of arterial hypertension in workers exposed to high noise levels. J. Hypertens. **5**, 463–466 (1987)
10. S.B. Venkateshiah, N.A. Collop, Sleep and sleep disorders in the hospital. Chest **141**(5), 1337–1345 (2012)
11. S.P. Banbury, W.J. Macken, S. Tremblay, D.M. Jones, Auditory distraction and short-term memory: phenomena and practical implications. Hum. Factors **43**(1), 12–29 (2001)
12. G.W. Evans, R. Stecker, Motivational consequences of environmental stress. J. Environ. Psychol. **24**(2), 143–165 (2004)

13. A. Tarnopolsky, J. Morton-Williams, Aircraft noise and prevalence of psychiatric disorders. Research Report, Social and Community Planning Research, London, 1980
14. S.A. Stansfeld, M.M. Haines, B. Berry, M. Burr, Reduction of road traffic noise and mental health: an intervention study. Noise Health **11**(44), 169–175 (2009)
15. J. Blair, D. Mitchell, K. Blair, *Psihopat* (Naklada Slap, Zagreb-Jastrebarsko, 2008)
16. K. Momirović, B. Wolf, Z. Džamonja, *KON 6-Kibernetička baterija konativnih testova* (Centar za primenjenu psihologiju, Beograd, 1992)
17. D. Prasher, Is there evidence that environmental noise is immunotoxic? Noise Health **11**(44), 151–155 (2009)
18. G.E. Miller, E. Chen, E.S. Zhou, If it goes up, must it come down? Chronic stress and the hypothalamic-pituitary-adrenocortical axis in humans. Psychol. Bull. **133**(1), 25–45 (2007)
19. M.A. Crook, F.J. Langdon, The effects of aircraft noise in schools around London Airport. J. Sound Vib. **34**, 221–232 (1974)
20. A.L. Bronzaft, E. Deignan, Y. Bat-Chava, N. Nadler, Intrusive community noises yield more complaints. Hear. Rehabil. Q. **25**, 16–22 (2000)
21. S. Melamed, T. Najenson, T. Luz, E. Jucha, M. Green, Noise annoyance, industrial noise exposure and psychological stress symptoms among male and female workers, in *Noise 88: Noise as a Public Health Problem*. Hearing, Communication, Sleep and Non-auditory Physiological Effects, vol. 2, ed. by B. Berglund (Swedish Council for Building Research, 1988), pp. 315–320
22. L. Standing, G. Stace, The effects of environmental noise on anxiety level. J. Gen. Psychol. **103**, 263–272 (1980)
23. R. Gregurek, M. Gozmi, *Buka i zdravlje* (Akademija medicinskih znanosti Hrvatske, Zagreb, 2005)
24. M. Milenović, S. Živković, M. Veljković, Noise and strategies for coping with stress, in Proceedings of the 11th International Conference Management and Safety, The Society of European Safety Engineers, Vrnjačka Banja, 2016, pp. 317–324

Chapter 13
Estimation, Assessment and Effects of Workers Exposure to Physical Noxae

Sorin Simion, Angelica Călămar, Daniel Pupăzan and Izabella Kovacs

Abstract Physical hazards (noise and vibration) are generally present in the working environment, a large number of workers from the European Union suffering from problems caused by workplace exposure to high levels of noise or vibration. The issue concerning exposure of workers to physical hazards also affects employers from the economic point of view, generating a significant decrease in working capacity. Low efficiency caused by exhaustion and overworking, mistakes performed in various operations caused by distractions are generated by exposure to noise and/or vibration noxae. Starting from theory to practice, the paper objectively highlights, using noise and vibration measurements, the comparison of results obtained with threshold values specified by in force legislation and their implications for the health of workers. The paper also analyses the ways in which noise and vibrations generated by technological processes influence workers activity, in order to estimate the level of exposure to occupational risk factors present during usual work hours and the identification of main diseases caused by them.

S. Simion (✉) · A. Călămar · D. Pupăzan · I. Kovacs
National Institute for Research and Development in Mine Safety and Protection to Explosion, INSEMEX Petroşani, 32-34 G-Ral Vasile Milea Street, 332047 Petroşani, Romania
e-mail: sorinsimion@insemex.ro

A. Călămar
e-mail: angelacalamar@insemex.ro

D. Pupăzan
e-mail: danielpupazan@insemex.ro

I. Kovacs
e-mail: izabellaeisler@insemex.ro

N. Herisanu and V. Marinca (eds.), *Acoustics and Vibration of Mechanical Structures—AVMS-2017*, Springer Proceedings in Physics 198,
https://doi.org/10.1007/978-3-319-69823-6_13

13.1 Introduction

Industrial development has led to increased work power and speed for equipment used in production, thus bringing a contribution to diversification and increased number of physical noxae sources and implicitly to increased number of exposed persons.

Noise and vibration are the most important industrial noxae that pose hazards to workers' health representing harmful factors arising from the deterioration and/or malfunction of work equipment [1].

Occupational risk refers to the likelihood of worker's injury or illness as a result of exposure to workplace hazards. Occupational risk assessment aims at ensuring worker safety, contributing to enhancement of prevention and protection measures quality [2]. Objective assessment of specific risks and quantification of level of noxae specific to each activity carried out by industry workers leads to an increase in occupational health and safety.

Improving working conditions in industry by applying the latest occupational health and safety technologies contributes to increasing the business's economic benefits (reducing the number of accidents and illnesses, namely payment of damages and medical care) as well society's benefits, as a whole. Occupational accidents and illnesses can cause serious financial losses for a business but also for society by occurrence of occupational illnesses that require the allocation of funds for their treatment and healing.

By harmonizing national legislation with European legislation, respectively by adopting health and safety at work directives and recommendations, the employer is solely responsible for the health and safety of workers. As the steps to be taken to ensure minimum health and safety at work conditions imposed by in force regulations, besides an organizational effort also require a sustained financial effort, knowledge of noxae levels respectively professional risk, leads to cost effectiveness of prevention measures taken to achieve a high level of security. The purpose of reducing exposure to physical noxae is to achieve optimal workplace comfort at a convenient price.

Exposure to physical noxae favours a decrease in working capacity through the emergence of a state of tension, a decrease in concentrating attention, in coordination of professional technical movements, in the ability to assess certain situations as a result of occupational stress. Voice masking effect, respectively masking of various sound and verbal signals during work processes leads to special, exhausting efforts concerning attention, lowering work efficiency.

13.2 Case Study

Within the research activities carried out by INCD INSEMEX Petrosani, a number of noxae measurements were carried out at the workplaces of economic traders with activity in the steel and extractive industries in accordance with the national legislation in force. In order to quantify the level of workers exposure, measurements of noise (according to GD no. 493/2006) and vibrations (according to the GD no. 1876/2005) levels were carried out in several workplaces considered typical, for establishing the technical and organizational measures to be taken (according to Law 319/2006).

13.2.1 Equipment Used

Bruel & Kjaer type 2250 integrative sonometers, Bruel & Kjaer type 4231 calibrators and the BZ 5503 software for viewing measurements were used for noise measurements. To ensure the quality of results, the sonometers and calibrators operating standards are checked in accordance to requirements of legislation in force. Acoustic pressure measurement to determine the A-weighted sound exposure level and/or equivalent continuous A-weighted sound pressure levels was carried out with the microphone placed in the position(s) usually occupied by the worker's head.

Measurement of vibration levels was performed by using a 4447 type vibration analyzer, type 4515-B-002 and type 4520-002 triaxial accelerometers, Bruker-Kjaer type 4294 calibrator, in accordance with in force legislation. Measurement of vibrations transmitted through the hand-arm system was performed for devices that had to be held with both hands, for each hand. The exposure reported was the highest value of the two. Assessment of the hand-arm vibration exposure level is based on the calculation of the daily exposure value for an 8-h reference period A (8), expressed as the mean square root (msr) (absolute value) of weighted acceleration frequency values, measured on the orthogonal axes a(hwx), a(hwy), a(hwz) [3].

13.2.2 Results and Discussions

Table 13.1 presents results of measurements performed at the main noise and vibration generating equipment used in the steel and extractive industries.

An estimate of the overall measurement uncertainty was included in reporting the results of measurements, taking into account the influence of factors such as:

- measuring apparatus;
- microphone/accelerometer location;

Table 13.1 Results of noise and vibration measurements

No.	Place of measurements	Measured noise level LEX, 8 h dB (A)	Exposure limit value LEX 8 h dB (A)	Exceeding of noise exposure limit values dB (A)	Measured vibration level (m/s^2)	Exposure limit value (m/s^2)	Exceeding of vibration exposure limit values (m/s^2)
Steel industry STEEL PALNT							
1	Electric-arc furnace platform	115	87	28	1.94	1.15	0.79
2	LF type steel refining plant	103	87	16	1.29	1.15	0.14
3	Bug re-bricking stand	97	87	10	5.54	5	0.54
4	Bug preparing shop	92	87	5	5.87	5	0.87
5	Distributor preparing shop	95\	87	8	1.59	1.15	0.44
6	MTF installation	98	87	11	5.75	5	0.75
Extractive industry SMU exploitation							
7	Hammer drill P 90	115.2	87	282	17.53	5	12.53
8	Pick hammer CA 14	99.6	87	12.6	9.63	5	4.63
9	Pick hammer CA 12	95.2	87	8.2	9.07	5	4.07
10	Pick hammer CA 10	93.5	87	6.5	8.39	5	3.39
11	Hammer drill PR8	90.7	87	3.7	12.65	5	7.65

- number of measurements;
- variation in time and space of noise and/or vibration source.

In accordance with HG 493/2006 on workers exposure to risks of noise pollution, exceeding of exposure levels have been noted, in the range of 3–28 dB (A). As shown in Table 13.1, the most noisy workplaces are also associated with high vibration levels, for example the electric-arc furnace platform (steel industry) and the perforation by using PR90 hammer drill (extractive industry).

Analysis of noise and vibration sources in technological processes in the iron and steel industries showed that most appropriate noise reduction solutions are:

- Implementation of active and reactive attenuators specific to equipments used;
- Installation of acoustic screens near the noise source and changing control elements;
- Limiting the exposure time and use of personal hearing protection;

To reduce the effects of noise and vibration on workers, without involving large financial costs, technical and organizational measures may be applied:

- Combating noise and vibration at source;
- Collective protective measures, including work management;
- Use of personal hearing protection.

After measuring and quantifying vibration and noise levels to which workers are exposed, measures may be applied to reduce occupational exposure to these noxae.

Avoiding and/or reducing exposure of workers to noise and vibration risks may be achieved by:

- choosing the right work equipment, ergonomically designed to produce vibration and noise exposure below the maximum limits allowed by in force legislation;
- provision of additional equipment that reduces the risk of injury caused by vibrations;
- appropriate maintenance programs for work equipment, workplace and workplace systems;
- adequate information and training of workers for the correct and safe use of work equipment to minimize exposure to vibration and noise;
- limiting duration and intensity of exposure;
- appropriate work schedules with adequate rest periods;
- provision of hearing protection equipment and clothing to protect exposed workers against cold and moisture (factors that aggravate the symptoms of vibration-related illnesses).

13.3 Effects of Exposure to Physical Noxae on the Worker and Health

Noise

Noise not only has a pathological action upon the auditory organ, causing temporary or permanent loss of hearing, but also has other harmful effects on the body in general. Exposure to high noise levels produces certain disorders such as increased heart rate, breathing rate, changes in blood pressure. Exposure to noise causes changes in the body's physiological balance. The general symptom is a feeling of tiredness, weakness; sometimes not even sleep or a few days of rest does not allow for a full recovery. Disorders can go as far as dizziness, syncope, permanent migraine and headaches, loss of appetite, profound weakness, and anemia.

Knowledge of the real harmful effect of noise exposure is limited by the fact that the investigation of short periods of exposure doesn't describe the real effects on the worker. Studying the long-term effects of noise on workers affected by noise exposure is hindered by the fact that most workers that start having auditory problems did not perform audiograms showing the route of exposure to noise [4].

Noises produced in the technological processes have numerous consequences on the human body, affecting the working and response capacity [5]. Depending on noise level, there are several categories of harmful effects on humans, namely: the masking effect, auditory fatigue, acoustic trauma, acute hearing loss; occupational deafness.

Vibrations

Vibrations that act upon the human body produce discomforts to physical and intellectual activity as a result of fatigue, mechanical deterioration of some body elements and subjective phenomena. The harmfulness of vibrations depends on their characteristics and conditions of transmission to the human body, depending on contact area with the vibrating object and exposure time. Most common lesions occur in bones, hand, shoulders and elbows joints. Mechanical damage may occur if accelerations are high enough and manifest by bone damage, lung damage, intestinal wall injuries, skull lesions, cardiac damage, peripheral vascular reactions (Raynaud's disease). Environmental factors also cause worsening of vibration diseases symptomatology, most important etiological favouring factors being high humidity and low temperature. The first signs of hand-arm vibration exposure are numbness and pain in upper limb phalanges, a disease called Raynaud's syndrome (white fingers disease) that translates into neurological and mental disparate or general syndromes, such as progressive muscle atrophies.

The purpose of reducing the effects of workers exposure to noise and vibration noxae is to increase safety and health at work. In order to achieve this goal, employers have at their disposal a wide range of applicable measures and normative acts that can improve working conditions, measures materialising into increased work capacity and improvement of workers health.

13.4 Conclusions

Following the analysis of noise and vibration levels in the extractive and steel industries, the following conclusions were derived:

- Risk assessment is a dynamic process (through refurbishment or reorganization of the technological process) allowing businesses and organizations to implement a proactive risk management policy at work. For these reasons, it is

important for businesses where the risk of exposure to noise and vibration is present, to carry out regular assessments of workplaces.
- The case study presented showed that in workplaces where there are high exceeding for noise noxae, exceeding of the maximum admissible vibration limits are also present, implying the need for identification and application of some technical—organizational measures that are effective for diminishing both noxae, in order to reduce workers exposure to both pollutants.
- Bringing forward some general technical and organizational measures to meet the employer needs is not enough; these measures need to be customized according to equipment, workplace, assessed risks, etc.; although exposure levels in extractive industry are similar in value to those in the steel industry, the sources of noise are different implying the need for noise reduction measures specific to each activity.
- Because of specific nature of the activities carried out in steel and extractive industries, all injurious conditions (noise, vibrations) for the worker can not be completely eliminated. The effects of these noxae can be reduced by using personal protective equipment, reducing exposure time, refurbishing, etc., to prevent occupational diseases that occur over time.

References

1. A. Călămar, M. Kovacs, L. Toth, Occupational health and safety regarding the exposure to noxious of workers from the steel industry. Metal. Int. **18**(2), 147–153 (2013)
2. G.D. Vasilescu, A. Draghici, C. Baciu, Methods for analysis and evaluation of occupational accidents and diseases risks. Environ. Eng. Manag. J. **7**(4), 443–446 (2008)
3. G.D. Vasilescu, E. Ghicioi, A. Draghici, N. Mija, Risk assessment of whole-body vibrations generated by industrial activities with environmental impact. Environ. Eng. Manag. J. **13**(6), 1453–1458 (2014)
4. A. Călămar, L. Toth, D. Pupazan, *Environmental Protection and Risk Factors Analysis* (Focus, Petrosani, 2016)
5. S. Simion, C. Vreme, M. Kovacs, et al., Exposure of workers to noise in mining industry, in *12th International Symposium Acoustics and Vibration of Mechanical Structures*, vol. 430 (2013), pp. 281–284

Chapter 14
Analysis of Acoustic Response and Treatment of a Rehabilitated Lecture Room

Diana Cotoros, Robert Cotoros and Anca Stanciu

Abstract Lecture room acoustics plays an utmost important role in the educational process, as all attendees should be able to listen in the best conditions to the information delivered by the teaching person. Many universities have old lecture rooms, some of them recently rehabilitated or renovated without taking into consideration the acoustic properties of the room. As many students sitting at the farther end of the room complain of problems regarding the lecture acoustics, the team selected one of the renovated lecture rooms in order to analyze the acoustic response in all locations and propose solutions for acoustic treatment.

14.1 Introduction

With the advent of technology and the possibilities of sound enhancement via sound reinforcement systems, the importance of room acoustics in teaching spaces has been more and more overlooked, as time went on, even though it is a key feature in the proper functioning of such a facility. Great emphasis must be placed on creating a vibrant and thriving environment for students to learn and be encouraged to learn further in. Therefore, it is important that every word that comes out of the lecturer's mouth should reach the ears of every single student in a clear fashion, in order to establish an efficient student-professor communication, in which most students can learn in a similar pace, without difficulty or confusion.

Therefore, this paper will approach all of the acoustic issues that could cause major problems in lecturer unintelligibility, including noise ambient levels,

D. Cotoros (✉) · R. Cotoros · A. Stanciu
Transilvania University of Brasov, Brasov, Romania
e-mail: dcotoros@unitbv.ro

R. Cotoros
e-mail: rcotoros94@gmail.com

A. Stanciu
e-mail: ancastanciu77@yahoo.com

N. Herisanu and V. Marinca (eds.), *Acoustics and Vibration of Mechanical Structures—AVMS-2017*, Springer Proceedings in Physics 198,
https://doi.org/10.1007/978-3-319-69823-6_14

reverberation time and room modes, ultimately calculating the RT60 and room modes of a seminar hall at Transilvania University and proposing several solutions to deal with or completely eliminate the problems that might occur.

14.2 Theoretical Aspects

Although it might not come as a big surprise, exterior ambient noise can be a large contributor to the overall sound quality of a teaching room. Whether it is an outside source, like the wind or people walking on the streets, or an inside source, like running computers, lights, vents and ducts, any surplus ambient noise, can easily interfere with the sound source that the students need to pay attention to and affect the clarity with which they understand their material.

Conventionally, the audio engineering community has elaborated a standard for such an issue, opting for a signal to noise ratio of +10 dB, in any given space or application [1].

Spaces through which external sound could leak into a room are usually poorly isolated areas, thin surfaces through which sound can easily pass, such as doors or windows. Therefore, it would be indicated to opt for quality double glazed windows, with a small amount dead space between the two sheets of glass. This would greatly improve the acoustic quality of any room, as not only would it keep less noise from coming in the room, but it would also stop noise from coming out. A similar strategy could be applied to teaching room's doors too.

Another issue concerns the reverberation time which is a fundamental building block in acoustics, as it is the one number that can reveal exactly what the characteristics of a given room will be. This number determines the role of a room, as it can dictate how intelligible speech will be in a certain room, and therefore, appropriate for studying, what frequencies will be boosted when listening to a sound source, or how suitable the hall is to be an amphitheater or concert hall [2, 3].

As a clear definition, reverberation time, or RT60 as it is conventionally named by acousticians, is the time it takes for a sound to decay to inaudible levels, or more specifically 60 dB, in a given room. This decaying phenomenon occurs because of the behavior of sound in closed spaces. With any surfaces, sound can either be absorbed or reflected. These behaviors are influenced by the absorption coefficient of the surface that the sound is hitting, which determines how much of the sound is absorbed into a material and how much is being reflected. The process of acoustically treating a room is, essentially, balancing out these ratios in accordance with the purpose of the room is and what established acousticians have decided upon as a convention.

Acousticians agree that the ideal reverberation time for speech and music is between 0.5 and 1 s. For larger spaces, a bigger reverberation time is required, as additional reflections help a voice travel farther. Therefore, buildings such as amphitheaters and even club music venues generally accept using a reverberation time between 1.5 and 2.5 s. Beyond 2.5 s, intelligibility starts to take a big hit and it is well indicated that no room, except for large halls and churches, should ever pass the 2.5 s mark.

As for the case of the seminar room in question, it is firmly believed that a reverberation time of 0.5–1 s would be most appropriate, as intelligible speech is the most important element. The room's measurements were taken and its length was of 13 m, its width was of 8 m and its height was of 3 m. To make sure that the RT60 goal was right, after calculating the room's volume it can be plugged into the Stephens and Bates formula for determining the ideal RT60 of a room. The equation is as follows:

$$\mathrm{RT60} = \mathrm{n}\left(0.012\mathrm{V}^{1/3} + 0.107\right) \tag{14.1}$$

In this context, the RT60 goal resulted in 0.62 s. This is the RT60 that should exist within the seminar room with the given dimensions. To get the full picture, the real RT60 of the room needs to be calculated, taking into consideration all of the materials in the room, as well as all of the chairs and desks that might absorb some of the reflected sound. The most often used formula for finding the real RT60 of a room was discovered by acoustics pioneer Wallace Clement Sabine, a formula which was referred to as the Sabine equation, leading to the following expression: $T = 0.161\ V/A$. In this expression, T is the RT60, V is the room volume in cubic meters and A is the equivalent absorption area in square meters. A is expressed as $A = \alpha \times S$, where α is the absorption coefficient and S is the surface of the absorbing area in square meters. The full A is obtained by calculating the sum of all the differing surfaces and their absorption coefficients as follows: $A = (\alpha 1 \times S1) + (\alpha 2 \times S2) + (\alpha 3 \times S3) + (\alpha 4 \times S4) + \ldots (\alpha n \times Sn)$. If $\alpha = 1$ then the surface is completely absorptive and if $\alpha = 0$ then the surface is completely reflective. Material and absorption coefficients for all the things in the seminar room were considered from [4, 5].

After plugging the material data, as well as the volume in the Sabine equation, the results provided a much more dramatic change to the original expectations. For each of the frequencies, the actual reverb time proved to be quite larger than the RT60 goal. In Fig. 14.1, there is a graph showing how the RT60 behaves in each of the main delimiting frequency bands.

The light blue line represents the RT60 goal, while the dark blue line represents the Sabine equation results. Most of the results are higher than expected, but there is an obvious resonating point at 250 Hz. It can be assumed that the material from which the walls are made of is to blame here, despite the reasonable size of the room, as concrete is known to be a highly reflective surface.

It is worth noting that no formula for reverberation time calculation is entirely accurate. Apart from the widespread Sabine equation, another very popular one is the Fitzroy equation, which is known to give dramatically different results.

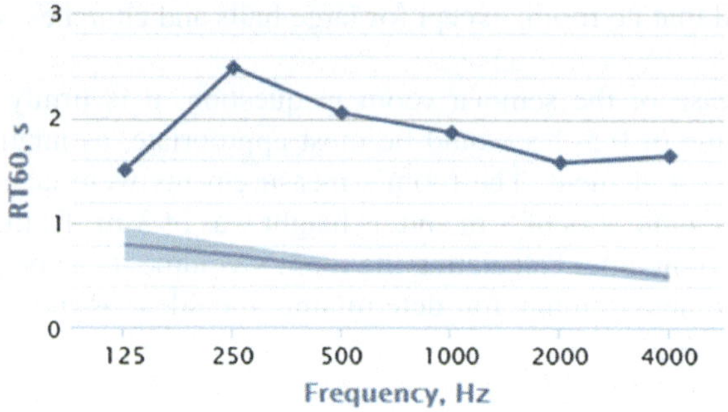

RT GOAL AND RESULTS

	125	250	500	1000	2000	4000
RT Goal, s	0.8	0.7	0.6	0.6	0.6	0.5
◆ Sabine, s	1.5	2.5	2.1	1.9	1.6	1.6

Fig. 14.1 Sabine equation results

14.3 Experimental Results

The only way to measure the RT60 of a room as accurately as possible is by going inside the room itself to measure it, which is what the team did, in order to verify these calculations. As the audio engineering domain developed at a quick rate, today it is possible to buy a simple reverb plug-in and use it to simulate and analyze an actual room, by using an impulse response. This type of reverb is called convolution reverb. Programs such as Logic's Space Designer and Audio Ease's Altiverb allow for the user to be able to put any audio track in the space of famous rooms and halls, such as the Royal Albert Hall, for example. Convolution also allowed for the existence of today's guitar amplifier simulators, which are used worldwide with great popularity at the moment. The same exact technique is what will be used here, to determine the RT60 of a teaching room.

Convolution reverb is done with the help of impulse responses. Impulse responses essentially contain all of the information with regards to a room's acoustical characteristics [6, 7]. Impulse responses are captured by recording how a room responds to a full range of frequencies, usually between 20 Hz and 20 kHz, which is the frequency range of the average human hearing. There are several techniques that can be used to record an impulse response. The measurements that were done here were finished by using a very popular method called the "frequency sweep" (see Fig. 14.2). This method involved using two high quality frequency measurement microphones (shown in Fig. 14.3), placed on either side of the room, in a spaced-pair pattern, with both capsules parallel to each other, while placing a large speaker on the opposite side of the room, in the middle, creating an equilateral triangle between the microphones and the speaker. The speaker will proceed to

Fig. 14.2 Experimental setup

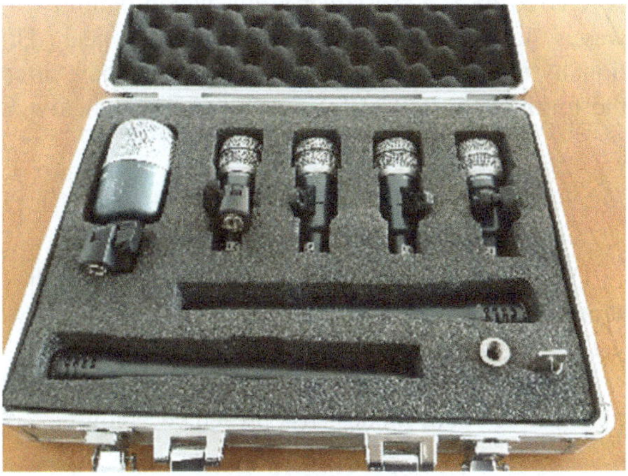

Fig. 14.3 Microphones kit

continuously play all of the frequencies which are audible to the human ear, in order, from 20 to 20,000 Hz. This sweep is then recorded and processed in a program called Space Designer, by Apple, which turns the entire sweep into a single transient impulse that can be used to create the convolution reverb. After analyzing the reverb from the seminar room, it was quite surprising to see that the results were almost identical to the Sabine equation.

It would seem that the seminar room requires some significant acoustic treatment. As large ceilings are known to be the main culprits of reflections, a good suggestion would be to cover the ceiling in Sonnex foam patches or even acoustic tiles.

14.4 Determining Room Modes

Apart from RT60, there is another element of acoustics which is elusive, yet very important to understand and correct, namely the room modes. A room mode, or standing wave, is created when the wavelength of a sound matches the dimensions of the room and the characteristics of how these room modes react is mostly based on the position of the sound source. Something which is especially common with loudspeakers, if a surface is highly reflective and a speaker pushed sound pressure into it, this pressure will bounce back. If the reflected sound arrives at the speaker just as it is about to do its second push, then the second push will be combined with the reflected sound and pushed back against the wall, amplifying the sound. This can build up and cause unwanted effects in the room. The standing wave can build up until the sound pressure level is equal to the amount of pressure that is absorbed by the walls in the room. Due to their length, lower frequencies tend to resonate stronger than higher frequencies. With standing waves, there is a frequency which is referred to as a crossover point, which delimits the frequency at which standing waves stop to occur and flutter echoes appear. This crossover point is conventionally referred to as the Schroeder Frequency, after the physician that discovered the formula for it. The formula can be seen below.

$$Schroeder = 2000\sqrt{\frac{RT_{60}}{V}} \tag{14.2}$$

Using this formula, it was discovered that the Schroeder frequency for the seminar room is 89 Hz. This means that standing waves will occur below 89 Hz, yet, after that, only flutter echo will appear.

The following formula is found on the sengpielaudio forums [4] and can be used to calculate room modes. It's presented below.

$$f = \sqrt{\frac{c}{2}\left(\frac{n_x}{L}\right)^2 + \left(\frac{n_y}{B}\right)^2 + \left(\frac{n_z}{H}\right)^2} \tag{14.3}$$

In this formula f is the frequency of the mode in Hz, c is the Speed of sound 343 m/s at 20 °C (68 °F) n_x is order of the mode of the room length, n_y is the order of the mode of the room width n_z is the order of the mode of the room height and L, B and H are the length, width, and height of the room in meters.

Using this formula within an online program called amroc, the room modes for the seminar room were calculated. It confirmed the Schroeder frequency, as well as show that standing waves will not cause huge problems in the room, as frequencies below 89 Hz are very unlikely to appear. By using a couple of well placed acoustic panels, the flutter echo that will occur can be broken up. In Fig. 14.4, there is an illustration from amroc for all of the room modes that occur on the frequency spectrum.

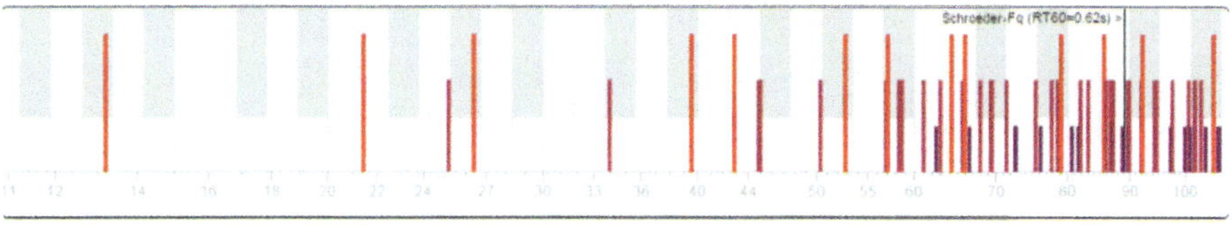

Fig. 14.4 Room modes

14.5 Conclusions

In conclusion, it's safe to say that there are teaching rooms in learning facilities that do not meet the acoustic requirements needed for a full learning experience for the students. Results have shown that ambient noise has little issue in the teaching room. The main problem that is affecting the room is the very reflective nature of concrete, which raises the RT60 more than it is needed. The over-reflective nature of these walls can cause the students to mishear words; certain voices might have an overall unpleasant quality to them, all of which contributes to a lesser experience in the teaching room.

However, this issue can be solved easily, by covering the ceiling either in Sonnex foam patches or even acoustic panels and placing a few well-thought out panels along the walls as well, to break flutter echo.

References

1. P. White, The sound reduction index, s.n.: s.i. downloaded from University of South Wales' Blackboard (2001)
2. Duke University|Department of Physics, Example of a Reverberation Time Calculation using the Sabine Equation. Available at: http://www.phy.duke.edu/~dtl/89S/sabineexample.html. Accessed 12 Mar 2016 (n.d.)
3. F. Alton Everest, *The Master Handbook of Acoustics*, 5th edn. (McGraw-Hill, New York, 2009)
4. Tontechnik-Rechner–sengpielaudio, Absorption Coefficients α of Building Materials and Finishes. Available at: http://www.sengpielaudio.com/calculator-RT60Coeff.htm. Accessed 12 Mar 2016 (n.d.)
5. JCW Acoustics, Absorption coefficients of common building materials and finishes. Available at: http://www.soundproofyourhome.com/absorption-coefficient-chart. Accessed 14 Mar 2016 (date not stated)
6. R. Nave, Reverberation time. Available at: http://hyperphysics.phy-astr.gsu.edu/hbase/acoustic/revtim.html. Accessed 14 Mar 2016
7. V. Nair, Recording impulse responses, designing sound. Available at: http://designingsound.org/2012/12/recording-impulse-responses/. Accessed 20 Apr 2016 (2012)

Chapter 15
An Appraisal of the Sound Field Generated by Helical Gear Pairs with Different Helix Angles, Manufactured by Various Technologies

Zoltan Korka, Lidia Filip, Bogdan Clavac and Aurel Bara

Abstract This paper aims to present the influence of the gear manufacturing technology on the sound pressure level (SPL) of a single stage helical gearbox. For this purpose, three gear pairs, having different helix angles (9°, 11° and, 13°, respectively) have been manufactured by three distinctive technologies. The first manufacturing alternative consisted of machining the gear teeth by milling with an accuracy grade 8, according to ISO 1328. The second alternative was designed to increase the quality class of the gears by grinding them with an accuracy grade 6 (according to ISO 1328), while the third manufacturing version involved applying a special fluoropolymer coating on the milled gear teeth. The gear pairs, which had different helix angles, and were manufactured by different technological variants, were tested, in various operating conditions, on an open energy test stand. The degree to which the different manufacturing technologies had an influence on the sound field generated by these helical gears was assessed, and the corresponding advantages and disadvantages of each manufacturing alternative were concluded. It was established that the best results in term of SPL were obtained by the gears which have been grinded, respective the gears with the special fluoropolymer coating. In fact, these two investigated manufacturing technologies lead to very similar results regarding the intensity of the generated sound field.

Z. Korka (✉) · L. Filip · B. Clavac · A. Bara
Department of Mechanics and Materials Engineering,
"Eftimie Murgu" University of Resita, Resita, Romania
e-mail: z.korka@uem.ro

L. Filip
e-mail: l.filip@uem.ro

B. Clavac
e-mail: b.clavac@uem.ro

A. Bara
e-mail: a.bara@uem.ro

N. Herisanu and V. Marinca (eds.), *Acoustics and Vibration of Mechanical Structures—AVMS-2017*, Springer Proceedings in Physics 198,
https://doi.org/10.1007/978-3-319-69823-6_15

15.1 Introduction

Besides high load capacity, failure-free operation and high efficiency, the main requirements for gear transmissions also include a low sound emission. With increasing environmental awareness, the sound field generated by gear systems, often referred to as noise, has become an important aspect in assessing the quality of the gear system. The limits for the generated sound field are established by existent standards, European regulations [1], or by the customers.

The mechanical components of a gearbox, including bearings, shafts and the gears themselves, are internally excited, mainly by the gear meshing, but also by unbalance or poor manufacturing accuracy, generating vibrations, which are then transmitted to the housing. The housing surface generates, in turn, pressure variations in the surrounding air, which are perceived as noise.

The internal excitation caused by the gear mesh process can typically be the result of following three factors:

- the transmission error, which is defined as "the difference between the actual position of the output gear and the position it would occupy if the gear drive were perfectly conjugate" [2];
- the elastic deflection under load, which causes mesh interference at the beginning and the end of the teeth engagement, also known as engagement impact;
- the variation of the mesh stiffness along the line of action.

As noise pollution produced by industrial equipment has lately become an important issue, the phenomenon of reducing the vibrations and noise in gear systems has received increased attention from research in the recent past years. Investigations focused on a wide range of aspects related to vibration and noise reduction, such as: influence of the transmission error [3], optimizing profile and tip relief of the gear teeth [4], application of special coatings on the gear teeth surface [5], or better gear manufacturing and finishing techniques [6].

15.2 Case Study

The study presented here is concerned with the influence of the gear teeth manufacturing technology on the sound field generated by helical gears having three different helix angles: 9°, 11° and, 13°, respectively. The main geometrical data of the examined gears pairs are shown in Table 15.1.

The first investigated technology consisted in manufacturing the gear teeth by milling with an accuracy grade 8, according to ISO 1328. Each of the manufactured gear pairs was measured on a Computer Numeric Control (CNC) gear tester of type PNC 150, produced by Klingenberg, Germany. The measurement results are shown in Table 15.2.

Table 15.1 Main geometrical data of the gear pairs

Geometrical data	Symbol [M.U.]	Value		
Centre distance	A [mm]	125		
Number of teeth	z_1/z_2	17/43		
Module	m_n [mm]	4		
Gear width	b [mm]	40		
Normal pressure angle	α_0 [°]	20		
Helix angle	β [°]	9	11	13
Addendum mod. coeff.	x_1/x_2	0.484/0.398	0.399/0.342	0.303/0.181
Refence diameters	d_1/d_2 [mm]	68.85/174.14	69.27/175.22	70.40/178.07

Table 15.2 Measurement results of the gears manufactured by milling

Gear type	Symbol [M.U.]	Pinion	Gear	Pinion	Gear	Pinion	Gear
Helix angle	β [°]	9		11		13	
Profile form error	f_f [μm]	19	20	20	22	20	19
Profile angle error	$f_{H\alpha}$ [μm]	11	14	4	12	14	14
Longitudinal form error	$f_{f\beta}$ [μm]	8	8	7	9	5	7
Total alignment error	F_β [μm]	25	10	24	9	25	15
Radial run out	F_R [μm]	38	12	29	36	30	35

Table 15.3 Measurement results of the gears manufactured by grinding

Gear type	Symbol [M.U.]	Pinion	Gear	Pinion	Gear	Pinion	Gear
Helix angle	β [°]	9		11		13	
Profile form error	f_f [μm]	4	2	3	3	4	2
Profile angle error	$f_{H\alpha}$ [μm]	2	1	2	3	2	2
Longitudinal form error	$f_{f\beta}$ [μm]	4	2	6	4	4	5
Total alignment error	F_β [μm]	6	6	9	5	7	6
Radial run out	F_R [μm]	16	10	18	12	16	16

The second manufacturing technology was designed to increase the quality class of the gears by grinding them with an accuracy grade 6, according to ISO 1328. The results of the gear profile measurements are shown in Table 15.3.

The third investigated manufacturing alternative consisted in applying two layers of Xylan 1052 coating (manufactured by Whitford Plastics Ltd. UK), each layer having a thickness of 12–20 μm. This fluoropolymer contains a mixture of MoS_2 (molybdenum disulfide) and PTFE (polytetrafluoroethylene), and was developed to aid dry film lubrication and anti-galling. The application technology is comprehensively described in [5].

15.3 Experimental Setup

This research was performed on an earlier developed test rig [5], which is shown in Fig. 15.1. The test stand is an open energy flow circuit, a hydraulic pump being used as a break. For the noise measurement, a Brüel & Kjaer microphone of type 4145 was placed above the gearbox housing.

The gear pairs, having different helix angles, and manufactured by different technological variants, were tested at an input speed of 1500 [rpm] and, in turn, at the following output loads: M_{t2} = 23, 26, 29, 32, 35 and 38 [Nm].

As a first step, the gearbox was equipped in turn with the three gear pairs having different helix angles and the teeth manufactured by milling. The noise measurement was performed for each teeth inclination, and for each of the six upper mentioned operating conditions, respectively.

In the next stage, the second gear manufacturing alternative has been tested. Thus, new pinions and gears, with the three different helix angles and the teeth manufactured by grinding have been mounted, one after the other, in the gear housing. Noise measurements were performed at the same operating conditions, for each helix angle.

For the last step, the gear pairs manufactured by milling were subjected to fluoropolymer coating procedure, consisting of two layers of Xylan 1052. Before starting the tests, a backlash measurement of the three gear pairs was performed. It was noted that the coating layers have decreased the gear backlash with values

Fig. 15.1 General view of the gear test rig

between 0.025 and 0.035 mm, which doesn't influence in a significant way the gear meshing conditions. Noise measurements were performed for the same operating conditions as for the previous stages.

15.4 Results and Discussion

Measurements of the sound pressure level (SPL) for the gear pairs operating in the before mentioned conditions were performed. The results are presented in Table 15.4 and Fig. 15.2.

Data analyses revealed that the highest noise emissions have been generated by the gears having the teeth manufactured by milling with accuracy grade 8, according to ISO 1328. Furthermore, the increase of the gear helix angle lowers the sound pressure levels for each of the investigated gear manufacturing technologies.

By comparison, the other two manufacturing technologies (grinding the gear teeth and applying two layers of Xylan coating), lead to very similar results, which refer to lower SPL levels.

As far as the advantages and disadvantages of these two technologies which resulted in lower noise levels are concerned, the fluoropolymer coating option presents, on the one hand, a good adhesion, an excellent resistance to chemical and environmental factors, even at operating temperatures up to 285 °C, is much cheaper and easy to use, compared to the teeth grinding. On the other hand, as specified by the manufacturer, the coating layers support a maximum contact stress of 343 N/mm^2 and thus it is applicable for noise reduction, only in transmissions where the upper mentioned contact stress is not achieved (i.e. in cinematic transmissions or for gear pumps). Therefore, even if it involves special technological facilities and higher manufacturing costs, teeth grinding represents the only applicable option for lowering the sound field generated by high loaded gear transmissions.

Table 15.4 SPL measurements on the milled gears (M), grinded gears (G) and coated gears (C)

SPL [dB]										
Helix angle β		9°			11°			13°		
Technology		M	G	C	M	G	C	M	G	C
Output torque M_{t2} [Nm]	23	76.44	71.95	72.25	75.27	72.63	72.74	74.17	70.94	71.15
	26	77.77	73.75	74.05	75.62	72.98	72.99	75.07	71.38	71.58
	29	78.71	75.05	75.31	76.57	73.55	73.71	75.85	72.15	72.25
	32	79.95	75.55	75.71	78.05	74.70	74.85	76.15	72.77	72.85
	35	80.40	76.75	76.80	78.35	74.80	74.95	76.50	73.52	73.85
	38	80.45	77.45	77.63	78.65	75.00	75.25	76.89	73.88	73.95

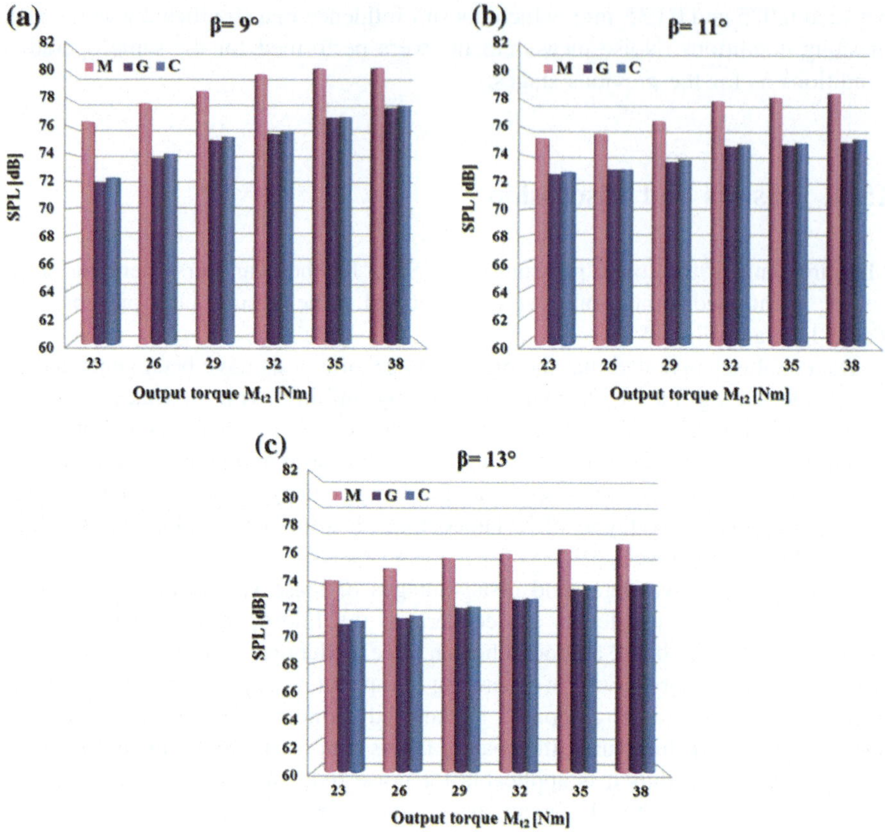

Fig. 15.2 Influence of the gear manufacturing technology, torque and helix angle on the noise of the gearbox

15.5 Conclusion

In this study, three different gear manufacturing technologies have been assessed, regarding the sound field generated by helical gears with different helix angles. Results show that the highest noise emissions have been generated by the gears having the teeth manufactured by milling with an accuracy grade 8, according to ISO 1328. The other two technologies, grinding the gear teeth in order to achieve an accuracy grade 6, and applying a special fluoropolymer coating on the gear teeth manufactured by milling respectively, lead to superior results in terms of sound pressure levels. Finally, the last two mentioned manufacturing technologies have shown very similar results regarding the intensity of the sound field generated by the helical gear pairs.

References

1. G.R. Gillich, N. Gillich, C.P. Chioncel, F. Cziple, Legal aspects concerning the evaluation of pollution effects due to vibrations in urban areas. J. Environ. Prot. Ecol. **9**(2), 465–473 (2008)
2. D.B. Welbourn, Fundamental knowledge of gear noise—a survey, in *Conference Proceedings of Noise and Vibrations of Engines and Transmissions*, Cranfield, UK (1979), pp. 9–14
3. S. Cho, J. Choi, J.H. Choi, S. Rhim, Numerical estimation of dynamic transmission error of gear by using quasi-flexible-body modeling method. J. Mech. Sci. Technol. **29**(7), 2713–2719 (2015)
4. S.S. Ghosh, G. Chakraborty, On optimal tooth profile modification for reduction of vibration and noise in spur gear pairs. Mech. Mach. Theory **105**(11), 145–163 (2016)
5. Z.I. Korka, G.R. Gillich, I.C. Mituletu, M. Tufoi, Gearboxes noise reduction by applying a fluoropolymer coating procedure. Environ. Eng. Manag. J. **14**(6), 1433–1440 (2015)
6. S. Jolivet, S. Mezghani, J. Isselin, M. El Mansori, Experimental and numerical study of tooth finishing processes contribution to gear noise. Tribol. Int. **102**(10), 436–443 (2016)

Chapter 16
Micro Turbo Engine JetCAT P80 Acoustic Evaluation

Grigore Cican, Marius Deaconu, Adina Toma and Adrian Gruzea

Abstract The purpose of this work is knowing the JetCAT P80 acoustic performances in order to know how to integrate him into the laboratory infrastructure and evaluate his noise impact on staff, students and on the entire campus, such that his exploitation can be performed without interfering with the other university courses. A series of noise measurements at different operating regimes were planned to evaluate and obtain the JetCAT P80 acoustic field. The paper will present inside and outside laboratory room noise measurements, correlated with JetCAT P80 speed in order to assess the staff and students noise exposure. Based on these measurements, the maximum running time will be calculated such as the exposure level to be kept under the legal noise limits. Taking into account the worst scenario (most loud operating regime), measurements results will be used to predict the noise contours in the campus area. The noise contours will be calculated using noise mapping commercial software and will help to schedule the faculty program. The obtained noise results are input for future experimental and numerical simulation work to reduce the jet noise by using different types of noise reduction gas dynamic method.

G. Cican (✉) · A. Gruzea
Faculty of Aerospace Engineering, Politehnica University of Bucharest, Bucharest, Romania
e-mail: grigore.cican@upb.ro

A. Gruzea
e-mail: gruzea.adrian@gmail.com

M. Deaconu · A. Toma
National Research and Development Institute for Gas Turbines COMOTI, Bucharest, Romania
e-mail: marius.deaconu@comoti.ro

A. Toma
e-mail: adina.toma@comoti.ro

N. Herisanu and V. Marinca (eds.), *Acoustics and Vibration of Mechanical Structures—AVMS-2017*, Springer Proceedings in Physics 198,
https://doi.org/10.1007/978-3-319-69823-6_16

16.1 Introduction

The micro turbo engines are generally used for Radio Comand models [1] but also for didactic purposes. The JetCat P80 is a micro turbo engine specially equipped with a test bench for laboratory usage [2, 3]. It is currently used by the Faculty of Aerospace Engineering from the Politehnica University of Bucharest for studies involving its gas-dynamic and aero-acoustic performance [4].

This study is the first step towards exploring and evaluating the acoustic characteristic of the micro turbo engine as well as investigating the effects, of the noise generated [5] in the university campus, on the students and on the teaching staff who are actively taking part in the laboratory experiments. To be more specific, the testing lab is located inside Faculty of Aerospace Engineering "Polizu" campus. The building hosting the JetCat P80 micro turbo engine consists of several other laboratories, various classrooms and professor's offices, and is surrounded in close vicinity by other university buildings containing more classrooms, laboratories and amphitheatres. This study is shaped to look into the micro turbo engine acoustic performance, to analyse the noise level in different places around "Polizu" campus, to calculate the noise contours for each individual building, to generate the campus acoustic map for various operating regimes, and to reveal the exposure time of the students/operating staff. The noise contours will be calculated using a commercial noise mapping software and will help schedule the faculty activity. The obtained results will be used as inputs for future experiments and numerical simulation work directed towards noise reduction using various gas-dynamics methods.

16.2 JetCAT P80 Acoustic Analysis

The bench test is presented in Fig. 16.1 and description is detailed in the US Patent, Jet Cat instruction manual [6].

The main objective of the acoustic measurements was to obtain the input data regarding the noise generated by JetCAT P80 (inside measurements) and also to validate the acoustic prediction results in campus area (outside measurements). Inside the laboratory, 3 random incidence 40AQ microphones were placed. Outside the laboratory, in various places on the campus (Fig. 16.2), 7 measuring points were placed, 6 free field 40AE microphones and 1 sound level meter type SOLO 01 dB. For noise measurements it was used the 01 dB ORCHESTRA multichannel acquisition system and the acoustic signals were processed with the dBFA software. The signals were recorded during the entire period of tests: start, idling, 89,298 rpm (77% from the Jet Cat P80 maximum speed recommended), and 109,830 rpm (96%).

Fig. 16.1 JetCatP80 test stand

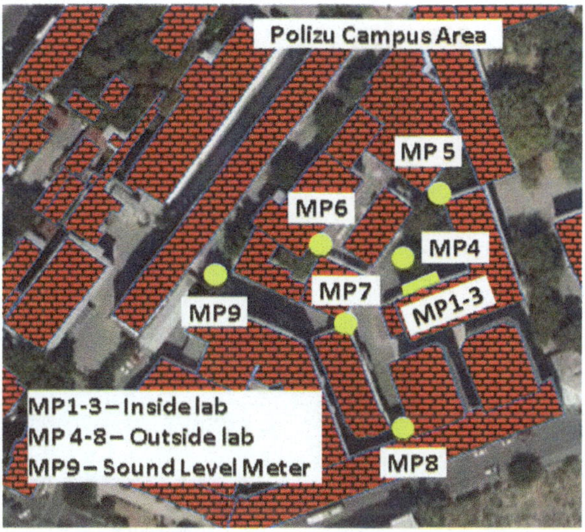

Fig. 16.2 Noise measuring points

Analysing the inside recorded values, the equivalent continuous sound levels were obtained for the entire operating period and for each stable regime, as is shown in Table 16.1. Also, a mean sound pressure level for the entire laboratory interior ($\overline{L_p}$) was calculated for each stable regime.

In order to eliminate the influence of the measurement duration, the sound exposure level was calculated and the values obtained for the entire functioning period are presented in Table 16.2. Another objective of the inside measurements was to determine the hourly exposure levels both for operating staff and for students. Taking into account the fact that one hour of laboratory it takes 50 min, the Lexp, 50 min was calculated and presented in Table 16.2.

Table 16.1 Interior sound levels measured for the entire period and for each stable operating regime in the three chosen position

Meas. point	Regime 1	Regime 2	Regime 3	Entire functioning period
	Idle	89,298 rpm	109,830 rpm	
	L_{Aeq} dB(A)			
MP1	104.3	107.4	110.4	103.3
MP2	106.7	110	112.9	105.8
MP3	105.5	108.6	112.3	104.8
$\overline{L_p}$	105.6	108.8	112.0	104.7

Table 16.2 Sound exposure levels inside the laboratory during entire operating period and hourly exposure levels

	MP1	MP2	MP3
SEL [dB(A)]	127.1	129.5	128.5
Lexp, 50 min [dB(A)]	92.2	94.7	93.7

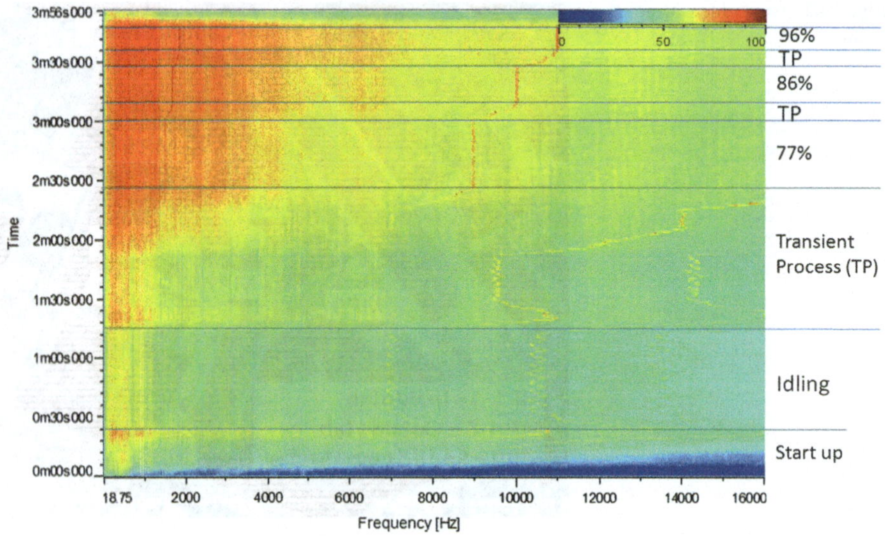

Fig. 16.3 FFT in time domain—point 4

In Fig. 16.3 is shown the results obtained in point 4 (near to the exit door of the laboratory) for Fast Fourier Transform Analysis (FFT) in time domain, and can be observed the spectral components of the micro engine speed, compressor, turbine and their variation in time. The measurement point 4 was chosen to be represented in such manner do the fact that is the nearest to the door, and the jet spectral component can be identified.

The outside equivalent sound levels are presented in Table 16.3 for the entire operating period of the turbo engine.

Table 16.3 Outside sound levels measured for the entire operating period

Points	P4	P5	P6	P7	P8	P9
L_{Aeq} dB (A)	101.7	75.5	76.0	70.1	66.3	60.8

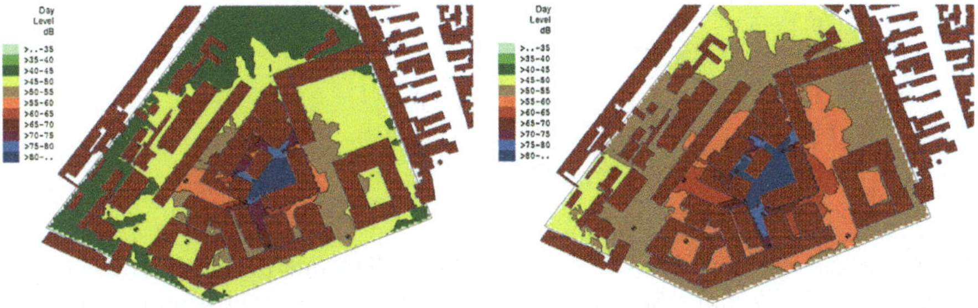

Fig. 16.4 Noise prediction −4 m height (regime 1, regime 3)

Table 16.4 Spectrum levels and overall sound level on each stable regime

Frequencies		16	31.5	63	125	250	500	1000	2000	4000	8000	Overall
Regime 1	L_p dB (A)	20.0	36.0	59.3	75.8	93.0	100.9	105.4	105.1	102.3	97.9	110.2
Regime 3		27.8	39.4	62.2	79.9	96.0	105.9	111.5	112.0	108.8	102.4	116.4

To take into account the acoustic attenuation given by the acoustic propagation in air on different frequencies, a frequency–dependent analysis was performed. These results will be taken into account in noise contours calculation step, were only the idling (regime 1) and the maximum regime (regime 3) are graphically represented (Fig. 16.4). So, in Table 16.4 are the 1/1 octave spectrum in point 4 and the overall sound level for each operating stable regime.

16.3 Prediction of JetCAT P80 Noise Contours

IMMI 2012 "IMMI-Noise Mapping and Noise Prediction Software" was used to realize the noise impact of JetCAT P80.

For the numerical model the recommendations from the national legislation Order 1830/2007 [7] based on the Directive 2002/49/EC were taken into account. As is specified in Order no. 864/2007, if there are no meteo data available, the average values for one year are considered −70% relative humidity and 10 °C for temperature. A special attention was given at buildings modelling, the attenuation due to reflection at the facades of the building was set at −1 dB.

In the model, the noise source was considered the area of the exit door (2.73 × 4.12 m) and it was model as a vertical area that generates noise.

Table 16.5 Comparative analysis measured-prediction

Meas point	Regime 1		Regime 2		Regime 3	
	Measured	Prediction	Measured	Prediction	Measured	Prediction
	L_{Aeq} dB(A)					
MP5	76.5	75.8	79.6	79	82.7	82
MP6	77.1	76.3	80	79.5	82.9	82.5
MP7	71	71	74.3	74.1	77.2	77.1
MP8	67.2	67.7	70.3	70.9	73.4	74
MP9	55.9	55.8	60.3	58.1	63.3	61.7

The virtual receiver points were modelled in the same positions as the microphones. In the next table the comparative sound pressure level from the measurement and the prediction are presented.

As can be seen from Table 16.5, there is a good agreement between the measured and predicted sound level, so the prediction can be declared as valid.

Accordingly with "Legislation regarding the acoustics in construction and urban area indicative C 125-2013" [8] the noise limits at the vicinity of the campus must be under 75 dB(A). Analyzing the values obtained in Fig. 16.4, we can conclude that even on maximum regime, this settlement is respected. Also for isolated actions characterized through high noise levels, a correction must be added depending on its duration, such that, considering a total of 25 min functioning time at maximum speed a time correction of −15 dB is imposed.

16.4 Conclusions and Future Work

As a result of this research we revealed the speed dependent acoustic characteristic of the engine, and more specific, the noise map for two operating regimes. High levels of noise may be observed inside the laboratory, so that occupational safety standards that require the use of hearing aids must be respected by both students and didactic staff.

Considering the low noise values predicted at the campus vicinity without the time correction of −15 dB to be applied, it can be concluded that the noise generated by the turbo engine does not raise noise issues.

Acknowledgements This work has been funded by University POLITEHNICA of Bucharest, through the "Excellence Research Grants" Program, UPB–GEX. Identifier: UPB–EXCELENŢĂ–2016 Research project title: "Study and research regarding aviation engines jet noise reduction using chevrons", Contract number: 81/26.09.2016.

References

1. K. Schereckling, *Gas Turbines for Model Aircraft* (Traplet Publications Limited, 2003)
2. W. Liou, C.H. Leong, Gas turbine engine testing education at Western Michigan University, in *45th AIAA Aerospace Sciences Meeting and Exhibit*, Reno, Nevada, 8–11 Jan 2007
3. C.H. Leong, J. Jacob, W. Liou, Development of a turbojet engine lab for propulsion education, in *40th AIAA/ASME/SAE/ASME Joint Propulsion Conference and Exhibit*, AIAA paper 2004-4085 (2004)
4. T.M. Raef, A. Elzahaby, S. Abdallahc, M.K. Khalil, S. Wagdy, Experimental and numerical investigations of noise from micro turbojet engine. Int. J. Sci. Eng. Res. **6**(8) (2015). ISSN 2229-5518
5. P.A. Savale, Effect of noise pollution on human being: its prevention and control. J. Environ. Res. Dev. **8**(4) (2014)
6. U.S. Patent 6216440, Jet Cat instruction manual (2008)
7. Romanian legislation, Order 1830/2007, Guidelines on the realization, analysis and evaluation of strategic noise maps (2007)
8. Romanian legislation, Indicative C 125-2013, Acoustics norm in construction and urban areas (2013)

Chapter 17
Procedures for Measuring and Modelling Underwater Acoustic Field Parameters

Delicia Arsene, Claudia Borda, Marinela Marinescu, Larisa Butu, Victor Popovici and Mihai Arsene

Abstract Considering the pollution acoustics increasingly acute underwater environment, it raises the need for an evaluation, structuring and mapping of underwater noise. It requires a harmonized methodological framework for data collection, field simulation of evolution acoustic characteristics for the modeling and plotting maps of underwater noise. This paper comes to make a contribution on procedures for measuring the characteristics of underwater acoustic field, by developing procedures for measuring and modeling the transmission loss and sound profile. Two scenarios are also presented for the production of underwater noise, that can be used for drawing maps of noise.

17.1 Introduction

In order to evaluate underwater acoustic environment pollution is required to issue action plans in this regard. These plans must be harmonized at European level. There has been progress in several directions: noise mapping and assessing noise pollution, action that is in an early stage and expansion; improving the ability to

D. Arsene (✉) · C. Borda · M. Marinescu · L. Butu · V. Popovici
University Politehnica of Bucharest, Bucharest, Romania
e-mail: delicia2008@yahoo.com

C. Borda
e-mail: ctistere@yahoo.com

M. Marinescu
e-mail: m_marinescuro@yahoo.com

L. Butu
e-mail: l_butu@yahoo.com

V. Popovici
e-mail: popovici_victor@yahoo.com

M. Arsene
SC AFICO SA, Bucharest, Romania
e-mail: arsenemihai@bksv.ro

N. Herisanu and V. Marinca (eds.), *Acoustics and Vibration of Mechanical Structures—AVMS-2017*, Springer Proceedings in Physics 198,
https://doi.org/10.1007/978-3-319-69823-6_17

compare strategic noise maps, including common indicators and the establishment of a database at EU level; identification of acute problems in some plans of action on the noise. The harmonized methodological framework will focus on setting noise maps and will consider specific issues, such as for example the data collection.

17.2 Measuring and Evaluating the Impact of Underwater Noise. Sound Field Program

The measurements described below were conducted in an investigation initiated to determine noise levels underwater ambient noise and sound transmission loss at a location in the Black Sea. Are presented below results of a sound field program. Program and measurements will be the subject of a database that will include acoustic field parameters, their variation, their modeling, useful too land that should be considered in evaluating the activities impacting on the marine environment.

17.2.1 Measurement of Transmission Loss

Acoustic properties of the environment can be characterized taking into account three parameters: source level (SL), the noise level received (RL) and transmission loss (TL). The source level of the sound characterize the power level near the source, before the signal suffer a loss of propagation. The measurements of these parameters must be specified reference distance used and the distance between source and receiver. Reference distance used for acoustic sources is 1 m. Transmission loss is the ratio of received sound level and source level.

Experimental background

In order to perform measurements was used a search vessel. It was recorded noise from the ship at anchor. On board the research vessel was installed a mobile laboratory for measurements of underwater acoustic field parameters.

Measurements were made in an experimental context that includes: a mobile sound recording system, consisting of a network of three hydrophones located at different depths, equipped with a float and ballast to keep the system at desired locations in a vertical position, research vessel, a motor boat to move, location/recovery the measurement system, mobile polygon displacement.

Locations implementations were numbered and measurement results were reported according to this numbering. Transmission loss measurements were made during several recordings. Measurements taken during implementation 5 of 19, provided the most comprehensive set of data on transmission loss compared to the range in which the measurements were carried out on a line between implementations locations 7 and 3. 5 implementation results have been compared with the modeling of the transmission loss on a line between implementations 7 and 3, with the source positioned at the real location [1].

Data collection

The recording system was sent with a motor boat-at 08:40:25, at the location 4:7° 55′ 50.640″N, 69° 31′ 30.180″W. Several narrow band tones with a duration of five seconds, with 1/3-octave band center frequencies ranging from 200 Hz to 2 kHz was passed through a hydrophone in water and the pressure was recorded sound location. Source hydrophone was positioned at a depth of 5 m and tones were sent to six different locations throughout the day (Table 17.1).

Data processing

Sound transmission consisted of tone duration of 5 s at each frequency listed, with source levels as shown in Table 17.2. Acoustic source levels reported at 1 m from the source to each center frequency are also shown in Table 17.2. Resulting sound pressure level as a function of frequency, which was used for comparison with modeling [2]. Comparison between transmission loss measurement results and modeling transmission loss will be analyzed further.

Source level was measured using a hydrophone placed near the transmitter and is shown in Fig. 17.1. Figures 17.2, 17.3 and 17.4 shows the modeled and measured transmission loss in band 1/3 octave for each transmission location.

Agreement between measured transmission loss in band 1/3 octave and the model is generally good, with typical differences of 2–5 dB, and a few exceptions. Levels measured and modeled broadband are plotted in Fig. 17.4 versus distance.

Table 17.1 Locations broadcast transmission loss measurements tones, for the first and the second implementation

Tone transmission	Transmission start time	Latitude N (°)	Longitude E (°)	Distance to the registration system (m)
1	10:24:20	44° 09′ 17.267″	28° 37′ 21.523″	734
2	16:18:00	44° 09′ 07.127″	28° 37′ 27.4″	632

Table 17.2 Center frequencies (summary) and source levels used in measurements of the transmission loss

Frequency (Hz)	Source level (dB re 1 μPa)	Frequency (Hz)	Source level (dB re 1 μPa)
200	131.4	631	157.4
316	141.5	1000	167.0
398	146.2	1584	158.9

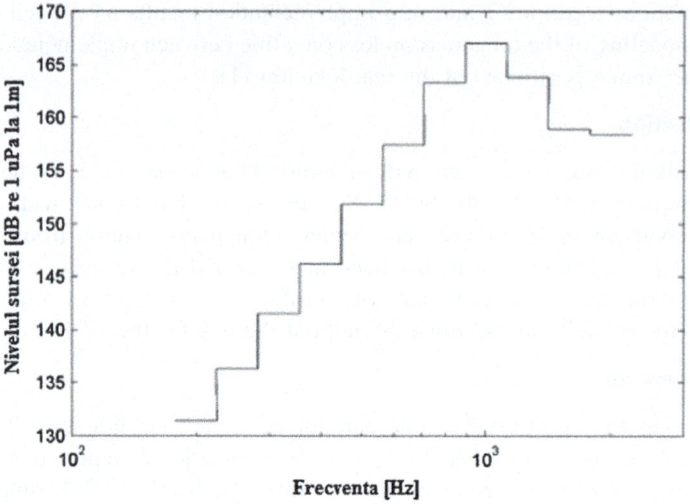

Fig. 17.1 Source level as a function of frequency for the transducer used in transmission loss experiments

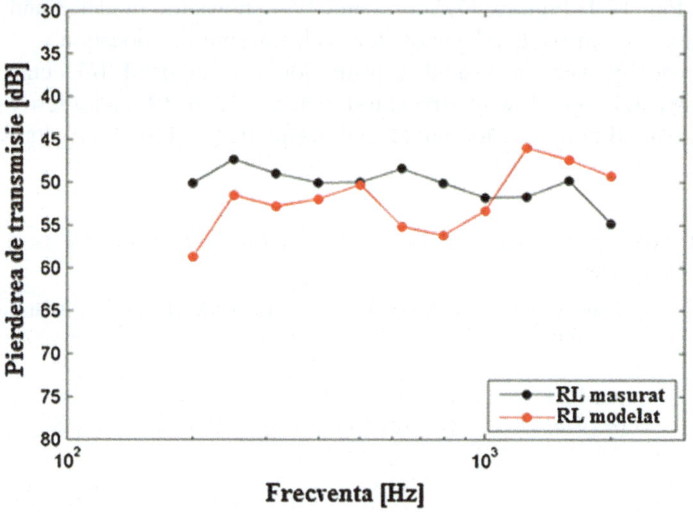

Fig. 17.2 Transmission loss (dB) according to the center frequencies of band 1/3 octave measurements and modeling of transmission 1 with source-receiver distance 834 m

Broad band levels are measured and modeled in agreement with a maximum difference of 3.5 dB, and a mean difference of 0.9 dB, suggesting that the model can predict received sound pressure levels in this region.

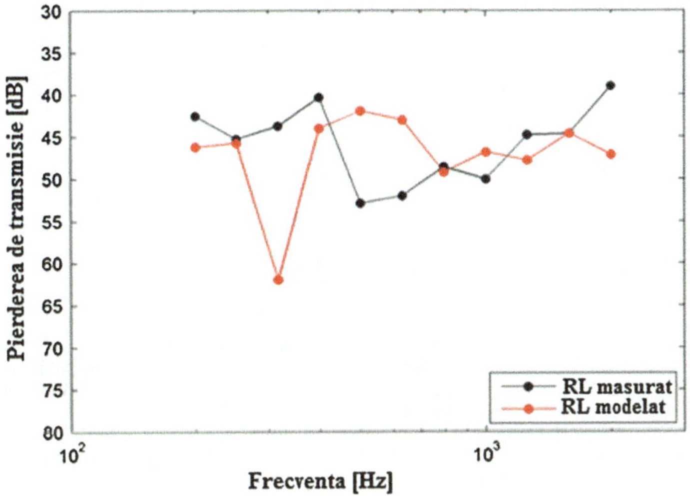

Fig. 17.3 Transmission loss (dB) according to the center frequencies of band 1/3 octave measurements and modeling of transmission 2 with source-receiver distance 532 m

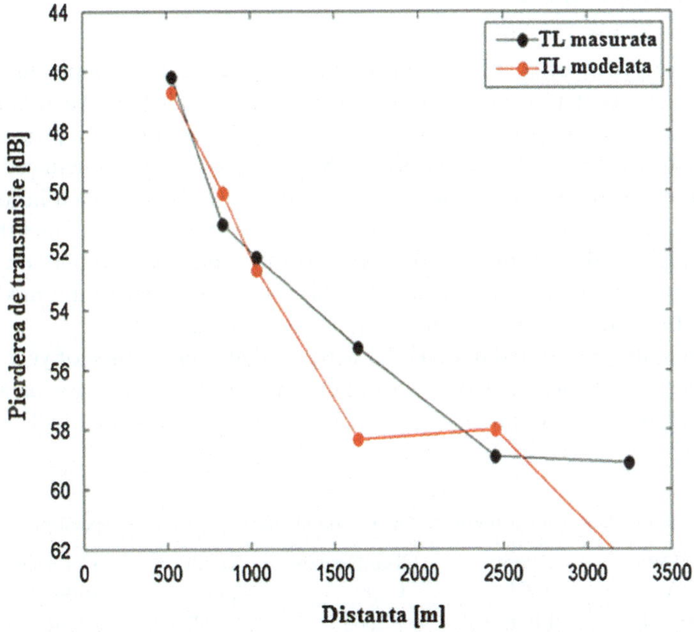

Fig. 17.4 Received broad band levels for measured and calculated transmission loss versus distance (170.7 dB re 1 μPa source level at 1 m)

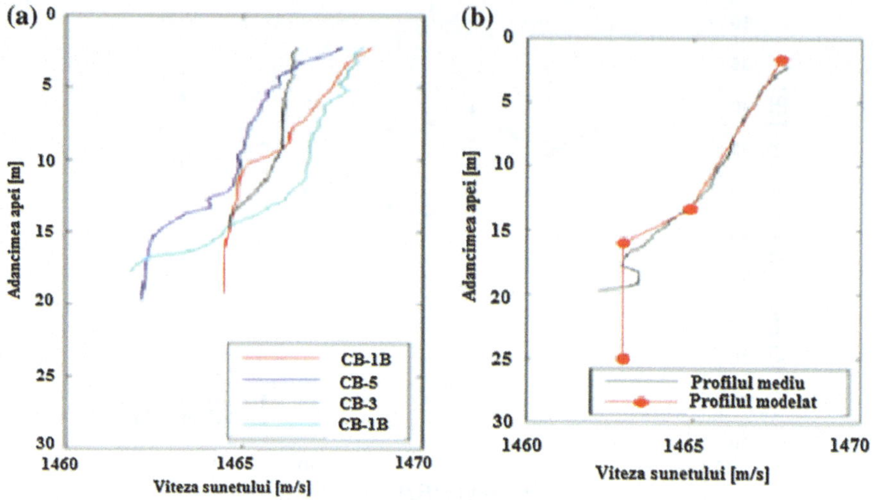

Fig. 17.5 The speed of sound according to depth from four CTD records, **a** average sound speed profiles; **b** approximate profile used as input for model

17.2.2 *Sound Speed*

Sound velocity was calculated based on the data from CTD equipment that measures conductivity, temperature and depth of water. CTDs have been launched to water in the area during transmission loss experiments. During the program in the field, CTD measurements were performed. Sound speed profiles calculated from the CTD data sets were examined and are presented below. This resulted in four sound velocity profiles shown in Fig. 17.5a, which were averaged to generate the profile shown in Fig. 17.5b. A representative profile of the velocity of sound, on the basis of the average sound velocity profile, has been used as input to the model of the sound field transmission loss scenarios presented in Fig. 17.5.

In order to develop sound field model will also consider the time of year you wish to evaluate the activity of acoustic location chosen. The necessary data plotting sound speed profile were purchased with CTD's and are presented in Table 17.3.

Table 17.3 Some data of measurements of the sound speed used for the speed plotting

No. crt.	Depth (m)	Celerity (m/s)	Temperature (°C)	Salinity (ppt)	Conductivity (ms/cm)
1	1.17	1490.53	17.065	15.242	21.74
3	5.07	1475.78	11.849	17.696	21.604
6	14.19	1453.4	6.003	17.836	18.612
7	16.24	1454.36	5.972	17.889	18.647
9	21.67	1454.33	6.160	17.964	18.816

The sound speed in seawater depends on the pressure (and hence the depth) temperature and salinity. Generally a change in temperature 1 °C produce a variation in the sound speed with 4 m/s and a change in salinity 1‰ bring a change in sound speed 1 m/s. As the temperature decreases with depth, while the pressure and salinity increases, sound speed profile depending on depth, presents a decreasing as depth increases. The curve shape calculated with the values that CTD purchased corresponds to with the resulting curve shape from the available data for the period of the year when measurements were carried.

The average levels of ambient noise broadband were determined from the measured data during periods of low noise, that during the noise flux was at a minimum and there were no identifiable sources of noise made by humans, such as maritime transport nearby.

17.3 Scenarios for Producing and Development of Underwater Acoustic Pollution

Developing scenarios pollution underwater acoustic environment is a necessary action to assess noise levels achievable during normal transit, and in the course of construction and operation of industrial projects which impact on the marine environment.

17.3.1 Scenario of the Transit Vessel for Transporting Containers

The scenario for transit of a container carrying vessel was modeled using sound speed data retrieved for late summer/early autumn, although the transit takes place throughout the year. Noise measurements performed on the carrier were carried out in the area entry into port. Planned operational procedures have included a tug escort, which was brought as a source of additional noise scenario modeling [3].

The source levels of the carrier, in 1/3 octave band, have been calculated using Eq. (1). Band source levels in band 1/3-octave for the two vessels are presented in Table 17.4.

$$SL = 163 + 10lgBD^4N^3/f^2[dBre\mu Pa] \qquad (1.)$$

The area inside each contour level and the average distance to each contour are presented in Table 17.5. Ship noise is classified as a continuous sound, so that the outline of 120 dB is relevant impact criteria, depending on limitations assumed by the Marine Directive (Marine Strategy Framework Directive of the EU) [4].

The sound level drops below 120 dB for intervals of greater than 1.5 km away from the ship and the area where noise levels can exceed 120 dB is 9 km^2.

Table 17.4 Levels of the source band 1/3 octaves for two ships

The carrier (calculated)			Tug (measured)
Broadband source levels (dB re 1 μPa la 1 m)			
168.6			174.4
The depth of the source (m)			
7			2,8
The center frequence (Hz)		Source level (dB re 1 μPa la 1 m)	
1	10	161.6	121.4
3	16	161.6	116.2
4	20	161.6	117.0
10	80	161.6	128.2
11	100	161.6	115.7
20	800	140.6	138.9
23	1600	135.5	132.2
24	2000	131.6	134.4

Table 17.5 Results of the dish carrier transit scenario

The level on the contour (dB)	The area inside (km^2)	The average distance (m)
110	46	3900
120	9	1500
130	2	700

17.3.2 The Docking Scenario of the Carrier Vessel

The scenario of carrier docking ship was modeled autumn, using sound speed profile measured in autumn. The carrier was modeled at the location emplacement, accompanied by four tugs (one at each corner of the docking area). The calculated levels of the source used in the transit scenario of the carrier were used as a source for the carrier level. Also, the tug source levels were used as a source level for the docking scenario, in accordance with Table 17.4.

The ship noise is classified as a continuous sound, so that the 120 dB outline is the relevant impact criterion. Sound level drops below 120 dB for ranges greater than 0.6 km away from the vessel and the area where noise levels can exceed 120 dB is 1.8 km^2, Table 17.6.

Table 17.6 The results of the carrier docking scenario

Contour level (dB)	Inside area (km^2)	The average distance (m)
110	5	1300
120	1.8	600

17.4 Conclusions

Measurement and modeling of transmission loss is a required activity in the preparation of noise maps. Since the construction, industrial and mining in the Black Sea are becoming increasingly important as a contribution to ambient noise requires an accurate assessment of anthropogenic noise emission. This paper makes a contribution in this area by: development of an experimental context for transmission and reception of sounds under water, development of a procedure to establish concrete steps to measure transmission loss in marine; a contribution to drawing up of hydro-acoustic parameters database. Between noise level and noise modeled received was a difference between 0.9 and 3.5 dB. This difference is acceptable because an error of 3 dB has no a major influence on marine life. Level of disturbance of marine life to be over 80 dB is accepted. Mean levels of environmental noise broadband ranged between 89.6 and 101.6 dB re 1 μPa into the studied area and 104.0 and 108.7 dB re 1 μPa at the entry into port.

Transmission loss measurements were performed to estimate the potential impact of underwater noise through two scenarios as follows: docking of the carrier vessel.

The average distance ranges to which the criterion of 120 dB is applied, which is used to evaluate the impact of the transit carrier and of the carrier docking scenario, were as follows: 1300, 1600 and 600 m from the source, respectively.

Underwater acoustic noise modeling is a useful tool to determine if noise management measures should be applied in the area and witch is the mode of putting them into practice.

References

1. M. Arsene, Studii teoretice şi experimentale privind impactul undelor acustice asupra mediului marin (Teza de Doctorat, UPB, 2013)
2. M. Arsene, V. Popovici, D. Arsene, C. Borda, M. Marinescu, L. Butu, Transmission loss measurement in the black sea, in *The Annual Symposium of the Institute of Solid Mechanics*, SISOM, and Sess. of the Com. of Acoustics, Buc, May 2012. ISSN 1843-5459
3. V. Popovici, D. Arsene, C. Borda, M. Marinescu, L. Butu, M. Arsene, Virtual laboratory for underwater noise attenuation, in *TEHNOMUS XVII, 17th International Conference on New Manufacturing Technology*, Suceava, Romania, 17–18 May 2013, pp. 375–378. E-ISSN-224-6016
4. M. ARSENE, Study for reducing noise pollution of the marine environment by introducing noise at the acoustic source. U.P.B. Scientific Buleltin, Series C, Iss. 4 (2014), pp. 33–48

Part III
Biomechanics and Bioacoustics

Chapter 18
Experimental Human Walking and Virtual Simulation of Rehabilitation on Plane and Inclined Treadmill

Daniela Tarnita, Ionut Geonea and Alin Petcu

Abstract The paper presents the results of the authors concerning the experimental human walking and numerical simulation of human rehabilitation on a treadmill. Using Biometrics data acquisition system based on electrogoniometers, experimental measurements for ankle, knee and hip joints of right and left legs during walking on plane and inclined treadmill are performed. The human legs motion assistance for rehabilitation is proposed with an attached exoskeleton. The numerical simulation of a virtual mannequin walking with the attached exoskeleton on a plane and inclined treadmill is performed, using ADAMS virtual environment. A comparison between human experimental measurements and numerical simulations of a virtual mannequin with exoskeleton is presented.

18.1 Introduction

The human gait analysis and rehabilitation devices are well studied in the literature. Also the systems and sensors for human gait experimental analysis is presented and used by researchers in their experimental studies [1, 2]. The recent development in this field is represented by the wearable sensors [3]. According to studied rehabilitation literature, passive and active rehabilitation devices are developed [4].

Active rehabilitation devices can be grouped in five categories, according to the rehabilitation principle [4]: treadmill gait trainers, foot-plate-based gait trainers, stationary gait trainers, over ground gait trainers, stationary gait trainers and ankle rehabilitation systems, as presented in Fig. 18.1. A part of the existing rehabilitation systems is intended to a single joint rehabilitation, like ankle joint orthoses and knee

D. Tarnita · I. Geonea (✉) · A. Petcu
Faculty of Mechanics, University of Craiova, Craiova, Romania
e-mail: igeonea@yahoo.com

D. Tarnita
e-mail: tarnita.daniela@gmail.com

A. Petcu
e-mail: petcu.alin.i@gmail.com

N. Herisanu and V. Marinca (eds.), *Acoustics and Vibration of Mechanical Structures—AVMS-2017*, Springer Proceedings in Physics 198,
https://doi.org/10.1007/978-3-319-69823-6_18

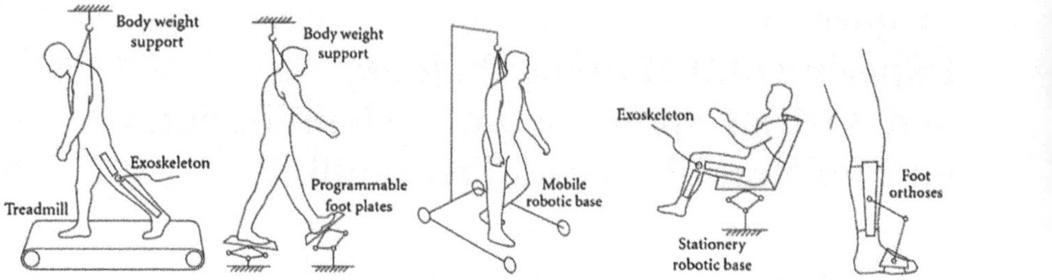

Fig. 18.1 Robotic system types for lower-limb rehabilitation

joint active orthoses [5]. Also are developed rehabilitation solutions to assist the entire leg joints. These solutions use motors for each joint and are expensive and difficult to implement on a large scale [6]. In the last years, cost-effective solutions of rehabilitation devices are developed, but only few of them assure anthropomorphic motion for leg joints [7, 8].

In the first part of present study, human motion characterization on plane and inclined treadmill using Biometrics Data Acquisition System is presented. In the second part, numerical simulation of human gait rehabilitation using an exoskeleton device for motion assistance is presented. The results obtained for human experimental walking and by numerical simulation in ADAMS environment are compared.

18.2 Human Gait Experimental Study on Plane and Inclined Treadmill

In addition to normal gait, one of the daily activities of a person is the walking on planes with different inclinations, which can cause a big difficulty for people with various lower limb disorders. The human gait experimental study is performed using the Biometrics Ltd. Data Acquisition system based on wearable electrogoniometers [9, 10]. The wearable Biometrics system allows capturing information during human gait and analyzing biomechanical data outside the laboratory [11–13]. For data acquisition two 8-channels DataLOG devices, at a frequency of 500 Hz, were used [14]. In Fig. 18.2a the block schema for data collection and processing in Biometrics is presented, while, in Fig. 18.2b the subject with mounted electrogoniometers and a DataLog device is shown.

Three tests are performed at a speed of 2.5 km/h, during 30 s each: (1) walking on plane treadmill (TM); (2) walking on a 3° inclined TM; (3) walking on a 6° inclined TM. During the three tests, knee and hip joint data was collected (flexion-extension angles (F-E) in sagittal plane and lateral rotation (L-R) angles in frontal plane) for the right and left lower limbs.

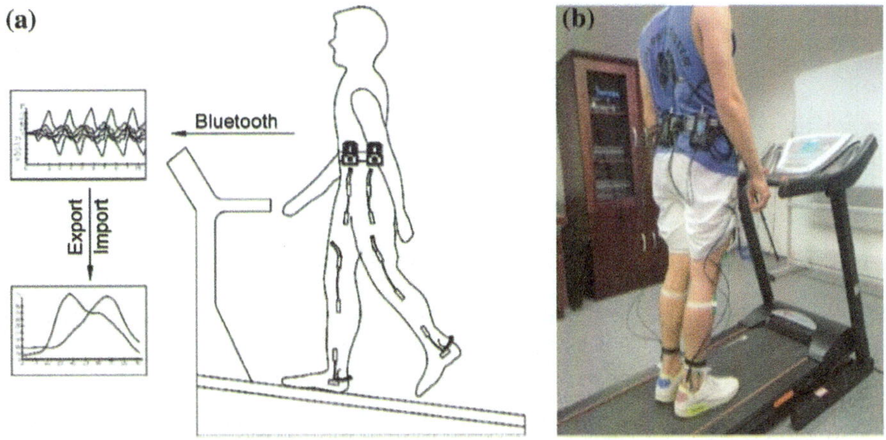

Fig. 18.2 Block schema for data collection using biometrics system (**a**), subject with goniometers and DataLog system (**b**)

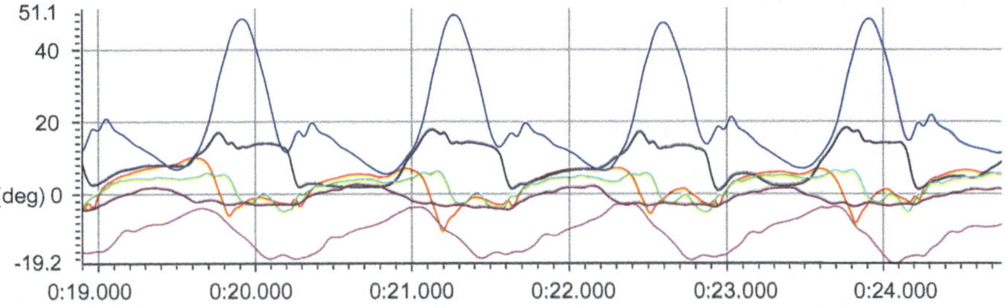

Fig. 18.3 Four consecutive cycles of F-E and L-R angles for right ankle, knee and hip joints

The experimental tests were approved by the Human Ethics Research Committee from the University of Craiova. They were performed by a healthy male subject in the Biomechanics Laboratory at INCESA Research Centre, University of Craiova.

The values of the subject anthropometric data are: Weight = 70 kg; Height = 170 cm; Hip–knee length = 42.5 cm; Knee–ankle length = 40 cm; Leg length = 82.5 cm.

In Fig. 18.3 four consecutive cycles of F-E and L-R for right hip, knee and ankle collected by using the Biometrics system, are shown. For this study, only the F-E angles present interest and they are considered in order to be processed and compared with numerical simulation results.

The F-E data files obtained for right and left knee and hip joints for the three different tests were exported from Biometrics to Simi Motion software, in order to obtain a normalized medium cycle. From the total number of performed cycles, for each F-E joint angle a number of 9 consecutive cycles were selected. The normalized curves corresponding to each of 9 cycles and the medium cycle were

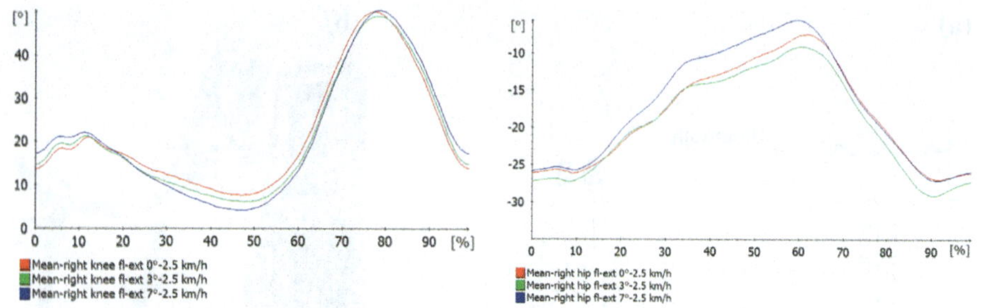

Fig. 18.4 Medium normalized flexion-extension experimental cycles corresponding to the right knee joint (left) and right hip joint (right) for all three tests

obtained. In Fig. 18.4 the three medium F-E cycles for the right knee and the three medium F-E cycles for the right hip are presented. Similar diagrams were also obtained for the left joints.

The flexion angle during the gait cycle revealed small differences with respect to flexion magnitude between the three tests. There were not big differences in the shape of the flexion angle.

18.3 Numerical Simulation of Human Rehabilitation with an Exoskeleton on Treadmill

One of the most used rehabilitation principle used in therapies consist in the use of a robotic rehabilitation device on a treadmill. This method has the advantage that the human disabled patient is suspended with belts to an upper frame, in this way maintaining his equilibrium. The exoskeleton used for rehabilitation has the structure presented in Fig. 18.5.

The exoskeleton design consists in a leg mechanism, with the structure composed from nine kinematic links (1…9), connected by 14 revolute joints. The link (5) of the exoskeleton is replicated by the human femur and the link (9) as tibia. The link (1) is the actuation element of the leg linkage. From motor (11) it is used a chain transmission (10) to transmit the motion to the two legs mechanism. The links (1) are fixed on the shaft (12) at 180°, to replicate the human gait stances. For this simulation model the exoskeleton is worn by a human mannequin designed with the anthropomorphic dimensions of the human subject. The simulation model is completed with a treadmill which has the possibility to adjust the inclination. The simulation model, presented in Fig. 18.5, with proper joints and ground contact is built in ADAMS simulation environment. To define the contact from the exoskeleton foot and treadmill it is used the ADAMS impact method, which is the most used in this case [7], and the used parameters are: contact stiffness, damping, contact force exponent and penetration depth. For this simulation the treadmill speed in specified to 2.5 km/h and the inclination angle is set to 0°, 3° and 6°. The

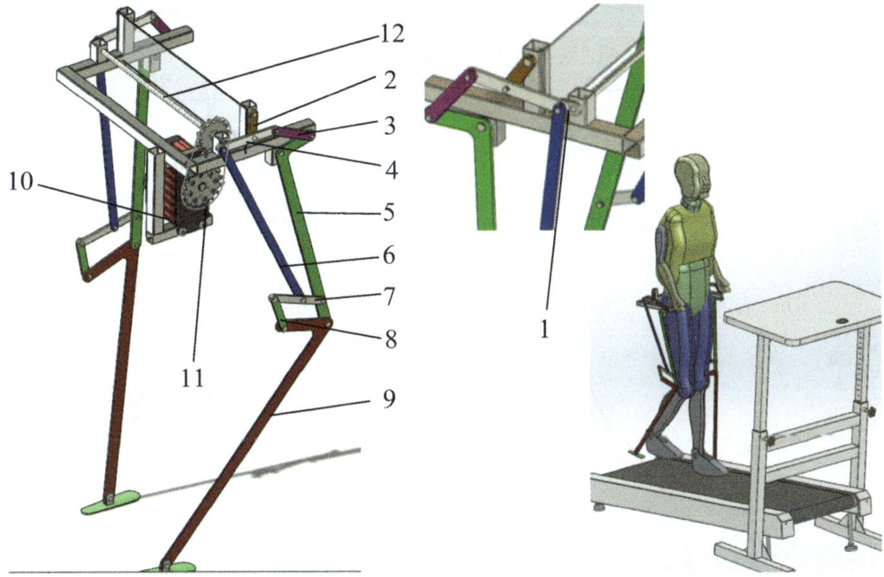

Fig. 18.5 Proposed exoskeleton design to assist human mannequin

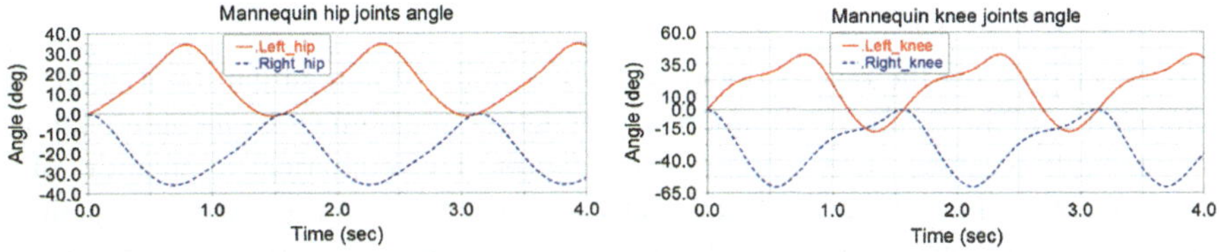

Fig. 18.6 Computed results of mannequin hip and knee joint angle variation

interest results are the mannequin hip and knee joints angular variations. The angular variation for the both hip joints and both knee joints are presented in Fig. 18.6. In Fig. 18.7 the computed vertical reaction forces for the exoskeleton knee joints during walking on inclined treadmill at 3° and 6° are shown.

The vertical force from the exoskeleton knee joints increase with the treadmill inclination. At 3° treadmill inclination the maximum value is 500 N and at 6° inclination it reaches 700 N. In Fig. 18.8, a comparison between medium cycle of the human and exoskeleton is presented. The knee joint angle of the human is 50° and for the exoskeleton is 60°. This difference appears because the exoskeleton model used for simulation is designed to achieve this angular amplitude for the knee joint. The angular amplitude of the exoskeleton knee and hip joint can be modified, adjusting the lengths of links (2), (3) and (4). For the hip joint, as can be observed from medium cycle comparison, the exoskeleton achieves 35° angular amplitude and human 30°.

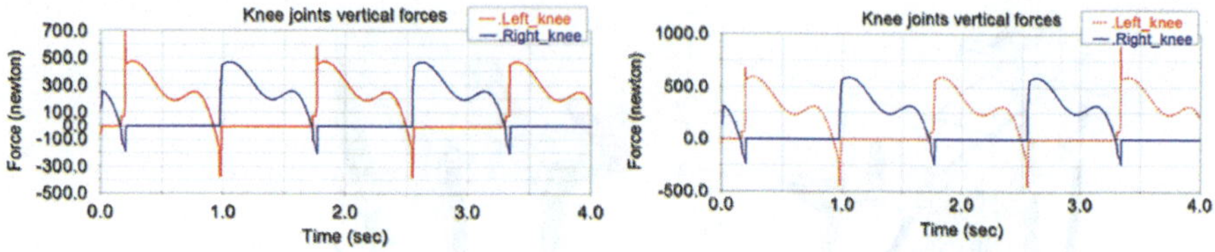

Fig. 18.7 Computed vertical forces of knee joint for 3° (left) and 6° (right) treadmill inclination

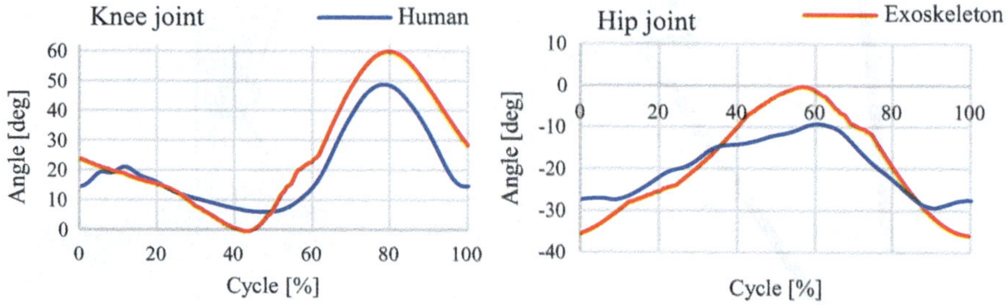

Fig. 18.8 Compared knee and hip joints angles achieved on 6° inclined treadmill

18.4 Conclusions

This paper presents the results obtained for the study of human gait on a treadmill compared with the results of a numerical simulation of a human mannequin walking on treadmill wearing an assistance exoskeleton. The numerical simulation of a virtual mannequin walking with the attached exoskeleton on a plane and inclined treadmill is performed, using ADAMS virtual environment. The obtained simulation results are used in order to validate the engineering feasibility of the proposed exoskeleton rehabilitation device and principle.

References

1. K. Anama, A.A. Al-Jumaily, Active exoskeleton control systems: state of the art. Procedia Eng. **41**, 988–994 (2012)
2. O. Ashkani, A. Maleki, N. Jamshidi, Design, simulation and modelling of auxiliary exoskeleton to improve human gait cycle. Australas. Phys. Eng. Sci. Med. (2016)
3. D. Tarnita, Wearable sensors used for human gait analysis. Rom. J. Morphol. Embryol. **57**(2), 373–382 (2016)
4. I. Díaz, J.J. Gil, E. Sánchez, Lower-limb robotic rehabilitation: literature review and challenges. J. Robot. (2011)
5. D.R. Louie, J.J. Eng, Powered robotic exoskeletons in post-stroke rehabilitation of gait: a scoping review. J. Neuroeng. Rehabil. **13**(1), 53 (2016). doi:10.1186/s12984-016-0162-5

6. T. Li, M. Ceccarelli, Design and simulated characteristics of a new biped mechanism. Robotica **33**(07), 1568–1588 (2015)
7. I. Geonea, M. Ceccarelli, G. Carbone, *Design and Analysis of an Exoskeleton for People with Motor Disabilities*. The 14th IFToMM World Congress, Taipei, Taiwan, 2015
8. I. Geonea, C. Alexandru, A. Margine, A. Ungureanu, Design and simulation of a single dof human-like leg mechanism. AMM **332**, 491–496 (2013)
9. W. Tao et al., Gait analysis using wearable sensors. Sensors **12**, 2255–2283 (2012)
10. D. Tarnita et al., Experimental measurement of flexion-extension movement in normal and osteoarthritic human knee. Rom. J. Morphol. Embryol. **54**(2), 309–313 (2013)
11. D. Tarnita et al., Experimental characterization of human walking on stairs applied to humanoid dynamics. Adv. Robot Des. Intellig. Control 293–301 (2016)
12. D. Tarnita, D.N. Tarnita, N. Bizdoaca, D. Popa, Contributions on the dynamic simulation of the virtual model of the human knee joint. Materialwissenschaft und Werkstofftechnik **40**(1–2), 73–81 (2009)
13. D. Tarnita, D.N. Tarnita, N. Bizdoaca, Modular adaptive bone plate for humerus bone osteosynthesis. Rom. J. Morphol. Embryol. **50**(3), 447–452 (2009)
14. http://www.biometricsltd.com/

Chapter 19
Prototype Device for Mastoid Obliteration in Pediatric Cholesteatoma Patients

Aurel Marin, Karoly Menyhardt, Alina Maria Marin, Ramona Nagy and Marioara Poenaru

Abstract Cholesteatoma represents a keratinization process of the middle-ear cleft or mastoid. This disease destroys the peripheral organs for hearing and balance, with possible intracranial consequences. The management of cholesteatoma is surgical and the primary aim is to remove the disease and prevent relapse. Mastoid obliteration with reconstruction of the bony external ear canal recreates the normal anatomy to avoid such morbidity. In this paper it is presented the prototype device, developed by the authors that will be inserted in the cavity. Design parameters were determined based on measurements conducted by the medical staff during mastoidectomy operations. The requirements of biomaterials used for middle ear reconstruction such as biocompatibility, surface energy, resistance to degradation and infection, mechanical stiffness, weight, bone deposition, function-appropriate design, easy to tailor and position, imaging artifact, were all taken into account and have been evaluated accordingly.

A. Marin (✉)
Pediatric Emergency Hospital, Cluj Napoca, Romania
e-mail: aurelmarin.pediatrie@yahoo.com

K. Menyhardt · R. Nagy
University Politehnica Timisoara, Timişoara, Romania
e-mail: karoly.menyhardt@upt.ro

R. Nagy
e-mail: ramona.nagy@upt.ro

A. M. Marin
Iuliu Hatieganu University of Medicine and Pharmacy, Cluj Napoca, Romania
e-mail: alinamaria.muntean@yahoo.com

M. Poenaru
Victor Babeş University of Medicine and Pharmacy, Timisoara, Romania
e-mail: marioara.poenaru@gmail.com

N. Herisanu and V. Marinca (eds.), *Acoustics and Vibration of Mechanical Structures—AVMS-2017*, Springer Proceedings in Physics 198,
https://doi.org/10.1007/978-3-319-69823-6_19

19.1 Introduction

A mastoid cavity resulting from a canal wall down mastoidectomy can result in major morbidity for patients due to chronic otorhea and infection, difficulty with hearing aids and vertigo with temperature changes. Mastoid obliteration with reconstruction of the bony external ear canal recreates the normal anatomy to avoid such morbidity. Few have studied the quality of life benefit that this procedure confers.

Cholesteatoma represents a keratinization process of the middle-ear cleft or mastoid. This disease destroys the peripheral organs for hearing and balance, with possible intracranial consequences. The management of cholesteatoma is surgical and the primary aim is to remove the disease and prevent relapse.

Reconstruction of the posterior canal wall and/or obliteration of the mastoid may be the best surgical treatment alternative for cholesteatoma.

19.2 Problem

A cholesteatoma is an abnormal growth of squamous epithelium in the middle ear and mastoid bone that can cause chronic middle ear infections and damage the hearing structures of the ear. It may progressively enlarge to surround and destroy the ossicles, resulting in conductive hearing loss. Hearing loss also may occur if the cholesteatoma obstructs the eustachian tube orifice, leading to middle ear effusion. Surgical therapy is required for most cholesteatomas. The extent and effectiveness of surgery depends upon the size of the cholesteatoma. Early diagnosis is crucial to an optimal outcome.

The surgical treatment of cholesteatoma frequently involved canal wall down mastoidectomy technique (CWDT) to provide complete removal of disease [1].

The complications associated with open cavity mastoid operations are the same to those possible in any procedure in which the mastoid bone is removed and structures in the middle ear are controlled. These include: (1) hearing loss, (2) facial paralysis, (3) vestibular symptoms, (4) cerebrospinal fluid leak, (5) infection, (6) recurrence, (6) aural drainage, (7) mucous retention and (8) recurrent cholesteatoma [2]. Obliteration methods are performed to resolve these problems. The obliteration is more efficient for anatomic and physiologic reason and can be executed by various techniques. Although these problems can be avoided using the canal wall up technique (CWUT), the incidence of cholesteatoma recurrence is higher with CWUT than with CWDT [3]. When patients have a destructive scutum or poor Eustachian tube function, surgeons usually choose to perform the CWDT to prevent the occurrence of retraction pockets, despite the disadvantages of this technique [4].

Yung and Karia [5] suggested mastoid obliteration as a routine procedure in all mastoid surgery. Mastoid obliteration should not be undertaken lightly, because it carries a risk of enclosing cholesteatoma within the cavity. Computed tomography (CT) may or may not be effective in detecting these residual. Overall, the clinical follow up is a valuable method in detection of recurrence.

19.3 Proposed Solution

Different synthetic materials can be used to obliterate and reconstruct the mastoid cavity. We want to emphasize the effectiveness of our device in mastoid obliteration and quality of life after the surgery in the pediatric patients. Dost et al. [6] were the first ones to use bioactive materials in reconstructive middle ear surgery.

Four major categories of surface-active biomaterials have been used throughout the years: dense hydroxylapatite (HA) ceramics, bioactive glasses, bioactive glass-ceramics, and bioactive composites [7]. The requirements of biomaterials used for middle ear reconstruction according to Jahnke et al. [8] are biocompatibility, high surface energy, high resistance to degradation and infection, high mechanical stiffness, low weight, no bone deposition, function-appropriate design, easy to tailor and position, no imaging artifact.

Bio-glass is a commercially available family of bioactive glasses, composed of SiO_2, Na_2O, CaO and P_2O_5 in specific proportions. High ratio of calcium to phosphorus promotes formation of apatite crystals; calcium and silica ions can act as crystallization nuclei, while phosphorus ions don't. Generally, there are four classes of bio-glasses [9]:

- 35–60 mol% SiO_2, 10–50 mol% CaO, 5–40 mol% Na_2O: bioactive, bonds to bone, some formulations bond to soft tissues,
- <35 mol% SiO_2: non-bioactive, nearly inert, gets encapsulated with fibrous tissue,
- >50 mol% SiO_2, <10 mol% CaO, <35 mol% Na_2O: bioactive, resorption within 10–30 days,
- >65 mol% SiO_2: technologically impractical.

For our implant we needed a material with biocompatible properties. Among the best candidates were Fluoroplastics, which represent a group of plastics where the molecules contain carbon and fluorine. The plastic polythene is a molecule consisting of a carbon chain with hydrogen atoms attached. PTFE is similar but the hydrogen atoms are replaced with fluorine atoms. The replacement of the hydrogen atoms with fluorine atoms dramatically changes the properties of the material, and fluoroplastics therefore tend to have special useful properties such as non-stick characteristics very high resistance to chemical.

Several biocompatible materials were considered, taking into account their chemical reaction, water absorption. The most suitable materials considered for our device are shown in Table 19.1, the finalist being high density Polyethylene (HDPE).

Table 19.1 Considered materials for implant device

	Nylon 6/6 GF30	PPS GF-40	PEEK	PVDF	Noryl	TPFE	PP	UHMWPE	HDPE
Density $(g/cm)^3$	2.8	1.94	2	1.72	1.06	0.8	0.9	0.928	0.958
Water absortion (%)	1.2	0.03	0.04	0.015	0.07	0.01	0.05	0	0.1
Tensile strength (Mpa)	32	43	53	70	63	30	40	40	20.3
Elongation (%)	120	20	30	20	25	40	100	300	10.4

Polyethylene is a thermoplastic polymer with variable crystalline structure and an extremely large range of applications depending on the particular type. It is one of the most widely produced plastics in the world. As a readily available alternative, high-density polyethylene (HDPE) has a large strength-to-density ratio, with stronger intermolecular forces and tensile strength and was used for fabrication.

After consideration of usability and maneuverability, a device was developed and subjected to simulation with finite element method and mechanical tests (Fig. 19.1). The parts were studied as a linear elastic isotropic model, with tensile strength of 22.1 N/mm^2, having a high-quality standard solid mesh, resulting in a response of Stress (von Mises Stress) of maximum 1.77041 N/mm^2.

Roland and Leach [10] reported that most biologic tissues have a tendency to atrophy over time, and some otologists recommend overfilling the cavity to compensate for the expected loss of the volume of the obliterating tissues over time. Also Yung stated that because the final shape and size of the obliterated mastoid

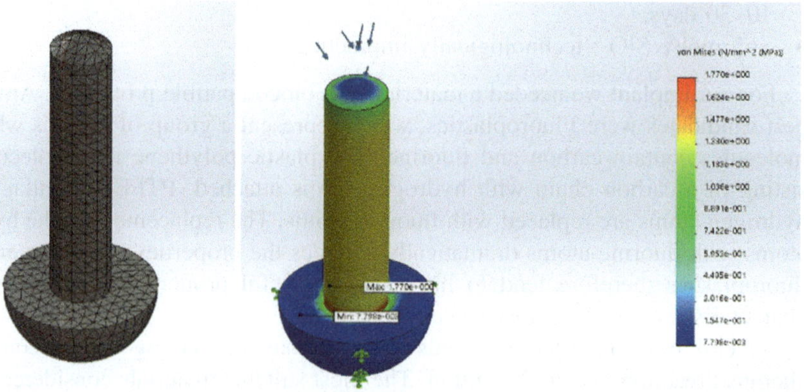

Fig. 19.1 Center pin—finite element method model

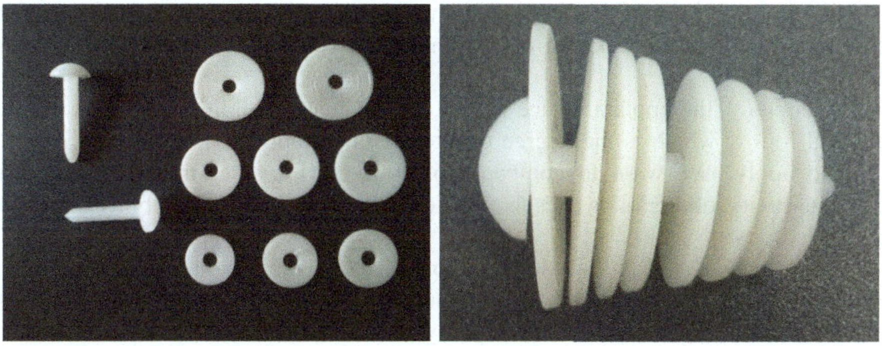

Fig. 19.2 Fabricated mastoid obliteration implant device

cavity cannot be predicted, using biologic tissues for obliteration can be a drawback. Experience has shown us that with long-term follow up, free muscle, and fascia flaps undergo atrophy, retraction, and fibrosis, acting differently from the synthetic materials which have very little chances over time [11].

The shape of the implant assembly (Fig. 19.2) was developed considering the variables met in the operation room such as size and geometry of the cavity, ease of material tailoring, easy positioning and assembly, respectively the post operatory interventions. Thus, clinical follow ups done after 1, 2, 3, 6, 12 months will permit a straightforward access to the investigated area.

19.4 Conclusions and Discussions

Ridenour [12] reported that after using the bioactive materials in mastoid cavity obliteration, three patients required revision surgery to remove bioactive ceramic material form 99 patients; it is our hope that using this prototype the success rate of this operation will rise.

Nyrop and Bonding [13] compared the long-term results of three different surgical techniques (modified radical mastoidectomy, Canal wall up mastoidectomy and atticotomy and atticoantrotomy). They reported a recurrence rate of 70% for cholesteatoma in cases treated with canal wall up surgery, as compared to 15% in patients undergoing canal wall down mastoidectomy with obliteration.

In this paper, we presented the prototype device, developed by the authors that will be inserted in the cavity. Design parameters were determined based on measurements conducted by the medical staff during mastoidectomy operations. Based on these values, a prototype device was created and tested for best fit. This device is very easy to apply and remove if inflammatory reactions occur after the surgery.

During the development phase we compared several materials used for mastoid cavity obliteration and also the effectiveness and compatibility with the patient's body.

In our study, the implant device will be used as mastoid obliteration material and we will follow which is the degree of fate for obliteration. This system it is very easy to apply. We want to compare this device with other materials used for mastoid cavity obliteration and also the effectiveness and compatibility with the patient's body.

A prospective trial study with this approach will be conducted for 1 year to determine if mastoid obliteration with the developed prototype device following mastoidectomy overcomes the open cavity problems or not.

As possible candidates, forty eight patients will be identified with either primary (first time open cavity) or secondary (opened cavity after radical and modified mastoidectomy in case of cholesteatoma) from 2017 till 2018.

Under general anesthesia, endaural or postauricular incision, 5 mm behind the sulcus, will be done in our cases. Open techniques will be used in this study. The posterior canal wall will be removed on the complete mastoidectomy. The cholesteatoma sac, pathologic mucosa, incus, malleus head, and tensor fold will be removed if needed. This will be followed by reconstruction of tympanic membrane and obliteration of the mastoid. A revision mastoidectomy is completed if required to exclude any diseased mastoid cells and to clean.

References

1. M.J. Ramsey, S.N. Merchant, M.J. McKenna, Postauricular periosteal-pericranial flap for mastoid obliteration and canal wall down tympanomastoidectomy. Otol. Neurotol. **25**(6), 873–878 (2004)
2. S.Y. Ho, J.F. Kveton, Acoustic neuroma—assessment and management. Otolaryngol. Clin. North Am. **35**(2), 393 (2002). doi:10.1016/S0030-6665(02)00004-X
3. R. Charachon, B. Gratacap, C. Tixier, Closed versus obliteration technique in cholesteatoma surgery. Am. J. Otol. **9**(4), 286–292 (1988)
4. D.H. Lee, B.C. Jun, C.S Park et al., A case of osteoma with cholesteatoma in the external auditory canal. Auris Nasus Larynx **32**(3), 281–284 (2005)
5. M.M.W. Yung, K.R. Karia, Mastoid obliteration with hydroxyapatite—the value of high resolution CT scanning in detecting recurrent cholesteatoma, Clin. Otolaryngol. **22**(6), 553–557 Published: Dec 1997
6. P. Dost, G. Lehnerdt, R. Kling et al., Surgical therapy for malignant melanoma of the external ear. HNO **52**(1), 33–37 (2004)
7. D.S. Finch, T. Oreskovic, K. Ramadurai et al., Biocompatibility of atomic layer-deposited alumina thin films. J. Biomed. Mater. Res. **87A**(1), 100–106 (2008)
8. K. Jahnke, P. Dost, M. Schrader, Biocompatibility studies of implants for reconstructive middle ear surgery—scientific basis. Trans. Implants Otol. Proc. 41–46 (1996)
9. M. Neo S. Kotani, T. Nakamura et al., A comparative-study of ultrastructures of the interfaces between 4 kinds of surface-active ceramic and bone. J. Biomed. Mater. Res. **27**(2), 281–281 (1993)

10. P. Roland, J. Leach, Reconstruction of the posterior ear canal, in *Middle Ear and Mastoid Surgery*, vol. 12, ed. by Haberman R II (Thieme Medical Inc., New York, 2004)
11. D.A. Moffat, R.F. Gray, R.M. Irving, Mastoid obliteration using bone pate. Clin. Otolaryngol. **19**(2), 149–157 (1994)
12. J.S. Ridenour, D.S. Poe, D.W. Roberson, Complications with hydroxyapatite cement in mastoid cavity obliteration. Otolaryngol. Head Neck Surg. **139**(5), 641–645 (2008)
13. M. Nyrop, P. Bonding, Extensive cholesteatoma: long-term results of three surgical techniques. J. Laryngol. Otol. **111**(6), 521–526 (1997)

Chapter 20
Comparative Analysis of the Noise and Vibration Transmitted to the Operator by a Brush Cutter

Elena Sorică, Valentin Vlăduţ, Petru Cârdei, Cristian Sorică and Carmen Brăcăcescu

Abstract Noise is one of the priority issues of environment, especially in crowded urban zones. A low level of acoustic pollution leads to the environmental and consumer health protection satisfaction level and thus increases the quality of marketed products. The Directive 2000/14/EC it requires manufacturers to apply to each equipment introduced on market the guaranteed value of the acoustic power, in this respect the user being more and more informed in order to choose some reliable products with low noise level and reduced vibrations. The paper consists in determination of the acoustic power level, respectively the vibrations transmitted to the user by a brush cutter, in various functioning regimes.

20.1 Introduction

Noises can have a dangerous action on human body, action that depends of more factors, namely: noise intensity level, its spectral component, duration and distribution of noise exposure during a working day, the total lifetime exposure.

The achieved researches have shown that a continuous noise that repeats itself at closed intervals is more annoying than a fleeting noise of short duration, which

E. Sorică (✉) · V. Vlăduţ · P. Cârdei · C. Sorică · C. Brăcăcescu
National Institute of Research-Development for Machines and Installations Designed to Agriculture and Food Industry, INMA, Ion Ionescu de La Brad Blv. No. 6, Sector 1, Bucharest, Romania
e-mail: postelnicu.elena@yahoo.com

V. Vlăduţ
e-mail: valentin_vladut@yahoo.com

P. Cârdei
e-mail: petru_cardei@yahoo.com

C. Sorică
e-mail: cri_sor2002@yahoo.com

C. Brăcăcescu
e-mail: carmenbraca@yahoo.com

N. Herisanu and V. Marinca (eds.), *Acoustics and Vibration of Mechanical Structures—AVMS-2017*, Springer Proceedings in Physics 198,
https://doi.org/10.1007/978-3-319-69823-6_20

repeated at long intervals. Noises consisting mainly in high frequency components are more harmful than those of low frequency. Also, machines noises produced by some technological processes are more annoying than natural ones, having equal intensities.

2000/14/EC directive belongs to improving air quality category and is part of the chapter—"Air quality, targeting emissions management outside buildings". Air quality is an environment priority issue and any kind of is recorded in regulations.

2000/14/CE directive establishes provisions (noise limits, public information regarding to noises produced by equipment, conformity assessment procedures etc.) related to equipment sound emissions for using outside the buildings. The directive follows in a general manner the principles and concepts of the modules for different conformity assessment procedures stages and rules, applying EC conformity marking [1].

It is clear that markings of the produces which indicate a guaranteed level of the acoustic power, will allow uses to choose the product after an appropriate information.

20.2 Materials and Methods

The sound power level of a noise source is determined using *acoustic pressure* (p). This represents a fluctuating pressure due to the presence of sound presence overlapped on the static pressure and is expressed in Pascals (Pa).

Sound pressure level (L_p) is ten times the logarithm to the base 10 of the ratio between the square of the measured sound pressure and the square of the reference sound pressure (20 μPa). Sound pressure level is expressed in dB.

Sound pressure level averaged over the measurement surface is given by [2]:

$$\overline{L'_p} = 10\lg\left(\frac{1}{N}\sum_{i=1}^{N} 10^{0,1L'_{pi}}\right), (dB) \tag{20.1}$$

where N is the number of measuring points;
L'_{pi}—sound pressure level measured in the i position of the microphone.

Sound pressure level averaged over the measurement surface and corrected is:

$$\overline{L_{pf}} = \overline{L'_p} - k_1 - k_2 \tag{20.2}$$

where k_1 is the correction for background noise;
k_2—the correction for the reflected sound (it applies to enclosed spaces).

The sound power level is given by:

$$L_w = \overline{L_{pf}} + 10\lg\left(\frac{S}{S_0}\right), \text{ (dB)} \tag{20.3}$$

where S is the measurement surface area;
$S_0 = 1\ \text{m}^2$.

In view of determining the acoustic power level is used modern equipment that in fact measures the acoustic pressure on measuring surface, makes the necessary corrections and then calculates the acoustic power level by the above formula. The measurements have been achieved on shoulder motor mower (Fig. 20.1), at INMA place.

Shoulder motor mower is designated to mower operations of the gardens and parks herbs etc., on plate and unlevelled land or in less accessible places. The cutting device consists in wires or discs head.

The main components of the motor mower are: engine, tubular rod handle, belt, mudguards, wires or discs head.

The main technical characteristics of the product are presented in Table 20.1.

Fig. 20.1 Shoulder motor mower

Table 20.1 Technical characteristics

Crt. No.	Characteristic	UM	Values
1	Engine type	–	2 stroke
2	Starting	–	Manually
3	Fuel mix, gasoline	ml oil/L	20
4	Engine power	HP	3
5	Engine capacity	cm^3	45.6
6	Engine speed at idling	rpm	2700
7	Rated engine speed on the driving shaft	rpm	12,300

20.3 Equipment Used for Testing

Measuring devices used are: roulette, 4000 Testovent anemometer type, DH 50 Thermo hygrometer, 2237 sound level meter, electronic tachometer, triaxial accelerometer and integrator system type PULSE with 12 microphones.

The equipment used to determine the sound power level is a measurement and analysis system based on PC— "System Type 3569 C PULSE multi-analysis" produced by Bruel and Kjaer, which consists in 12 microphones with preamp, amplifier and signal conditioning module with 12 measuring channels, assisted by a notebook computer and software required for the acquisition, processing, interpretation and presentation of data in tabular form. Also, includes a calibration module type 4231 which generates on the frequency of 1 kHz, a noise level of 94 or 114 dB The calibration value of 114 dB is used when measurements are made in a noisy environment (noise level >50 dB) [3–5].

Calibration of measuring channels is performed at the beginning of each set of measurements required to measure a noise source.

In order to determine the sound power level, the following operations were performed, necessary for preparing the product for tests:

- noise source dimensions were measured;
- measuring the surface area was calculated [4].

In Fig. 20.2 is presented the positioning of the 6th microphones on measuring surface according to SR EN ISO 22868:2011.

Coordinates of the 6 microphones positions are shown in Table 20.2.

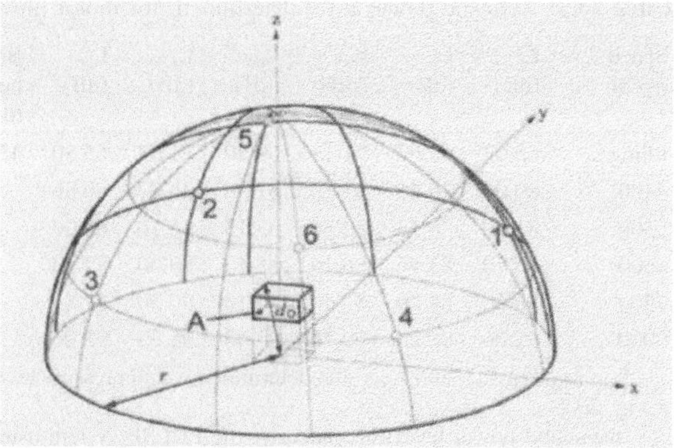

Fig. 20.2 The 6th microphones positions for measuring the hemispherical surface [6]

Table 20.2 Coordinates of the 6 microphones positions for r = 4 m

r = 4 m	x	y	z
1	·2.6	2.6	1.52
2	−2.6	2.6	1.52
3	−2.6	−2.6	1.52
4	2.6	−2.6	1.52
5	−1.12	2.6	2.84
6	1.12	−2.6	2.84

20.4 Results and Discussions

Environment conditions of shoulder motor mower testing were:

- temperature: 22.2 °C;
- humidity: 43.8%;
- wind speed: 0.8 m/s.

In Tables 20.3 and 20.4 is presented the average acoustic pressure level for each of the 6 microphones and acoustic power level, at idling and various speeds, for shoulder motor mower endowed with mowing head, respectively with metal tool.

The graphical representations of acoustic pressure level, depending on microphones positions on the measuring surface and the speed of the motor, are presented in Fig. 20.3.

The results for acoustic pressure level measured at user's ear are presented in Tables 20.5 and 20.6. The measurements were performed in working condition, at 10 cm from the user ear, using the sound level meter. The background noise was measured before every repetition, using the same device.

Table 20.3 Acoustic power level determined, for motor mower endowed with mowing head

Speed (rpm)	L_{p1} (dB)	L_{p2} (dB)	L_{p3} (dB)	L_{p4} (dB)	L_{p5} (dB)	L_{p6} (dB)	Background noise (dB)	Acoustic power level L_w (dB)
Idling	66.97	63.07	71.13	69.20	71.67	65.50	45.1	89
4400	80.06	76.46	81.10	81.36	83.83	80.86		101
5300	81.50	78.53	82.86	83.13	85.20	82.06		103
6600	86.70	83.96	86.10	85.73	89.00	87.53		107
7500	86.60	84.20	87.40	86.10	90.30	87.83		108
9000	92.06	88.26	92.76	92.43	95.33	93.06		113

L_{p1}, L_{p2}, L_{p3}, L_{p4}, L_{p5} and L_{p6}—the determined sound pressure level in each of the 6 measurement points in Fig. 20.2

L_w—the sound power level calculated by the PULSE system using the formula (20.3)

Table 20.4 Acoustic power level determined, for motor mower endowed with metal tool

Speed (rpm)	L_{p1} (dB)	L_{p2} (dB)	L_{p3} (dB)	L_{p4} (dB)	L_{p5} (dB)	L_{p6} (dB)	Background noise (dB)	Acoustic power level L_w (dB)
Idling	74.47	73.20	75.77	75.27	76.63	74.43	45.1	89
4400	78.30	75.43	79.73	82.33	82.90	78.60		101
5300	81.73	79.83	83.46	83.96	84.36	83.00		103
6600	87.76	86.73	88.93	88.96	91.66	88.10		109
7500	89.90	86.80	90.13	90.63	91.60	90.10		110
9000	92.06	90.23	94.20	94.50	98.50	93.66		115

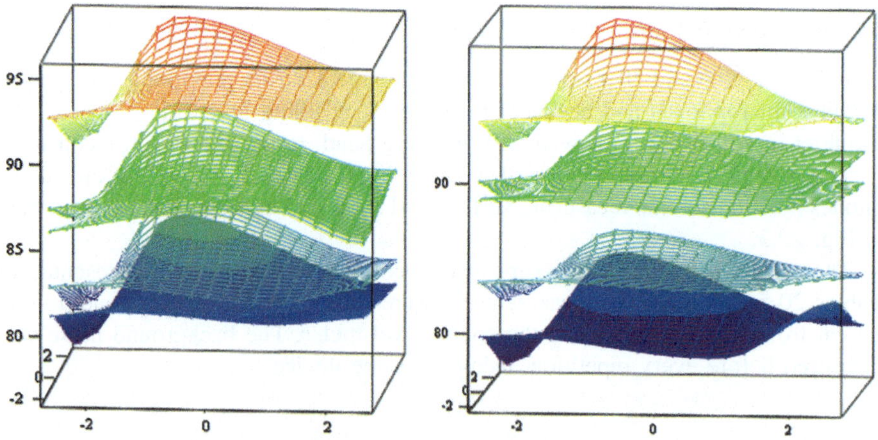

Fig. 20.3 Graphical representations of acoustic pressure level for the motor mower endowed with mowing head (left) and with metal tool (right)

Table 20.5 Acoustic pressure level at user's ear produced by motor mower with mowing head

Crt. No.	Measuring time interval (min)	Average background noise (dB)	Continuous acoustic pressure, equivalent at user's ear (in work) L_{eqA} (dB)
1	5	44.6	94.3
2			94.5
3			94.1
Average			94.3

L_{eqA}—continuous acoustic pressure level, A—weighted equivalent

Table 20.6 Acoustic pressure level at user's ear produced by motor mower with metal tool

Crt. No.	Measuring time interval (min)	Average background noise (dB)	Continuous acoustic pressure, equivalent at user's ear (in work) L_{eqA} (dB)
1	5	44.6	95.9
2			95.4
3			95.3
Average			95.2

Table 20.7 Vibration values at the left and right handle for the motor mower with mowing head

Frequency weighted acceleration a_{hw} (m/s^2)			
Left		Right	
Minimum speed (*idling*)	Maximum speed	Minimum speed (*idling*)	Maximum speed
2.5	5.2	2.1	4.3

Table 20.8 Vibration values at the left and right handle for the motor mower with metal tool

Frequency weighted acceleration a_{hw} (m/s^2)			
Left		Right	
Minimum speed (*idling*)	Maximum speed	Minimum speed (*idling*)	Maximum speed
1.9	4.1	1.6	3.1

There were determined the vibration values at the left and right handle, both for the shoulder motor mower with mowing head and with metal tool, according to SR EN ISO 22867: 2012 [7] (Tables 20.7 and 20.8).

20.5 Conclusions

The acoustic pressure level and acoustic power level increase when the speed of the motor increases. Also, it was found that upon a certain speed (6600 rpm in our situation), the acoustic power level is higher for the shoulder motor mower endowed with metal tool in comparison with the shoulder motor mower endowed with mowing head. The acoustic power level, at idling, was 89 dB, both for motor mower with mowing head and motor mower with metal tool.

Continuous acoustic pressure, equivalent at user's ear (in work), for motor mower with metal tool was higher than the one measured for motor mower with mowing head.

The vibration values at the left handle are higher than the right handle in both situations: with mowing head or metal tool. Also, the vibration values measured for the shoulder motor mower with metal tool are less than the vibration values for the shoulder motor mower with mowing head.

Although this equipment is not subjected to noise limitation, only noise marking according to 2000/14/EC directive, it is recommended that workers operating with such equipment to wear noise protection equipment (ear muffs).

Markings indicating the guaranteed acoustic power level of the products will enable consumers and users to choose the product knowingly.

References

1. Directive 2000/14/EC of the European Parliament and of the Council of 8 May 2000 on the approximation of the laws of the Member States relating to the noise emission in the environment by equipment for use outdoors. Official J. L **162,** 0001–0078 (2000)
2. SR EN ISO 3744:2011—acoustics. Determination of the acoustic power and energy levels for noise sources, using acoustic pressure. Technical methods of closed conditions of those of an empty field above a reflecting plane
3. Sound Level Calibrator type 4231—technical documentation
4. E. Postelnicu, V. Vladut, C. Sorica, P. Cardei, I. Grigore, Analysis of the sound power level emitted by portable electric generators (outdoor powered equipment) depending on location and measuring surface. Appl. Mech. Mater. **430**, 266–275 (2013). ISBN 978-3-03785-877-6
5. PULSE Multi-analyzer system type 3560—technical documentation
6. SR EN ISO 22868:2011—Forestry and gardening machines. Acoustic testing code for portable machines, equipped with an internal combustion engine. Expertise method (accuracy class 2)
7. SR EN ISO 22867:2012—Forestry and gardening machines. Vibration test code for portable hand-held machines with internal combustion engine. Vibration at the handles

Chapter 21
Fluid Dynamics in Simplified pre and post-Stented Coronary Bifurcation

Alin-Florin Totorean, Iuliana-Claudia Hudrea and Bernad Sandor

Abstract Coronary bifurcations are among the most challenging atherosclerotic lesion subsets for percutaneous interventions, even in the era of drug-eluting stents. Bifurcations represent specific regions in the coronary system where blood flow disturbances occur, phenomenon that leads to increased susceptibility to atherosclerosis. Flow patterns in bifurcations are inherently complex, including the creation of recirculation zones associated with low and oscillating wall shear stress, conditions that favorize atherosclerotic plaque formation. Percutaneous coronary intervention with stent placement is thought to induce further hemodynamic alterations due to both the procedure technique and the protrusion of stent struts into the lumen, which are the main determinants of restenosis. Although flow behaviour at the bifurcation sites and its role in promoting atherosclerosis has been widely acknowledged, to the moment few studies have focused on the description of local blood flow environment created after stent implantation, for commonly used bifurcation stenting techniques. The aim of this paper is therefore to describe local blood flow distributions for a normal, stenosed and stented coronary bifurcation, using 2-D idealised configurations, with correlation to the anatomical and pathological conditions. The main hemodynamic parameters included in the study and their potential role in determining in-stent restenosis were: Wall Shear Stress and recirculation length. The global effect of the presence of stenosis, respectively stent

A.-F. Totorean (✉)
Department of Mechanics and Strength of Materials, Politehnica University of Timisoara, Timisoara, Romania
e-mail: alin.totorean@upt.ro

I.-C. Hudrea
Victor Babes University of Medicine and Pharmacy Timisoara, Timisoara, Romania
e-mail: claudia_hudrea@yahoo.com

I.-C. Hudrea
Cardiology Department, Institute for Cardiovascular Diseases, Timisoara, Romania

B. Sandor
Center for Advanced Research in Engineering Sciences, Romanian Academy, Timisoara Branch, Timisoara, Romania
e-mail: sandor.bernad@upt.ro

N. Herisanu and V. Marinca (eds.), *Acoustics and Vibration of Mechanical Structures—AVMS-2017*, Springer Proceedings in Physics 198,
https://doi.org/10.1007/978-3-319-69823-6_21

is analyzed by pressure drop comparisons. The normal coronary bifurcation determines the increase of velocity in the apex zone and development of large recirculation regions in the vicinity of the bifurcation walls, with extended regions with Wall Shear Stress values below 0.4 Pa, considered to be critical and prone for atherosclerosis formation. Stenosis induces the highest pressure drop among all the investigated configurations, together with large recirculation regions. In spite of the fact that stent remodelled the arterial lumen and increased the vascular diameter to the normal configuration, the presence of the stent's struts favoured the initiation of multiple recirculation regions after each strut, which are responsible for in-stent restenosis on long term evolution. The results of this analysis may account as critical information that can help develop better strategies for improved stent designs and stenting techniques.

21.1 Introduction

Bifurcations are regions in the coronary system that are associated with higher risk of developing atherosclerotic lesions than the nonbifurcation areas. Flow patterns in bifurcations are inherently complex with the occurrence of blood flow disturbances, including the creation of recirculation zones associated with low and oscillating wall shear stress, conditions that favorize atherosclerotic plaque formation [1–4]. These hemodynamic alterations have been proven to modulate gene expression, promote of endothelial cell dysfunction, increase the uptake of lipoproteins and favourize leukocyte adhesion, thus contributing to the initiation and progression of atherosclerosis [1–4].

Approximately 20% of all percutaneous coronary interventions involve bifurcations lesions. Stent implantation for this specific lesion subset is considered technically challenging even in the era of drug-eluting stents, with high in-stent restenosis and thrombosis rates [4].

In the current medical practice there are several bifurcation stenting techniques, including the following most frequently used: Main vessel stenting ± side branch angioplasty, (Provisional) T-stenting, V-stenting, Y-stenting, the Crush or Culotte technique, Kissing stents. Most of these techniques require the use of two stents, and even though in many settings the use of two stents is unavoidable, it is recommended to use provisional strategy with single stent as often as possible [4].

Inserting a stent in a bifurcation site inevitably alters the local geometry and consequently creates disturbed flow patterns, conditions that can influence restenosis and can also facilitate the accumulation of plateles, leading to stent thrombosis. The presence of the stent struts at a bifurcation site, with frequent protrusions into the lumen, induces the creation of vortices and zones of low and oscillatory Wall Shear Stress adjacent to the vessel wall. These flow disturbances are strongly associated with early intimal thickening [1–4]. Even when proliferative responses to these altered hemodynamics are limited by drug-eluting stents (DES), restenosis may still occur. Moreover, abnormal flow patterns associated with delayed stent healing due to anti-proliferative medication, represent a possible cause of stent thrombosis [1–4].

Abnormal blood flow patterns at bifurcations promote in-stent restenosis through endothelial cell dysfunction, inflammation, cellular proliferation. The biological phenomenon that leads to restenosis and stent failure is partially related to and explained by local flow conditions. On this matter, the Computational Fluid Dynamics technique represents a special tool that offers the possibility to quantify and correlate several hemodynamic indicators (wall shear stress, zones of recirculation) with restenosis, thus providing additional knowledge that could improve the understanding of the phenomenon [4–7].

Although flow behaviour at the bifurcation sites and its role in promoting atherosclerosis has been widely acknowledged, to the moment few studies have focused on the description local blood flow environment created after stent implantation, for commonly used bifurcation stenting techniques. It is of high importance to better understand the hemodynamic changes that occur at the sites of stented bifurcations and their influence on restenosis and thrombosis, as these represent the major drawbacks of percutaneous interventions even to this date.

The aim of this paper is therefore to describe local blood flow distribution for a stented coronary bifurcation, compared to a non-stenosed and a stenosed one, in order to understand the impact of the presence of the stent on local hemodynamics and its correlation to in-stent restenosis and thrombosis. The main hemodynamic parameters analysed in the study and their potential role in determining in-stent restenosis and thrombosis were wall shear stress and recirculation length, using 2-D idealised configurations, with correlation to the anatomical and pathological conditions. The results of this analysis may account as critical informations that can help develop better strategies for improved stent designs as well as better stenting strategies and techniques.

21.2 Vascular Model

We considered for our study simplified 2-D geometrical configurations associated to native, stenosed and stented coronary bifurcations as shown in Fig. 21.1.

Stent configuration was chosen having geometrical characteristics associated to the investigated coronary bifurcation condition (Fig. 21.2).

Artery diameter was considered to be D = 4 mm and the bifurcation angle of 45°, The domain was meshed using an average value of 610,000 structured cells.

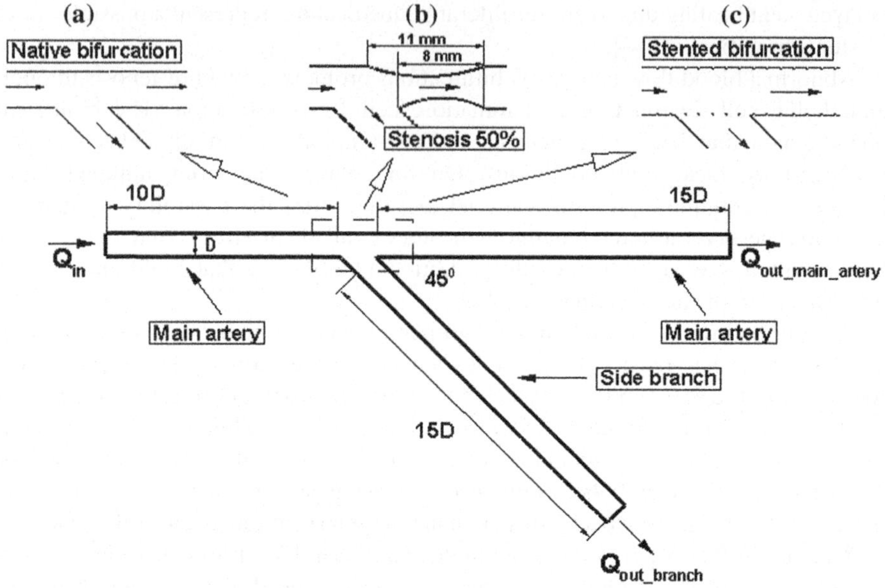

Fig. 21.1 Vascular simplified coronary bifurcation models used in the numerical analysis: **a** native; **b** stenosis with 50% diameter reduction (type of bifurcation stenosis: Medina 0-1-0); **c** stented configuration

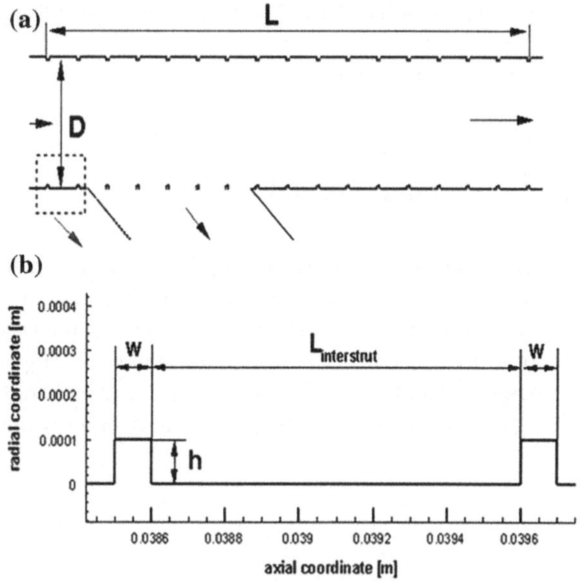

Fig. 21.2 Stent characteristics: **a** general view and position in the bifurcation; **c** geometrical description

21.3 Boundary Conditions

The boundary conditions used for numerical simulations are:

(a) Inlet: physiological flow rate associated to the coronary bifurcation Q = 225 ml/min.
(b) Outlet: outlet pressure is defined to be 0 Pa.
(c) Walls: nonslip and rigid.
(d) The fluid is assumed to be incompressible, similar to blood with density of 1050 kg/m^3 and dynamic viscosity of 0.00368 kg/ms. The physical properties remain constant and gravitational effects were negligible during all the analysis. The numerical simulations were performed using the commercial CFD software Fluent 6.3 [8], under steady-state and laminar conditions. The Navier-Stokes governing equations were solved iteratively until the residual of continuity and velocity reached the convergence criterion of 10^{-9}.

21.4 Results

In the presence of bifurcation large recirculation regions develop in all the investigated situations. In case of native bifurcation, there is a large recirculation region developed in the region with the apex (Fig. 21.3a), which is prone for atherosclerosis formation and evolution. Stenosis induces multiple recirculation zones situated in the vicinity of the bifurcation and the occlusion (Fig. 21.3b), whereas the presence of the stent lowers the extent of these regions, but on the other hand stent induces multiple recirculation regions situated in the vicinity of each strut (Fig. 21.3c).

The presence of recirculation regions are associated to zones with low values of Wall Shear Stress under 0.4 Pa. Wall Shear Stress was computed along the inferior wall of the coronary artery situated beside the bifurcation in all the investigated cases. It can be observed that the presence of stent induces a multiple high variation of the parameter with critical values under this pathological level of 0.4 Pa, which supports the idea that in time there could be developed in-stent restenosis. WSS variation is shown in Fig. 21.4.

Pressure drop along the coronary artery bifurcation is highly influenced by the geometrical characteristic (Fig. 21.5). The highest pressure drop is associated to the stenosed situation, whereas the presence of the stent improves this parameter. This could be correlated to the fact that the vessel diameter was biomechanically reestablished, and this way the resistance to the fluid flow was decreased.

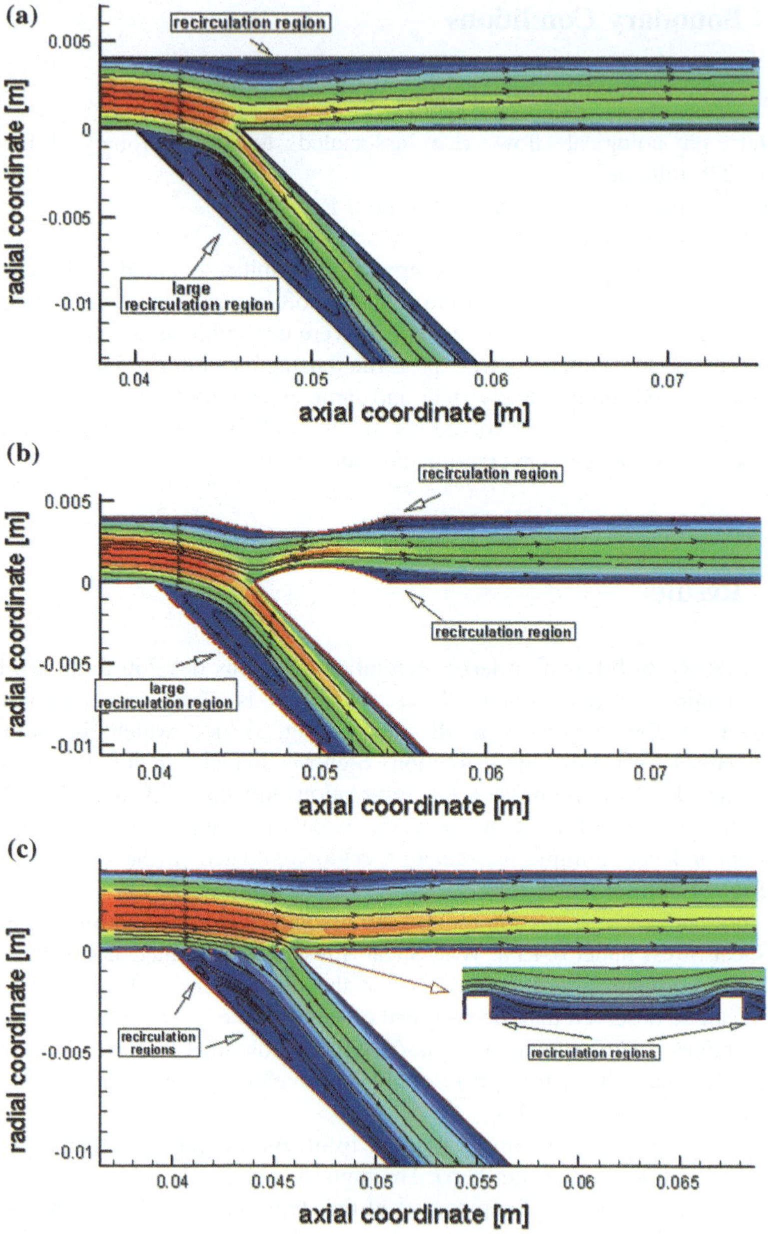

Fig. 21.3 Velocity field and streamlines view: **a** native coronary bifurcation; **b** stenosed bifurcation; **c** stented configuration

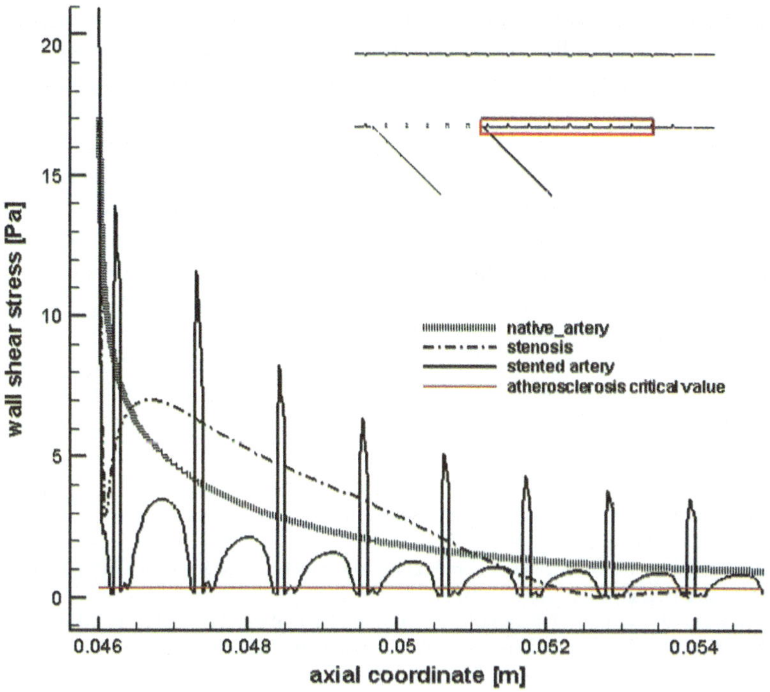

Fig. 21.4 Wall Shear Stress evolution along the wall in the vicinity of coronary bifurcation, in case of native, stenosed and stented configuration

Fig. 21.5 Pressure drop along the coronary bifurcation in case of native, stenosed and stented situation

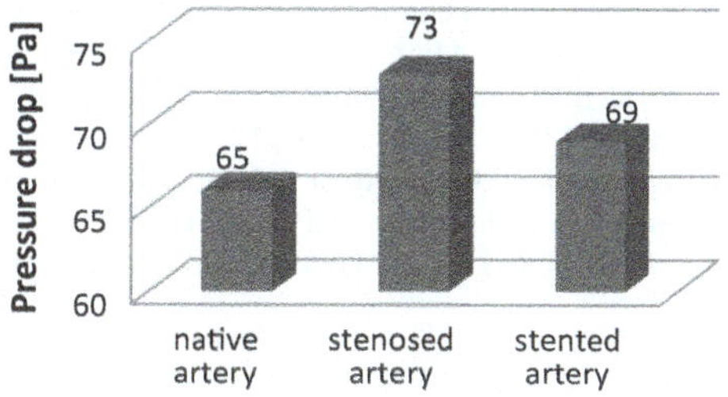

21.5 Conclusions

Both atherosclerotic and restenotic lesions are determined by blood flow alterations at the site of the bifurcations. For the non-stenosed coronary, the presence of the side branch determines the increase of velocity in the apex zone and development of large recirculation regions in the vicinity of the bifurcation walls, with extended regions with Wall Shear Stress values below 0.4 Pa, considered to be critical and prone for atherosclerosic plaques formation. In the case of the stenosed artery, the presence of the stenotic lesion was shown to induce the highest pressure drop among all the investigated configurations, together with large recirculation regions, drawbacks that supposedly should be solved by stent placement. In spite of the fact that the stent remodelled the arterial lumen and increased the vascular diameter to the normal configuration, the presence of the stent's struts favoured the initiation of multiple recirculation regions after each strut, zones associated with low Wall Shear Stress, which are responsible for the initiation and progression of in-stent restenosis.

Although the present study has limitations (the use of an idealized 2D model, the analysis of o single stenting technique model), the results of this analysis may account as critical information that can help develop better strategies for improved stent designs and stenting strategies and techniques.

References

1. J. García García, P. García Carrascal, F. Castro Ruiz, F. Manuel Martín, J.A. Fernández, Effects of bifurcation-specific and conventional stents on coronary bifurcation flow. An experimental and numericalstud. J. Biomech. (2017)
2. A. Antoniadis et al., Biomechanical modeling to improve coronary artery bifurcation stenting. JACC: Cardiovasc. Interv. **8**(0), (2015)
3. C. Chiastra, W. Wu, B. Dickerhoff, A. Aleiou, G. Dubini, H. Otake, F. Migliavacca, J.F. LaDisa, Computational replication of the patient-specific stenting procedure for coronary artery bifurcations: From OCT and CT imaging to structural and hemodynamics analyses. J. Biomech. **49**, 2102–2111 (2016)
4. D.G. Katritsis, A. Theodorakakos, I. Pantos, M. Gavaises, N. Karcanias, E.P. Efstathopoulos, Flow patterns at stented coronary bifurcations: computational fluid dynamics analysis. Circ. Cardiovasc. Interv. **5**, 530–539 (2012)
5. F. Migliavacca, C. Chiastra, Y.S. Chatzizisis, G. Dubini, Virtual bench testing to study coronary bifurcation stenting. EuroIntervention 11, V31–V34 (2015)
6. J.S. Raben, S. Morlacchi, F. Burzotta, F. Migliavacca, P.P. Vlachos, Local blood flow patterns in stented coronary bifurcations: an experimental and numerical study. J. Appl. Biomater. Funct. Mater. **13**(2): e116–e126 (2015)
7. E.S. Bernad, C.I. Hudrea, S.I. Bernad, A.F. Totorean, A.I. Bosioc. *Luminal flow alteration in presence of the stent.* AIP Conference Proceedings 1702, 080011, 2015
8. FLUENT 6.3 User's Guide, ANSYS-Fluent Incorporated (2006)

Chapter 22
Low Cost Adaptive Plantar Supporters for Soil Vibration Absorption

Barbu Braun and Mihaela Baritz

Abstract The paper describes a research stage concerning developing a solution for soil vibration absorption. The reason is to protect the human body from vibration having major negative effects in time, like low limbs numbness, atrophy of tissues and/or blood and lymphatic veins and arteries, loss of stability etc. A new method for adaptive plantar supporter prototyping was proposed and developed. It was applied and tested for two persons with plantar diseases, working in vibration conditions. It was proved that wearing such of special plantar supporters with vibration absorption and plantar conformation corrective role, would have a major positive influence on the body, in terms of vibration absorption and stability increasing.

22.1 Vibration Influence on the Human Body

Vibrations transmitted to the body (especially to the lower limbs) represent are encountered in many situations (working on mining of building (e.g. in case of rigs handling)) [1–3]. Exposure to the vibration may cause in short time some problems like numbness of the limbs, loss of stability, fatigue. Long time exposure to shocks and vibrations can lead to serious illness, like: atrophy of tissues and/or blood and lymphatic veins and arteries, local paralysis [4–6].

The soil vibration, especially felt by the employers who work on road maintenance (e.g. working with the rigs) may cause different and gradual problems like: lower limb numbness, plantar tissues atrophies, loss of body stability, lower limb joints deterioration [3, 4]. For this reason, environment condition evaluation, standpoint acoustic and vibration became more and more important for public

B. Braun (✉) · M. Baritz
Design Product and Environment Department, University Transilvania Brasov, Brașov, Romania
e-mail: braun@unitbv.ro

M. Baritz
e-mail: mbaritz@unitbv.ro

N. Herisanu and V. Marinca (eds.), *Acoustics and Vibration of Mechanical Structures—AVMS-2017*, Springer Proceedings in Physics 198,
https://doi.org/10.1007/978-3-319-69823-6_22

health [2, 5]. Thus means that the degree of vibration measurement, via different sensors and dedicated devices were strongly increased in the last years [7, 8]. Starting from the soft tissue model vibrations during running, different solution to improve the shoes performances have experienced a strong development [9]. The main aspect analyzing the shoes performances when standing, walking or running refers to loads on the main lower limbs joints [10, 11].

22.2 Proposed Method for Special Adaptive Foot Insole Prototyping

Considering the importance of the vibration on the body and the actual developed methods for their effect reduction on the lower limbs, our research focused to develop a new solution for low costs, foot insoles prototyping.

For this, a study regarding different materials composing the foot insoles it was done. To reduce the costs and especially for waste recycling, it was taken into account to re-use different types of elastic materials, providing from scrapped accessories. Different types of material samples were tested mechanical standpoint (degree of compression due to static and dynamic loads subjecting, for a certain period of time). Also the samples testing standpoint their behavior in saline solution at normal and high temperatures was performed (simulation the real condition when shoes wearing (e.g. feet sweet) [12]. Among the tested materials can be listed the following: textile coating, gel, active carbon, foam polyurethane, wool coating, plastic, rubber and silicone. It was proved that the rubber, silicone and foam polyurethane materials were the best, standpoint mechanical static and dynamic loads and saline solution.

In this order, for the plantar insoles prototyping the following steps have been established: (a) creating the core, containing several rubber and silicone items, encapsulated into a textile pocket, (b) disposing the core to a sock, made of silicone or rubber materials, (c) covering with a sandwich made of textile and polyurethane materials, for shocks absorption and antiperspirant role, (d) disposing the insoles into the shoes. An example of plantar insole prototyping (for the left foot) is presented in Fig. 22.1.

It was demonstrated that even the recycled materials made of silicone and rubber keep the elasticity and resistance properties, they could be successfully used for shoes insoles having shocks and vibration absorption role. The number of rubber or silicone items contained into the pocket strictly depends on the degree of vibration of the soil.

Once established the most adequate materials and the prototyping procedure, the method could be successfully adapted for several persons working in soil vibration condition. The most representative examples are two persons with plantar diseases: the first being diagnosed with flat feet and the second with pronounced arch. The reasons for which these have been considered the most representative examples for

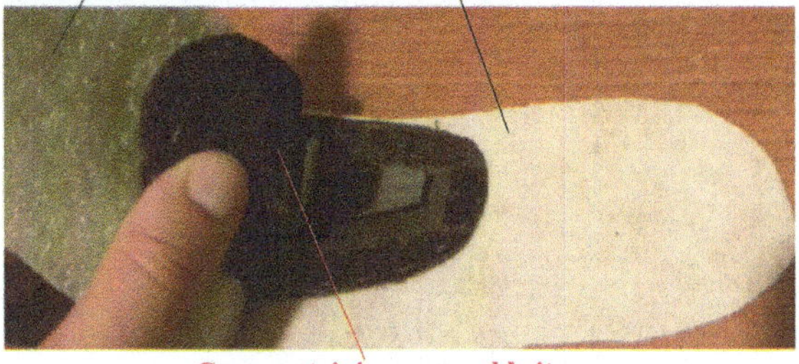

Fig. 22.1 Example of prototyping of foot insole for the left foot

the study are the following: in case of flatfoot, the vibration transmission is taken over the entire plantar surface, so that the vibrations are transmitted to a greater extent to the foot joints and to entire body. In case of pronounced arch, the vibration are transmitted only to the heel and metatarsal areas, at an increased intensity (up than two or three times greater). Thus could lead to atrophies and muscle pain.

In case of the second person, with pronounced arch, it was necessary to manufacture two cores (respecting the procedure described above), one for the heel area and the other for the metatarsal area.

22.3 Effects of Prototyped Foot Insoles Wearing

To evaluate the effects of the plantar supporters, standpoint vibration compensation, the following study was done: both subjects were tested in two distinct situations: wearing their regular shoes and, after that, wearing their shoes, including the prototyped insoles. The test invoked two steps.

The first one was to subject them to vibrations for a period of time of 15 min, due to a training equipment for toning the muscle mass of athletes before different competitions. Experimentally, it was proved that a period of time of about 15 min for vibration exposure is enough to have a visible effect in terms of body behavior.

The second testing step was to evaluate the body stability, referring to the center of mass displacement amplitudes in sagittal and lateral planes, using a Kistler force plate, containing four piezoelectric sensors (Fig. 22.2). The period of time for the stability evaluation was established of about 30 s, being enough to observe the center of mass displacement evolution [13, 14]. The tests were made in static conditions (when standing), being the most suggestive evaluation in terms of vibrations influence. In Fig. 22.2b is presented an example in which a person wearing shoes containing prototyped insoles was evaluated after a session of exposure to vibrations. For this, each person was disposed to a stand with the feet on a vibratory plate, progressive degrees, with hands on a fixed handle stand.

To determine the vibration influence on the body stability, a comparison between body stability before and after subjecting to vibrations was made, as shown

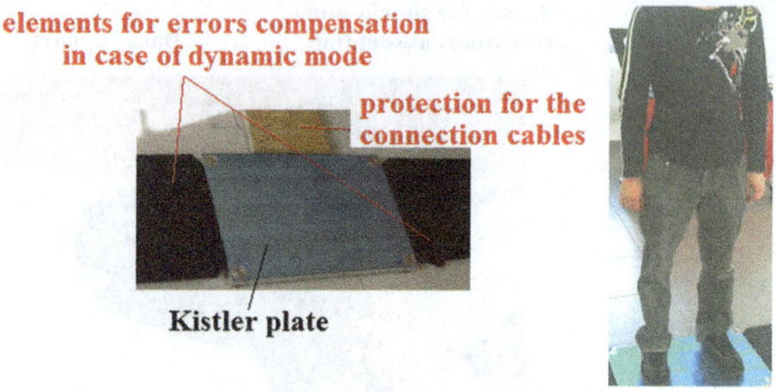

Fig. 22.2 Used equipment to evaluate the body stability

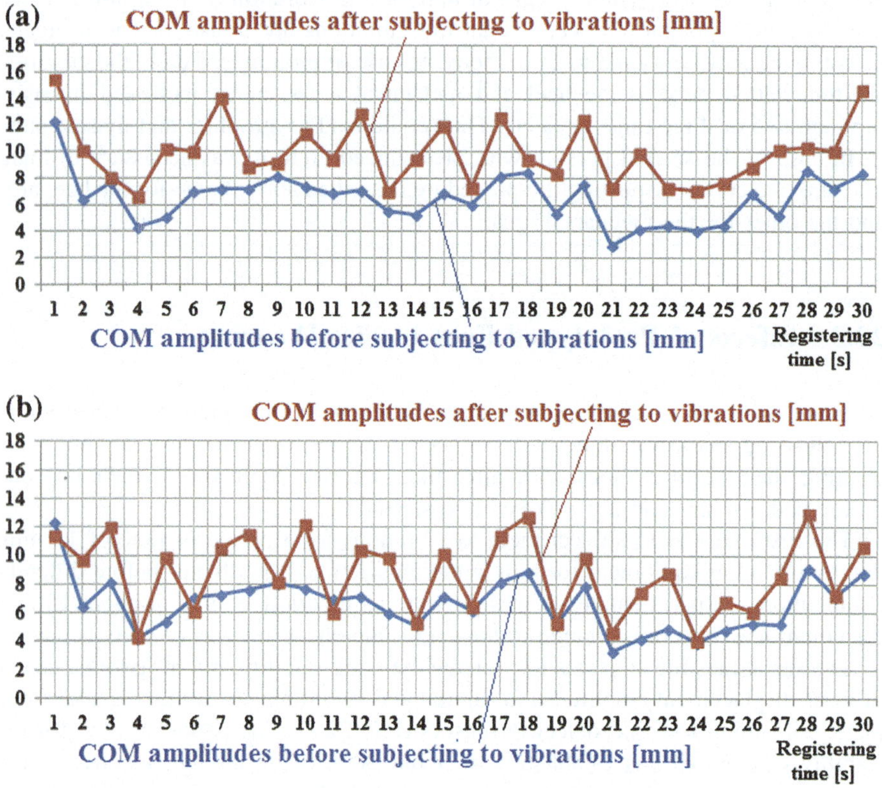

Fig. 22.3 Subjecting to vibrations effects in terms of body stability in case of the first tested person (with flat feet): **a** wearing regular shoes; **b** wearing the shoes containing the prototyped foot insoles, absorption role

in Figs. 22.3, 22.4, 22.5 and 22.6. In red is represented the center of mass displacement amplitude evolution during testing, before subjecting to vibrations and in blue, is displayed the evolution after vibration subjecting. The amplitudes were

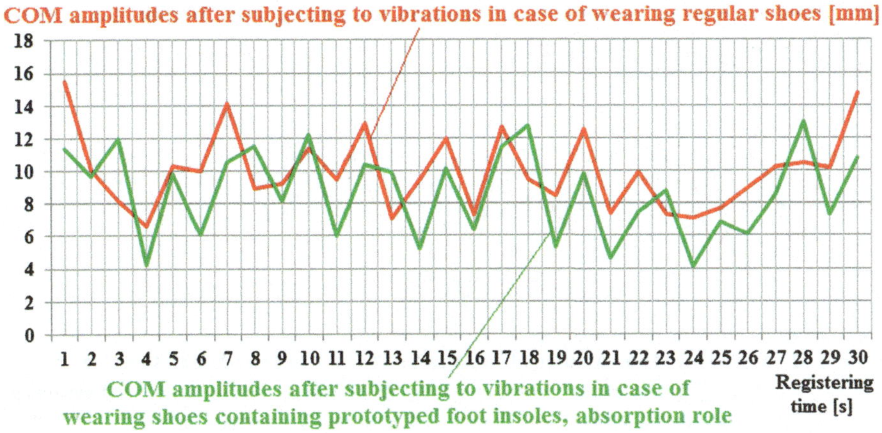

Fig. 22.4 Comparative results for the first person, on body stability after subjecting to vibrations in case of wearing regular shoes and in case of wearing shoes with prototyped insoles

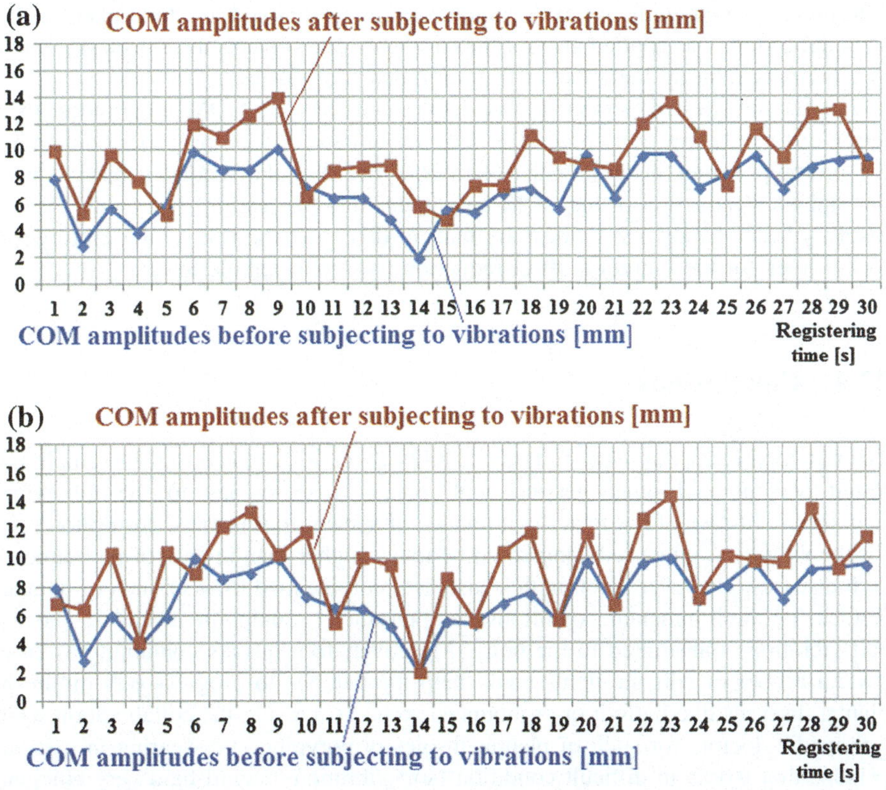

Fig. 22.5 Subjecting to vibrations effects in terms of body stability in case of the second tested person (pronounced arch): **a** wearing regular shoes; **b** wearing the shoes containing the prototyped foot insoles, absorption role

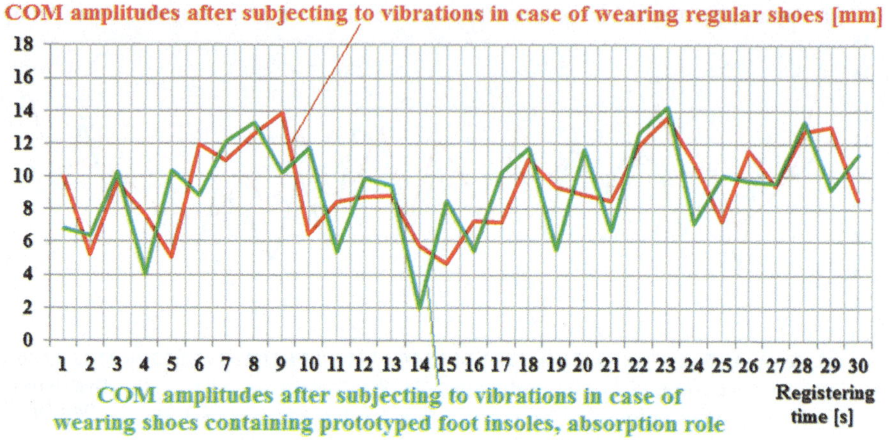

Fig. 22.6 Comparative results for the second person on body stability after subjecting to vibrations in case of wearing regular shoes and in case of wearing shoes with prototyped insoles

determines as average between the amplitudes in sagittal and in lateral planes. The same procedure was applied also for the case when the subjects have been subjected to vibrations wearing the shoes with prototyped foot insoles.

For the first tested person it was determined that the COM amplitudes after vibrations subjecting was about 20% reduced in case of wearing shoes containing the prototyped foot insoles, with absorption role. For the second subject it was determined a reduction of the COM amplitudes due to wearing shoes with special insoles of about only 1.2%, meaning that, in this case, the role of the prototyped foot insoles for the vibration effect reduction was considerable lower.

22.4 Conclusions

The obtained results in terms of body stability evolution, related to the exemplified tested persons and other persons with plantar diseases led to the following hypothesis: In case of flat feet, the influence of special foot insoles, absorption role for the vibrations effect reducing could be much higher than in case of pronounced arch diseases. However, even in this case, is recommended to wear such of plantar insoles for environmental conditions compensation, standpoint mechanic shocks and vibrations transmitted to the body. Moreover, foot insoles containing as core several rubber or silicone items could help the persons to progressively diminish plantar diseases like flatfoot or pronounced arch. As a result, the collaboration with orthopedic doctors for such of plantar insoles prototyping and wearing in case of persons that works in difficult condition (soil vibrations) could be a very efficient solution.

Acknowledgements This paper is part of the current researches in the Developing and Research Institute of TRANSILVANIA University of Brasov and due to the collaboration with the two persons.

References

1. Environmental Noise management, *Assessing Vibration: A Technical Guide*. Department of Environment and Conservation (2006), pp. 19–26
2. ILO Codes of Practice, *Protection of Workers Against Noise and Vibration in the Working Environment* (International Labour Office Geneva, 1984), pp. 18–25. ISBN 92-2-103604-9
3. O. Mary, R. Joseph et al., *The Industrial Environment. Its Evaluation and Control*. U.S. Department of Human Service (1973), pp. 333–336
4. M. Bovenzi, C. Hulshof, *Risks of Occupational Vibration Exposure*. VIBRISKS, FP5 Project No. QLK4-2002-02650, European Commission Quality of Life and Management of Living Resources (2006), pp. 12–21
5. V. Archer, *Whole-body Vibration: Causes, Effects and Cures*. NAPIT, Health and Safety (2007), pp. 22, 23
6. World Health Organization, *Occupational Exposure to Vibration from Hand-held Tools*. Teaching Materials, pp. 23–46
7. C. Druga, I.C. Rosca, S. Lache, in *Perception of whole body vibrations*. 2nd International Conference, Advanced Composite Materials Engineering, COMAT 2008, 9–11 October 2008, Brasov, Romania, 2008
8. M. Griffin, *Handbook of Human Vibration*. Human Factors Research Unit, University of Southampton (1990)
9. N. Asadi, A. Zadpoor, in *A model-based parametric study of soft tissue vibrations during running*. IEEE Benelux EMBS Symposium, pp. 49, 50 (2007)
10. H. Miller, J. Hamill, Computer simulation of the effects of shoe cushioning on internal and external loading during running impacts. Comput. Methods Biomech. Biomed. Eng. **12**(4), 481–485 (2009)
11. G.R. Johnson, The effectiveness of shock-absorbing insoles during normal walking. Prosthet. Orthot. Int., pp. 91–95 (1988)
12. B. Braun, M. Baritz, in *Correlative analysis of shoe insoles acting in moist and cold/warm environment*. Computing and Solutions in Manufacturing Engineering, COSME'16, pp 21, 22, 2016
13. B. Braun, Life style influence on body stability for young subjects. Bull. Transylvania Univ. Brasov **9**(58), (2016)
14. M. Baritz, D. Cotoros, L. Cristea, B. Braun, in *Analysis of stability asymmetries and of gait cycle used to determine the recovery degree for subjects with traumatic locomotion disorder*. 8th International Conference Optimization of Mechanical Engineering, Manufacturing Systems, Robotics and Aerospace, OPTIROB, pp. 534–539, 2013

Chapter 23
Influence of Sound Physical Characteristics on Some Human Physiological Parameters

Diana Cotoros, Ionel Serban and Anca Stanciu

Abstract The influence of certain types of sounds as health hazards for human beings is a constant preoccupation for researchers during the last decades. Not only it is likely to inflict auditory effects but it is proven to produce other physiological as well as psychological effects upon people subjected to noise. Of course, not all types of sounds are damaging for health, their influence is strongly related to their physical characteristics like: frequency, intensity or harmony. Each change in these physical characteristics may produce different effects upon human health that can be monitored by help of several physiological parameters. As many researchers focused upon the auditory effects and psychological issues under the influence of noise, our team headed towards a less investigated aspect related to the risk of cardio-vascular diseases for persons subjected to unhealthy types of sounds, a risk which was mentioned in a thematic issue of the European Commission in 2015 and also some influences regarding posture control.

23.1 Introduction

Noise pollution has become a constant and stressful component of the everyday modern life. Not only can it lead to harmful physiological effects like hearing loss, various types of diseases of the inner ear, vegetative disorders, risk of cardio-vascular diseases, etc. but it also has potential dangerous psychological effects which may affect at a certain extent people's performances and wellbeing [1]. Additionally, researches show that noise, especially at night is likely to create

D. Cotoros (✉) · I. Serban · A. Stanciu
Department of Product Design, Mechatronics and Environment, Transilvania University of Brasov, Brașov, Romania
e-mail: dcotoros@unitbv.ro

I. Serban
e-mail: serban_ionel1984@yahoo.com

A. Stanciu
e-mail: ancastanciu77@yahoo.com

N. Herisanu and V. Marinca (eds.), *Acoustics and Vibration of Mechanical Structures—AVMS-2017*, Springer Proceedings in Physics 198,
https://doi.org/10.1007/978-3-319-69823-6_23

blood vessels damage, increased heart rate and stress hormones, leading on long term to high blood pressure and many heart or circulatory diseases [2].

Of course not all types of sounds are harmful; on the contrary, it was proven that certain type of harmony, frequencies and intensities may be an effective form of therapy [3]. Music becomes more and more an effective complementary medicine that helps patients to recover from serious physiological and psychological impairments. Therefore there should be some kind of balance in dealing with sounds. Where is the limit between harmful and pleasant? And is this limit the same for all people?

Some types of work performances can be affected in a negative way, especially those requiring more concentration or complex analysis [4], some activities may become more pleasant if there is too much routine involved and productivity may be increased by help of adequate chosen music or some types of sounds go into the category of noise pollution and create unpleasant results which may or may not become serious illnesses, according to the individual sensitivity and personality.

Motor performance depends highly on the postural stability control [5] and as the auditory system is connected to the vestibular system, air disturbances caused by sound stimuli will affect the vestibular system and as well the posture control [6].

Some researchers proved that in certain situations a mobilization effect upon the postural control can be obtained by help of sounds [7].

Obviously people do not react the same way to sounds and also the sound physical characteristics are extremely important as far as their influence upon the human body is concerned. The purpose of our paper is to determine the influence of sound frequency while keeping the intensity constant, upon the heart rate and posture of some healthy subjects in order to be able to identify the pleasant and especially the unpleasant frequencies that might affect their future health and activity.

23.2 Materials and Methods

In order to obtain information regarding the individual influence of various types of sounds upon human beings our team selected a sample of 34 young, healthy subjects, aged 21–24, males and females, with no previous serious health issues and subjected them to different types of sounds, with various frequencies and an intensity of 100 dB. The intensity was maintained constant as we wanted to determine the influence of the frequency only for this experiment. After a few trials we decided that 100 dB was the appropriate intensity for the size of the laboratory. Their heart rate was measured before and immediately after the sound exposure in order to determine the short term implications upon the cardio-vascular system.

The subjects were instructed to listen to the sounds for 30 s and rate the annoyance produced by hearing various frequencies. They awarded 3 points for the most annoying sound, 2 points for the second place and 1 for the third place.

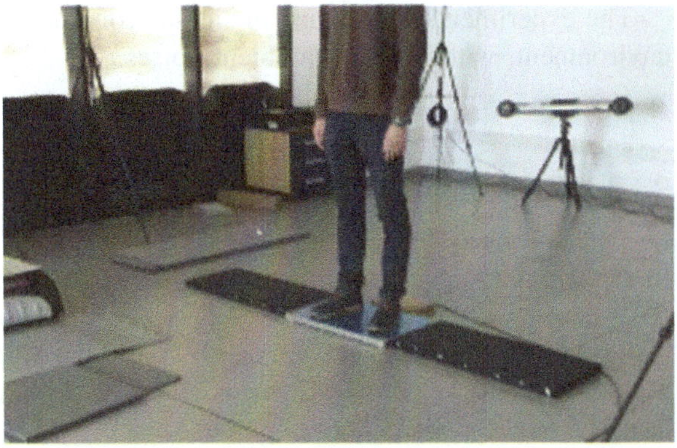

Fig. 23.1 Subject on the force platform

The results were recorded in an Excel database in order to be processed by statistical methods.

Also some of the subjects were instructed to stand on a force platform in a relaxed stance, intermediate base of support according to their comfort while their stability area and reaction forces to the ground were determined in silence and then subjected to unexpected noise stimuli (see Fig. 23.1).

One of the stimuli was common for all subjects, a sound of 440 Hz at 100 dB, characteristic for a loud concert sound, while the other was created at a random higher frequency. By help of the Bioware software the diagrams corresponding for each subject were recorded and analyzed.

The sounds were produced by a free sound generator software using a laptop attached to speakers which were hidden from the subject, in a location in front of the subject, the speakers oriented towards the studied person, as shown in Fig. 23.2. The software offers the possibility of generating sine wave sounds at any required frequency and intensity.

Fig. 23.2 Experimental setup

The experiments took place in similar conditions, around noon, in comfortable environment, with no other ambient noises.

23.3 Results and Discussions

In the beginning all subjects measured their heart rate and recorded the results. Then the persons were subjected at first to a sound of 20 Hz frequency, as this is the lowest threshold of human hearing. Of course not all people, even if their hearing is normal, are able to perceive these sounds. They were asked to write down if they can hear the sound or not. Thus, 20 subjects (58.82%) heard the sound and 14 subjects (41.18%) did not hear the sound.

Then they were subjected for 30 s to sounds with various frequencies, not necessarily in an increasing order, as follows: 164, 440, 800, 4186, 196, 8000, 1100, 16,000, 20,000, 293 Hz.

The subjects awarded 3 points of annoyance to the most disturbing sound, 2 points for the runner up and 1 point for the one on the third place from annoyance point of view and the results were analyzed for the entire lot. After summing up the points corresponding to each frequency, the following classification came up, see Fig. 23.3.

The most disturbing sound, occurring in all classifications on one of the first three places is the 4186 Hz sound, representing the highest frequency of a piano, then the 16,000 Hz sound situated close to the highest limit of human hearing and the 1100 Hz sound takes the third place.

Immediately after the experiments, the subjects were required to measure their heart rate again. The difference between their pulse in the beginning of the experiment and at the end of the experiment is shown in the following histogram (Fig. 23.4).

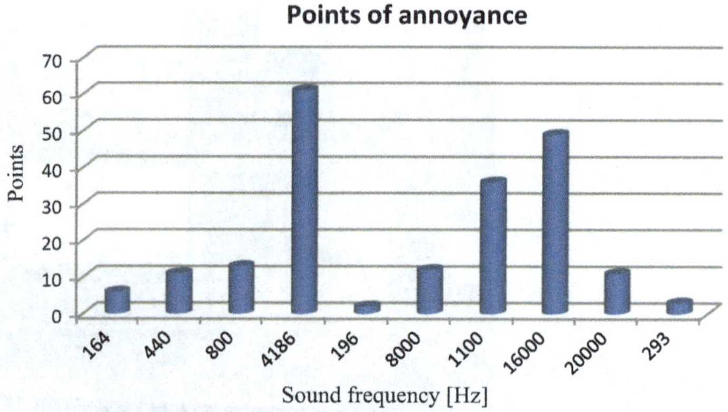

Fig. 23.3 Points of annoyance classification

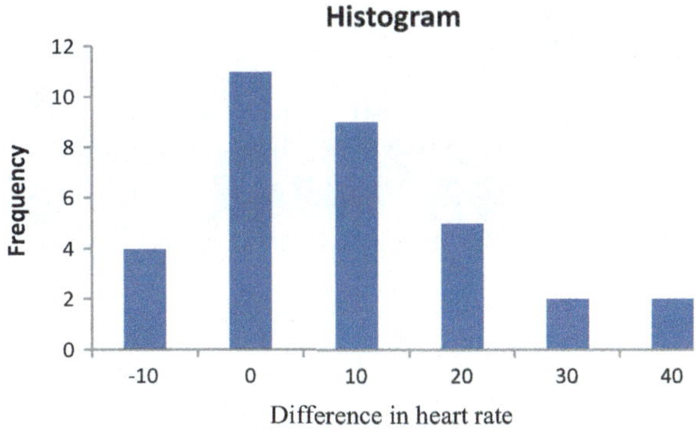

Fig. 23.4 Histogram of differences in the subjects' heart rates

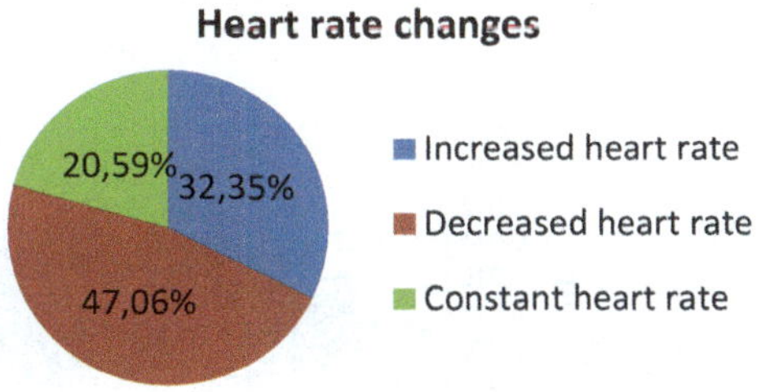

Fig. 23.5 Change in subjects' heart rate

Analyzing the histogram, it becomes obvious that most subjects recorded no significant increase of their heart rate, many of them had no obvious change, a small number even had a drop in the heart rate and around 10 persons were recorded with a significant increase of the heart rate (32.35%), as shown in the diagram in Fig. 23.5, which proves once again that not every person has the same way in dealing with noise pollution, some people react immediately changing their physiological parameters, but some people are not affected on short term.

Another lot of 14 subjects (6 males and 8 females), aged 21–24, in very good health conditions were required to stand for 30 s on a Kistler force platform, first in silent ambient, then subjected to a sound of 440 Hz after 15 s and then to a random chosen frequency sound in order to analyze their stability and posture when subjected to sound stimuli. The speakers were located in the front left and right of the subject and the sound oriented directly towards the subject, while the operator producing the sounds stayed hidden from their line of sight. Bioware software dedicated to the processing of the results obtained from Kistler force platform was further used for posture analysis.

When no sound stimuli was performed the reaction force Fz and the stability area (representing the projections of the body centre of mass on the support area) have regular shapes, the differences between the reaction forces are somewhere between 21 and 31 N, while the body sway in the stability area remains between 0.013 and 0.026 m with one exception, according to the Bioware statistics. An

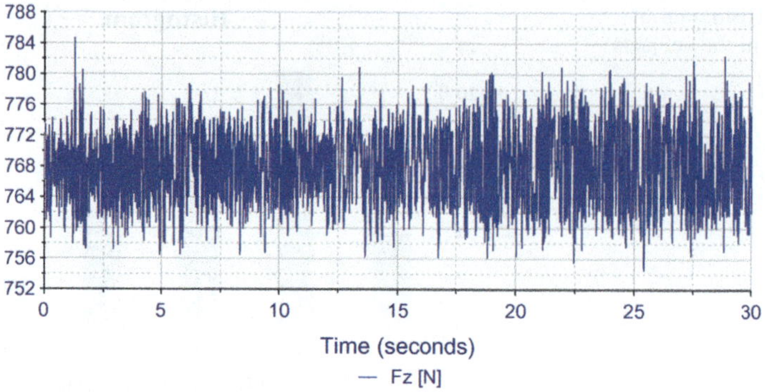

Fig. 23.6 Diagram of reaction force (no sound stimulus)

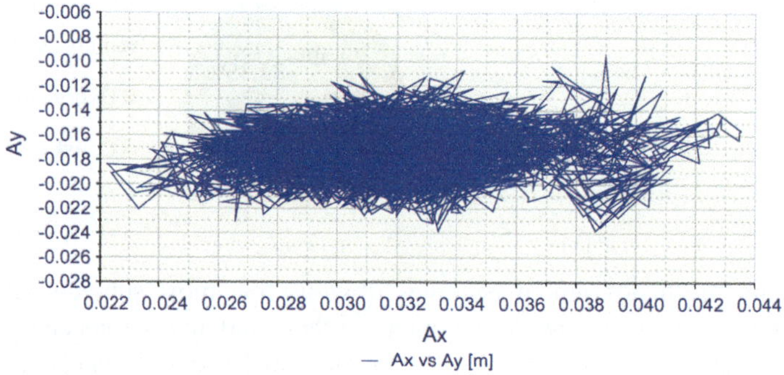

Fig. 23.7 Stability area (no sound stimulus)

example of diagram for the reaction force Fz and for the stability area with no sound stimuli is presented in Figs. 23.6 and 23.7.

The second experiment required the subjects to stand comfortably and relaxed on the force platform, while sometime around the 15th second, a sudden noise of 440 Hz corresponding to a concert sound was generated. The result was interesting to analyze because the reaction force showed a sudden decrease in variation after the 15th second, according to the software statistics and also the body sway was diminished for 8 subjects and increased insignificantly for the other 6 subjects. The diagrams in Figs. 23.8 and 23.9 are showing this phenomenon for the same subject as before.

Analyzing the results we find that the 440 Hz sound is mobilizing the body in maintaining the posture, though a small additional sway appears back and forth. This can be better understood following the graphical representations of the reaction force range compared for the situation with no sound stimulus and the one when a 440 Hz sound was generated, Fig. 23.10.

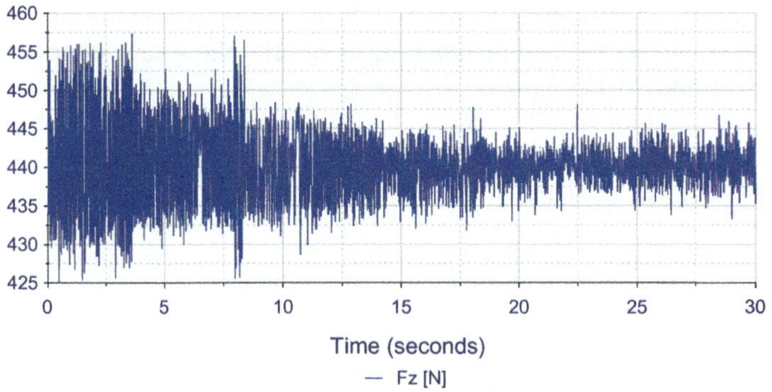

Fig. 23.8 Diagram of reaction force for 440 Hz stimulus

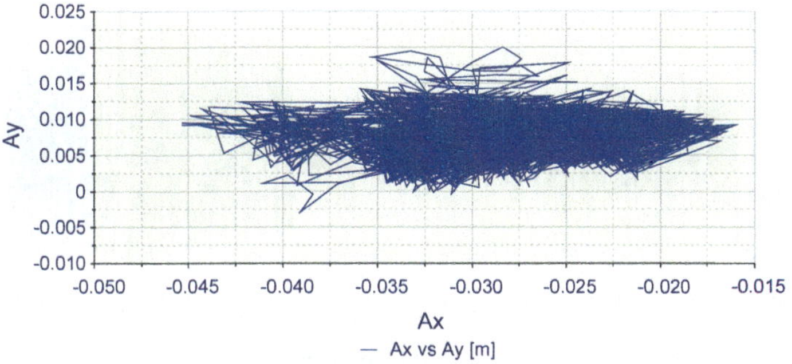

Fig. 23.9 Stability area for a 440 Hz stimulus

Fig. 23.10 Range of average force Fz (no stimulus compared to 440 Hz stimulus)

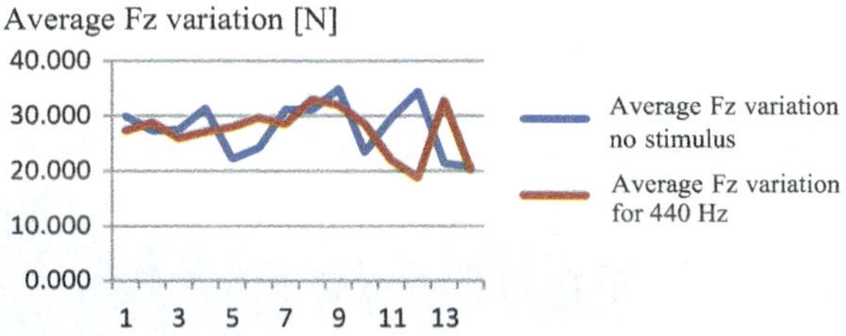

By analyzing the representation it becomes obvious that the difference in range for the two situations is insignificant, the force range even decreases sometimes during the sound stimulus.

The third part of the experiment required the subjects to stand on the force platform and at a given time one of the most disturbing sounds according to the classification of the annoyance points was generated. Tests were done for 4186 Hz and also for 16,000 Hz, but some lower frequency sounds were also tried.

Both diagrams representing the reaction force show an obvious increment at the moment the sound started and maintaining even after the sound stopped, as shown in Figs. 23.11 and 23.12.

The corresponding stability areas for 4186 and 16,000 Hz represented in Figs. 23.13 and 23.14 are showing a sway of 0.044 m and respectively 0.026 m, and also a small loss of stability towards the back, as though the subject was "hit" by the sound.

Even the forces along the axis Ox and Oy, whose range usually do not go over 8 N for Ox and 6 N for Oy, are reaching 10.83 N for Fx and 8.43 N for Fy when the 4186 Hz stimuli is applied, though the force variation does not exhibit changes at the moment of the sound stimuli, as shown in Figs. 23.15 and 23.16.

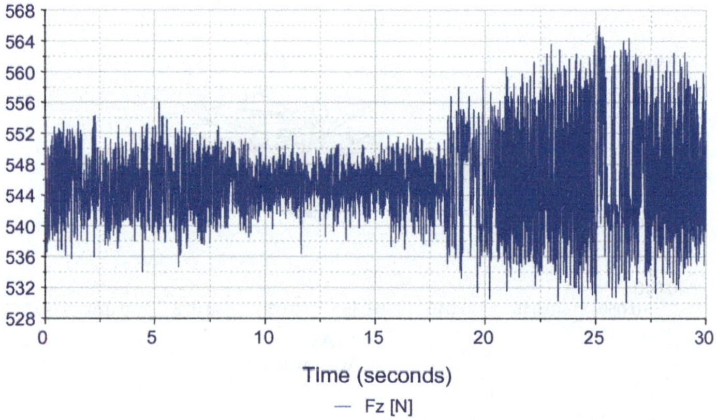

Fig. 23.11 Diagram of reaction force for a 4186 Hz stimulus

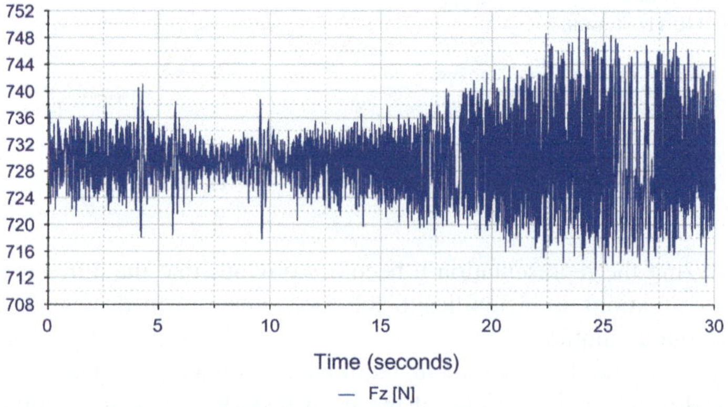

Fig. 23.12 Diagram of reaction force for a 16,000 Hz stimulus

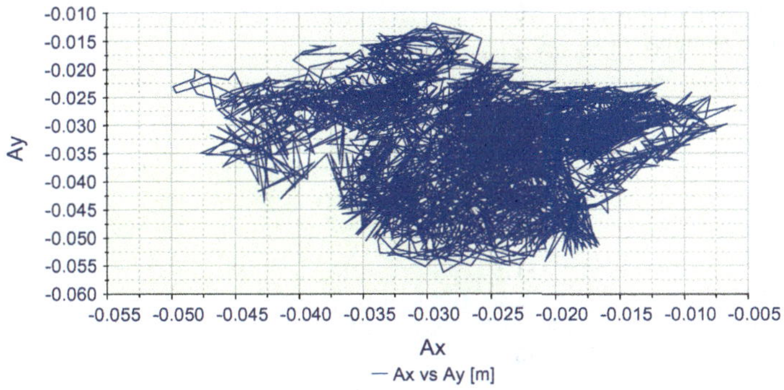

Fig. 23.13 Stability area for a 4186 Hz stimulus

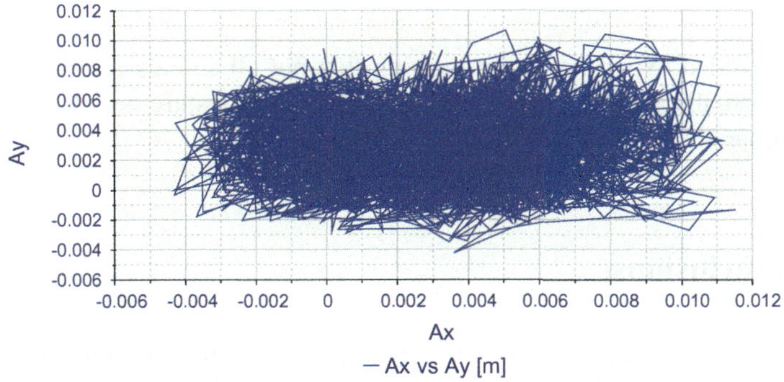

Fig. 23.14 Stability area for a 16,000 Hz stimulus

Fig. 23.15 Force Fx for a 4186 Hz sound stimulus

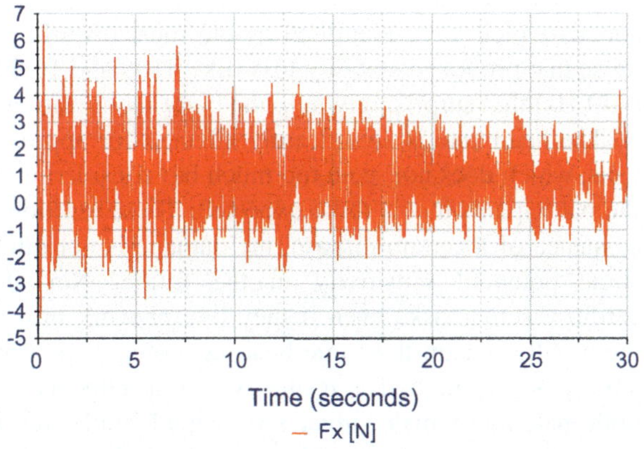

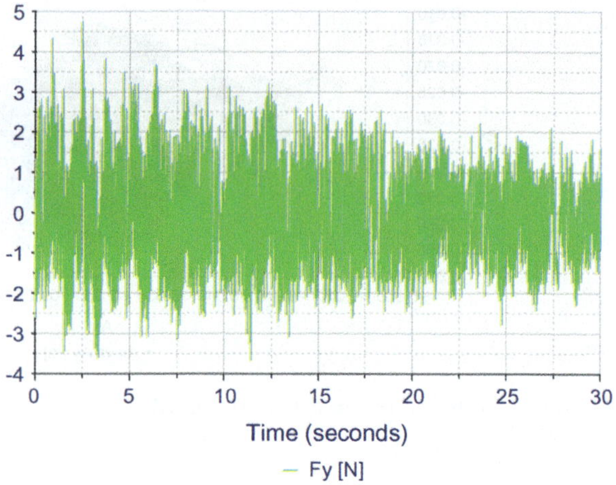

Fig. 23.16 Force Fy for a 4186 Hz sound stimulus

The performed experiments are proving that not all sounds produce unpleasant effects and usually the frequencies affecting the body posture are also those which produce subjective bothering effects.

23.4 Conclusions

Our environment is full of sounds but the exponential technological development lead also to some malfunctions in people's life, one of them related to noise pollution. Of course not every sound represents noise pollution; pleasant sounds in musical form may provide even therapy possibilities. The study performed in this paper aims at determining the subjective influence of different types of sound frequencies on healthy people and check them out with objective means. The sounds which were identified as the most disturbing by the participants are also those that provoked stability loss (like 4186 and 16,000 Hz). There are also sounds (440 Hz) that surprisingly help the body mobilization and improvement of balance.

The heart rate is significantly affected for approximately one third of the subjects, which does not seem too much but considering the age and the good health of the participants this is to be expected. These reactions do not necessarily mean the existence of a certain disease, the damage is not immediate but they may indicate some potential sensitivity in the future. Not all persons are subjected to cardio-vascular risks even if they live or work in a noisy environment, not all of them will develop the same hearing problems or vegetative diseases but the possibility is very high. The results were not influenced at all by the subjects' gender, both male and female subjects presented similar results in similar conditions.

For these reasons the studies aimed at individuals with different types of medical history, genetics, age, etc. subjected to various sounds, of different frequencies, and in future studies intensities, should further determine connections between the characteristics of the human beings and the physical characteristics of sounds that may help them to recover or trigger some kind of illness, in order to search for the first and avoid the second.

References

1. G. Jansen, E. Gros, *Noise Pollution* (Wiley, New York, 1986), pp. 225–236
2. *** European Commission, Science for Environment Policy, Thematic Issue: Noise Impacts on Health, Issue 47 (2015)
3. J. Gamache, *Effects of Sound on the Human Body* (Concordia University, Portland, 2015). https://www.slideshare.net/JustinGamache1/action-researchproposal-theeffectsofsoundonthehum-anbodyjustingamacheconcordiauniversity7272015
4. M. Shahid, H. Bashir, Psychological and physiological effects of noise pollution on the residents of major cities of Punjab (Pakistan). Peak J. Phys. Environ. Sci. Res. **1**(4), 41–50 (2013)
5. B. Siedlecka, M. Sobera, A. Sikora, I. Drzewowska, The influence of sounds on posture control. Acta Bioeng. Biomech. **17**(3), (2015)
6. M.R.M. Mainenti, L.F. De Oliveira, T. De Melo, M.A. De Lima, J. Nadal, Stabilometric signal analysis in tests with sound stimuli. Exp. Brain Res. **181**, 229–236 (2007)
7. M. Dozza, F.B. Horak, L. Chiari, Audio-biofeedback improve balance in patients with bilateral vestibular loss. Arch. Phys. Med. Rehabil. **86**(7), 1401–1403 (2005)

Chapter 24
Relative and Absolute Angles Computed from Pathologic Gait Data

Dan Ioan Stoia, Cosmina Vigaru and Lucian Rusu

Abstract The paper proposes a preliminary study upon the angular differences that occur between the sound and prosthetic lower limbs of a trans-tibial amputee patient. Based on landmark spatial coordinates measured during gait movement, the paper presents firstly the computation of the absolute and relative angles of the thigh and shank in all three anatomical planes. The relative angles were computed for knee and hip joints and not for the ankle joint due to the limited movement in the prosthetic limb. The spatial coordinates of the anatomical landmarks were measured using an ultrasound based motion system and further processed in order to obtain the angular parameters. At the end of the study, angles and angular velocities were presented for the sound and prosthetic limbs, underlining the differences that occur and the compensatory mechanisms.

24.1 Introduction

The increasing number of patients that undergo limbs amputations lead to a continuously growing market of prosthesis. Together with the technological development, the prosthesis advanced from rigid constructions to bionic limbs that are neurologically controlled [1, 2].

In order to develop and improve the prosthesis ability to behave as a natural limb, kinetic and kinematical analysis on unilateral and bilateral amputee's movements have to be made. Tae et al. performs dynamic gait analysis on

D. I. Stoia (✉) · C. Vigaru · L. Rusu
University Politehnica Timisoara, Timişoara, Romania
e-mail: dan.stoia@upt.ro

C. Vigaru
e-mail: cosmina.vigaru@upt.ro

L. Rusu
e-mail: lucian.rusu@upt.ro

D. I. Stoia · C. Vigaru · L. Rusu
Romanian Academy Branch of Timisoara, Timişoara, Romania

N. Herisanu and V. Marinca (eds.), *Acoustics and Vibration of Mechanical Structures—AVMS-2017*, Springer Proceedings in Physics 198,
https://doi.org/10.1007/978-3-319-69823-6_24

above-knee amputees in order to quantitatively evaluate the musculoskeletal system during walking and stair climbing. The aim of the study was to identify the differences in kinematical parameters and electromyographic response among different tasks. They identified a significant change in pelvic movements and hip abduction, while the other angles recorded in lower limb joints showed no statistic difference [3]. Schmalzet et al. conducted a study on transtibial and transfemoral amputees in order to evaluate their kinematics, kinetics and EMG condition in ascending and descending the stairs. Only the angles in the sagittal plane were recorded for hip, knee and ankle. The results indicated a qualitatively similar angular behavior among transtibial amputees and healthy subjects. But, a large difference in flexion moment was recorded. It seems that transtibial amputees compensate the loss of the foot and ankle with a high reduction of the flexion moment in initial contact phase of the gait [4]. The study of kinetic and kinematic gait parameters on patients having a magnetic prosthetic suspension was conducted by Eshraghi et al. The results of the new prosthesis suspension were compared with two other types of prosthesis. Comparisons between the sound leg and the prosthetic leg indicate a slightly reduced range of motion for the hip and knee of the prosthetic limb. However, no significant statistical differences were recorded [5]. Gaffney et al. investigate a lot of transtibial amputees, in order to assess the movement compensation that occurs in trunk and pelvis. They use computation of translational and rotational angular momentum based on motion data. Differences in translational angular momentum indicate unique movement patterns of healthy and amputee groups. The loss of ankle function in amputee patients prove to be compensated by higher rotational angular momentum of the trunk and pelvis [6].

In gait analysis there are a series of parameters that are generally accepted to characterize the motion. They are separated in two groups: spatial-temporal (swing and stance phase, step length, step width, walking velocity, stride time, and cadence) and kinematic (joint angles of the hip/knee/ankle, and thigh/trunk/foot angles) [7, 8]. The purpose of the study was first to compute the orientations of the thigh and shank and the joint angles of the hip and knee, and second to identify differences among the sound and prosthetic lower limbs of a transtibial amputee. The computing rely on spatial coordinates data acquired during subject's gait (walking) using an ultrasound measurement und system.

24.2 Materials and Methods

The measurements were conducted in the Motion Analysis Laboratory of Politehnica University Timisoara using Zebris ultrasound system. The system uses ultrasounds emission-reception principle in order to determine the spatial coordinates of the body landmarks. The body landmarks are physically related to the miniature microphones that are attached to the subject. The investigated patient undergoes a right limb transtibial amputation and at the moment of investigation was wearing a modular shank prosthesis that has one degree of freedom:

dorsal-plantar flexion of the ankle joint. The patient uses the prosthesis since 5 years, so he shows great confidence in movements.

After a short five minutes training time, the patient was equipped with the investigation markers and the anatomical landmarks were pointed. The patient was asked to walk on a flat surface with a self-selected, but as constant as possible, velocity. Five trials containing 3–4 steps each were recorded in order to obtain a statistically significant volume of data. The sampling rate of the recording system was selected to 25 Hz, a sufficiently high frequency for the patient's velocity.

The spatial coordinates of the anatomical landmarks recorded during gait trials were exported as time series for further processing. The Fig. 24.1 depicts the anatomical landmarks position for one limb, the other being symmetrical. It can be observed that X axis represents the direction of the body progression; Y axis is along the inferior-superior direction of the body and Z axis represents the lateral-medial direction.

The anatomical landmark for hip joint is defined only by one point, while for the knee and ankle joint two axes were defined. The articular axis of the knee is constructed from two points: internal and external, their position being known at any moment in time. The ankle joint is defined in the same way.

By using the landmark coordinates, the absolute and relative angles were computed. The absolute angles of the shank and thigh (Fig. 24.2) represent the orientations of the segment about the horizontal (X and Z) axes, while the relative angles represent the angles of the hip and knee joints (Fig. 24.3).

The absolute angles were computed using atan2 function (24.1), that returns the angle between the positive X or Z axis and the point defined by its coordinates (*ex. Xk*, *Yk*, *Zk*). Because the thigh and shank are always oriented in the positive half plane of the Y axis, the absolute angles computed are always positive. The tangent function was defined as a ratio between the Y and X coordinate differences (24.2).

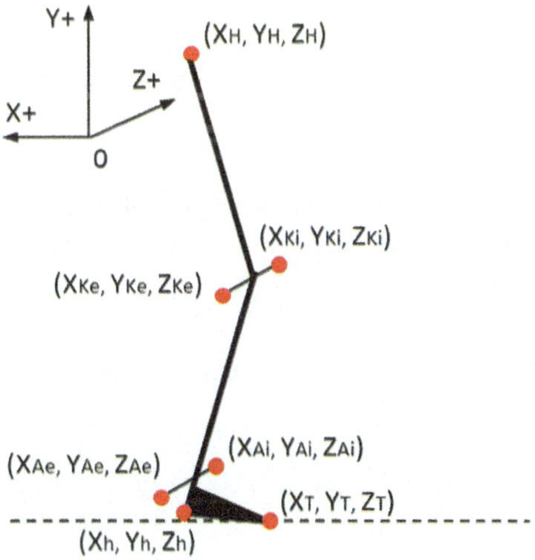

Fig. 24.1 Body landmarks' coordinates

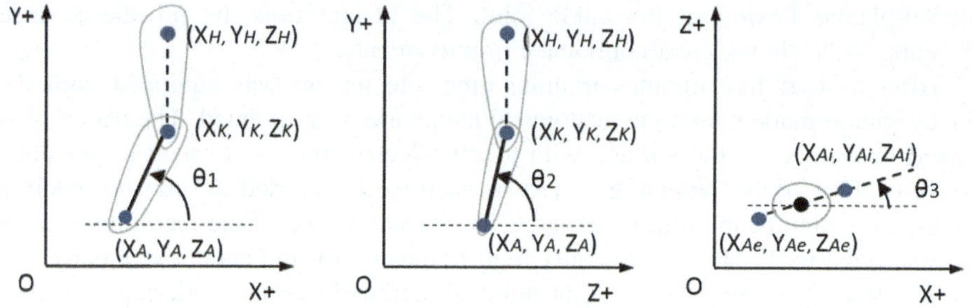

Fig. 24.2 Absolute angles measurement—shank orientation

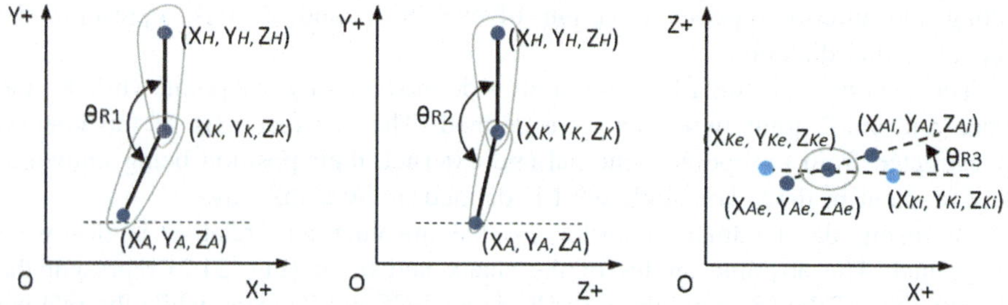

Fig. 24.3 Relative angle measurement—knee joint angles

In order to define the tangent function for the knee and ankle, the Euclidean segmental center was computed.

$$a\tan 2(y, x) = \begin{cases} \arctan\frac{y}{x}, & \text{if } x > 0 \\ \arctan\frac{y}{x} + \pi, & \text{if } x < 0 \text{ and } y \geq 0 \\ \arctan\frac{y}{x} - \pi, & \text{if } x < 0 \text{ and } y < 0 \\ \pm\pi/2, & \text{for } x = 0 \text{ and } y > 0; y < 0 \end{cases} \tag{24.1}$$

$$\theta_1 = a\tan 2\frac{Y_K - Y_A}{X_K - X_A}\ (\text{deg}) \tag{24.2}$$

The relative angles were computed using the cosine theorem (24.3), after the determination of the Euclidean distances between the points of known coordinates. The angular velocity in joints was calculated using the numerical differentiation (24.4).

$$\theta_{R1} = a\cos\frac{d_{HK}^2 + d_{KA}^2 - d_{HA}^2}{2 \cdot d_{HK} \cdot d_{KA}}\ (\text{deg}). \tag{24.3}$$

$$\theta_t = \frac{\theta_{t+1} - \theta_t}{\Delta t}(\text{deg}/s) \tag{24.4}$$

24.3 Results and Discussions

The results are presented in the form of angles and angular velocities as function of time, for both of the limbs. The representations (Figs. 24.4 and 24.5) depict slight amplitude and frequency modifications. The prosthetic limb generates a higher shank flexion in order to compensate the ankle function that the prosthesis is not providing (pronation-supination movement). In the transversal plane (*XoZ*), the angulation of the prosthetic limb exhibits a mechanical like movement, comparing to the sound leg. However, the dominant movement in gait (flexion-extension in *XoY* plane) shows no significant change from the qualitative point of view.

The angular velocity in knee joint exhibits higher values for prosthetic limb. This has to do with the confidence of the patient in standing on the prosthetic limb, and indicates a relatively lower stance phase. So, even if the angles seem qualitatively the same, there is an unbalanced situation of the stance phases of the two limbs (Fig. 24.6).

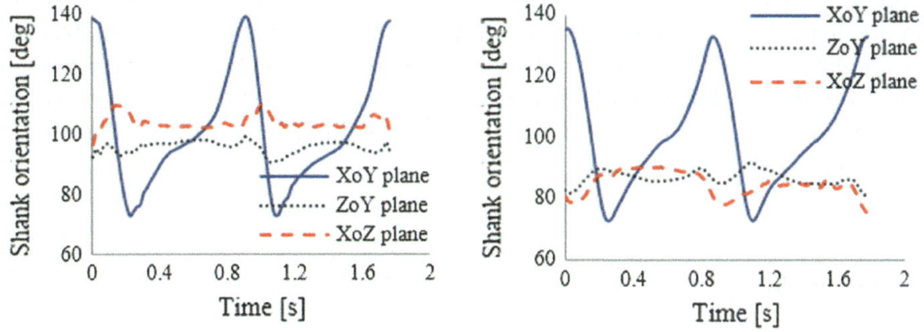

Fig. 24.4 Shank orientation of prosthetic (**a**) and sound (**b**) limbs

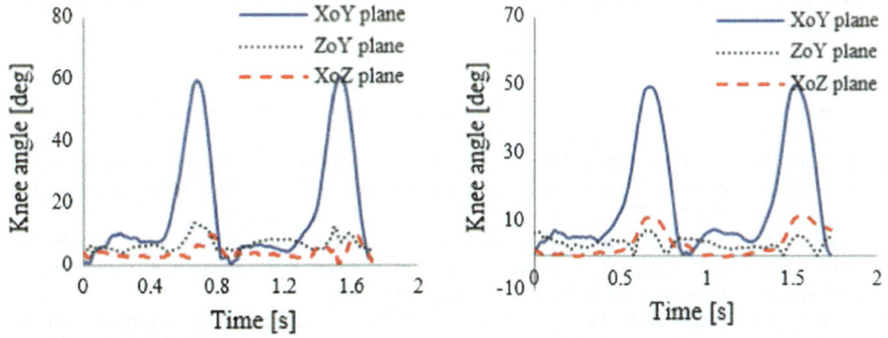

Fig. 24.5 Knee joint angle of prosthetic (**a**) and sound (**b**) limbs

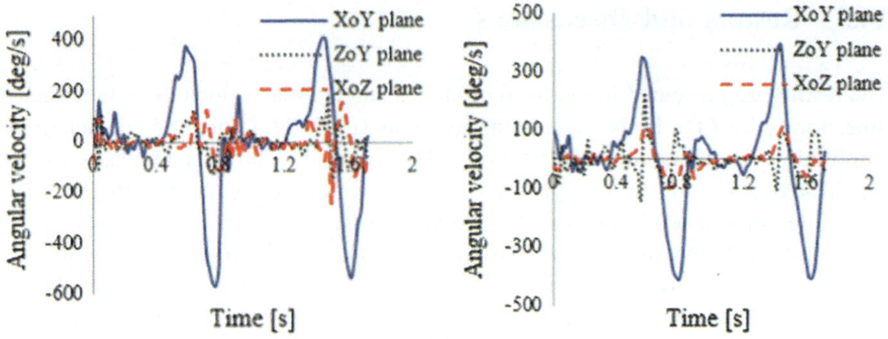

Fig. 24.6 Knee joint angular velocity of prosthetic (**a**) and sound (**b**) limbs

24.4 Conclusions

The paper presents a preliminary study of gait kinematics conducted on a transtibial amputee patient. The results have two characteristics: a qualitative aspect of the angles that indicates a good similarity between the sound and prosthetic limbs; and a quantitative aspect that shows differences in amplitude and frequency between the limbs, differences that lead to unbalanced gait phases. This type of data can be used in improving the designing of modular prosthesis, in order to achieve the natural functionality of the lower limb.

References

1. L.J. Hargrove, A.J. Young, A.M. Simon, N.P. Fey, R.D. Lipschutz, S.B. Finucane, E.G. Halsne, K.A. Ingraham, T.A. Kuiken, Intuitive control of a powered prosthetic leg during ambulation: a randomized clinical trial. JAMA **313**(22), 2244–2252 (2015)
2. E.H. Sinitski, A.H. Hansen, J.M. Wilken, Biomechanics of the ankle-foot system during stair ambulation: implications for design of advanced ankle-foot prostheses. J. Biomech. **45**(3), 588–594 (2012)
3. S.B. Tae, C. Kuiwon, H. Daehie, M. Museong, Dynamic analysis of above-knee amputee gait. Clin. Biomech. **22**, 557–566 (2007)
4. T. Schmalz, S. Blumentritt, B. Marx, Biomechanical analysis of stair ambulation in lower limb amputees. Gait & Posture **25**(2), 267–278 (2007)
5. A. Eshraghi, N.A.A. Osman, M. Karimi, H. Gholizadeh, E. Soodmand, W.A.B.W. Abas, Gait biomechanics of individuals with transtibial amputation: effect of suspension system. PLoS One **9**(5), e96988 (2014)
6. B.M. Gaffney, A.M. Murray, C.L. Christiansen, B.S. Davidson, Identification of trunk and pelvis movement compensations in patients with transtibial amputation using angular momentum separation. Gait & Posture **45**, 151–156 (2016)
7. C. Kirtley, *Clinical Gait Analysis, Theory and Practice* (Churchill Livingstone, Elsevier, London, 2006)
8. G. Robertson et al., *Research Methods in Biomechanics* (Human Kinetics, USA, 2004)

Chapter 25
Analysis of Temperature Variations in the Hand-Finger Assembly as a Result of Exposure to Vibration of Tools

Barbu Braun and Mihaela Baritz

Abstract Following the use and manipulation of vibratory tools, the human operator is exposed to a series of influences that can alter the anatomical structure of the hand-finger assembly (HFA) or trigger or aggravate certain pathological manifestations. An analysis of temperature and hand temperature variations is important from the perspective of personalized determination of behavior and maneuverability under these external actions. In the first part of the paper are reviewed several aspects related to the structure of the hand and the forms of manifestation of the effect of the vibrations on them. In the second part of the paper the experimental system is presented and the investigative procedure is developed. In the final part of the paper we analyze the results obtained from two subjects investigated by the projected procedure.

25.1 Introduction

According to the international human protection standards regarding its exposure to the effect of vibrations, the accepted level for different types of activities is determined according to the type of activity and duration of exposure. As the standardization system shows "ISO 10819 from 2013 specifies a method for the laboratory measurement, data analysis, and reporting of the vibration transmissibility of a glove with a vibration-reducing material that covers the palm, fingers, and thumb of the hand. ISO 10819:2013 standard specifies vibration transmissibility in terms of vibration transmitted from a handle through a glove to the palm of the hand in

B. Braun (✉) · M. Baritz
Design Product and Environment Department, University Transilvania Brasov, Brasov, Romania
e-mail: braun@unitbv.ro

M. Baritz
e-mail: mbaritz@unitbv.ro

N. Herisanu and V. Marinca (eds.), *Acoustics and Vibration of Mechanical Structures—AVMS-2017*, Springer Proceedings in Physics 198,
https://doi.org/10.1007/978-3-319-69823-6_25

one-third-octave frequency bands with central frequencies of 25–1250 Hz" [1]. Throughout its activity, the human operator is subjected to various forms and sources of vibrations in the environment, which manifests predominantly on the two components of the study: the manifestation of the vibrations in the entire human body and respectively the exposure of the arm-hand-fingers assembly to the vibrations generated by the manipulated objects. Daily exposure to the effect of vibrations, sometimes stretching over long periods of time, causes the appearance and development of physical damage to the human body, affects the circulatory and urological systems, or can cause long-lasting effects on the bone and limb joints. As shown in the documents of Canadian Center of Occupational Safety and Health "hand-arm vibration exposure affects the blood flow (vascular effect) and causes loss of touch sensation (neurological effect) in fingers. One of the earliest methods used for identifying the severity of these symptoms was the Taylor-Pelmear classification method" [2].

This classification method determines four exposure classes as follows: *1. Occasional, one or more finger; 2. Occasional, distal and middle finger; 3. Frequent attacks all/most fingers and 4. Same as 3 with skin change in finger tips.* The degree of severity of affecting the hand-fingers assembly by the vibration level also depends on another set of factors, such as physical, biodynamic, or specific (personalized). Usually, these physical damage results in the phenomenon of "white fingers syndrome", or damage at the joints and wrists and/or the elbow level.

White fingers syndrome (Raynaud syndrome), in advanced stages, is characterized by a white-violet color of the fingertips caused by damage of the arteries and nerves at the soft tissue level of the hand, from exposure to vibration [3, 4]. In the early stages of this event, symptoms in those fingers, affected by the effect of vibrations, are in the form of tingling, numbness and loss of sensitivity and control. Loss of sensitivity and finger control, even for short periods of time, can be a direct and immediate danger, especially when switching from vibration exposure to precision work. Damage to the wrist or elbow joints is often caused by long-term exposure to vibrations produced by percussion tools. This causes the joints and muscles of the forearm and is accompanied by reduced control and muscle strength in the forearm [5, 6].

Also, whole body vibrations can affect the perception of body orientation and the stability of the bipodal posture. As it is known, the level of vibration is dependent on the type of human body tissue or the area in which it measures their effect, to the area where the vibrations work. Thus, at the level of the arm (forearm) there is a frequency range of 5–10 Hz (16–30 Hz), compared to the hand (fingers) where this range increases to 30–50 Hz. At the level of the upper body of the human body this resonance frequency is between 20–30 Hz for the head and 20–90 Hz respectively for the eyes [5]. The hand, as the anatomical structure is a very important sub-assembly for various activities requiring precision, dexterity but also strength and resilience. By its structure, the hand may also suffer a series of dysfunctions that cause varying degrees of severity following the onset of Raynaud's syndrome. According to researchers [7], there are five levels of damage with different degrees of disfunctionality. The state of vasoconstriction in the hands and fingers as a result of exposure to a vibratory environment is determined by the structure of the palm circulatory system (the contact area with the vibratory tool), but also by the dimensions and shape of the outer/inner surfaces, according to the research [8] through different investigation protocols.

25.2 Experimental Setup

For the purpose of thermography analysis of temperature variations in the palm and fingers surface, in the case of vibration exposure, a specific methodology has been developed, consisting of an experimental system and a dedicated procedure.

The experimental system (Fig. 25.1) comprises a set of apparatus and devices (force plate-Kistler type, MediTutor gloves, HVM, vibration plate Sport, thermovision videocamera Flir type), which record the behavior of the subjects during the investigations, and the established procedure is specific to the analysis of the sample of subjects chosen to show this type of effect of the vibrations on the hand-finger assembly. The sample of subjects who participated in the investigations established by the procedure is composed of 10 subjects with an average age of 22.4 years, having good health and having anthropometric and physiological characteristics corresponding to the 75 percentile category. All of them gave written consent and were informed that they could withdraw from the experiment at any time.

The procedure for acquiring and analyzing the data obtained from the experiment follows a specific way of this investigation in the following steps: *1. Analyzing the environment in which the investigation will take place; 2. Analysis of the physiological and anthropometric data of subjects participating in the experiment; 3. The acquisition of initial phase data (hand-finger and whole body temperature, bipodal postural stability status, straight-hand dexterity) from subjects when they are resting and they are in a good metabolic state; 4. Exposing the vibration set to the hand-finger assembly using a drill and two different working materials; 5. The acquisition of data on vibration exposure established through the procedure data; 6. Data acquisition after the experiment of the same parameters*

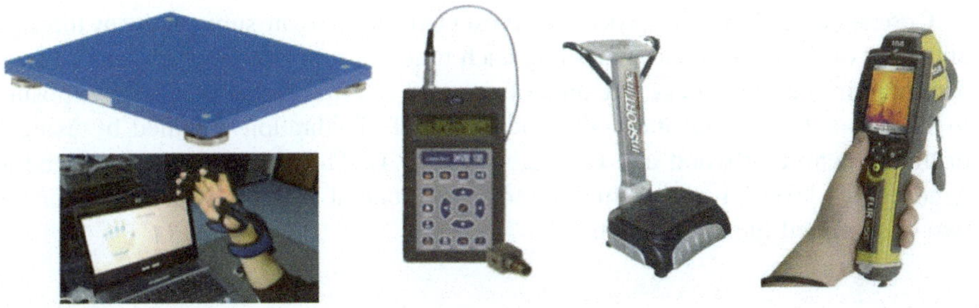

Fig. 25.1 The equipment used in the experimental setup

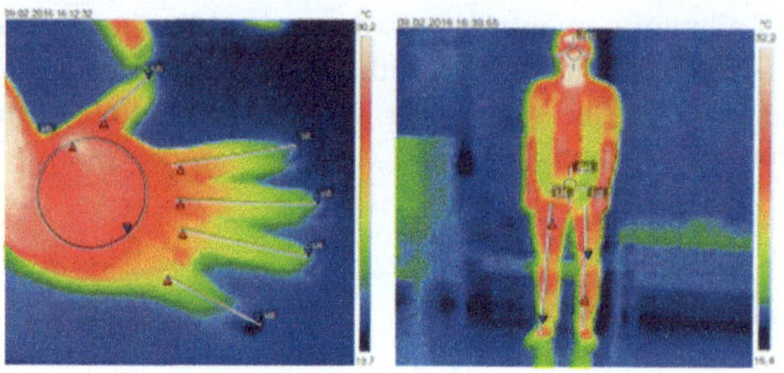

Fig. 25.2 Initial temperature of subject A (hand-finger and whole body)

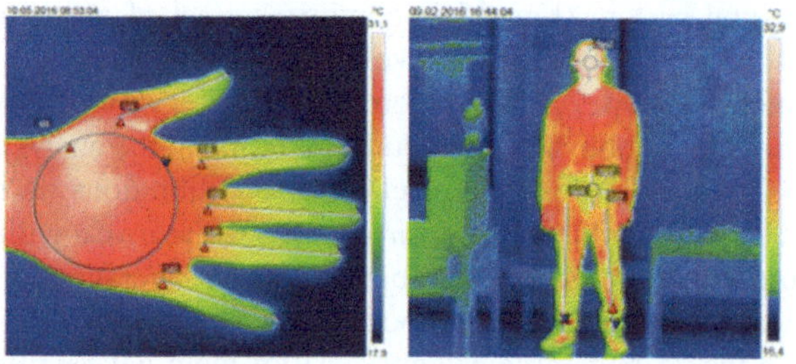

Fig. 25.3 Initial temperature of subject B (hand-finger and whole body)

(temperature, stability, dexterity); 7. Performing the procedures in section 4 if the subjects wear protective gloves against the effect of machine tool vibrations.

For example, two more special cases were selected from the sample of subjects, namely a male subject (A), with initial temperature of the hand-finger assembly 24.73 °C and a frail anatomical constitution (Fig. 25.2) and a second male subject (B), with an initial temperature of the same structure hand-finger of 26.88 °C and a more robust anatomical constitution, but part of the same percentile category (Fig. 25.3). From the point of view of the hand-finger temperature, the average parameters of the entire sample were 25.74 °C, corresponding to a lower average temperature than the average age; temperature of the environment (23 °C) and normal health status of the human subjects sample.

Correspondingly to the procedure presented, the human subjects went through all stages of the experiment, recording each time the temperature of the hand-finger assembly in the same areas, the area of postural stability before and after exposure to vibrations throughout the body, and the level of vibration obtained by using a drilling machine in wood and concrete (Fig. 25.4). The temperatures measured at hand-fingers level have been highlighted a substantial variation, especially for the two exemplified human subjects.

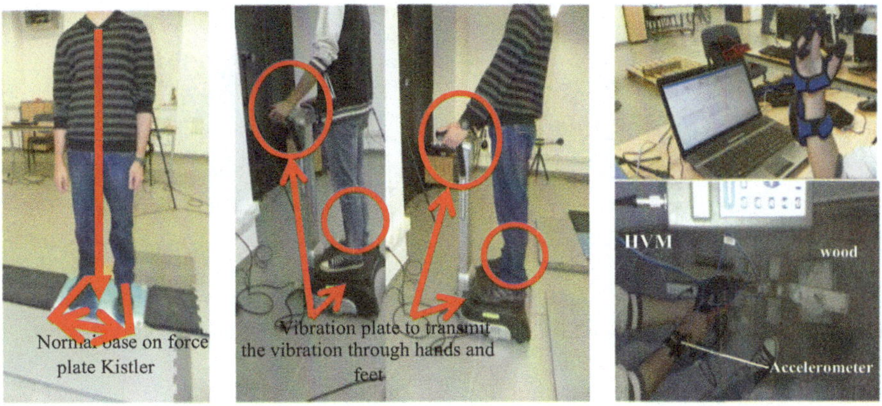

Fig. 25.4 Measurement of postural stability and inducing vibrations into human body

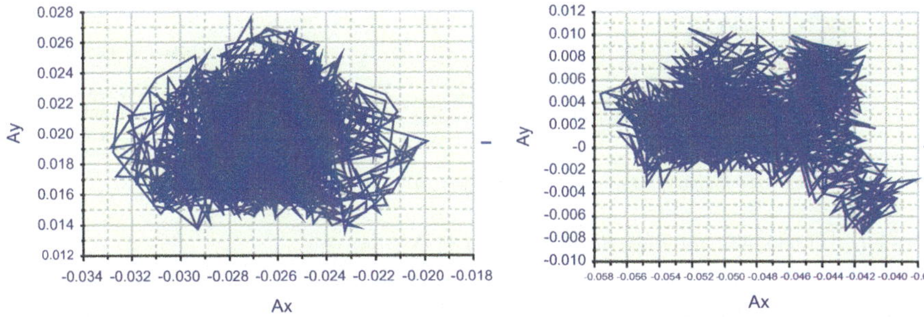

Fig. 25.5 Stability area [m × m] before/after vibrations exposure of human body subject *A*

25.3 Results

The results of the measurements obtained from the two subjects presented in the example will be compared in the following way: Initial results (Kistler force plate and body temperature) with the results after vibration exposure of the whole body (Figs. 25.5 and 25.6); Initial results (temperature and dexterity) with hand-arm vibration results on hand-arm; Vibration results of hand-fingers assembly without antivibration gloves with hand-arm vibration results with antivibration gloves; Results measured with HVM vibrator—wood comparison versus concrete with anti-vibration gloves versus without antivibration gloves, human subject A versus human subject B.

This will highlight the common aspects but especially the influence of hand-finger vibrations in relation to temperature and dexterity variation when performing maneuvers with or without antivibration gloves.

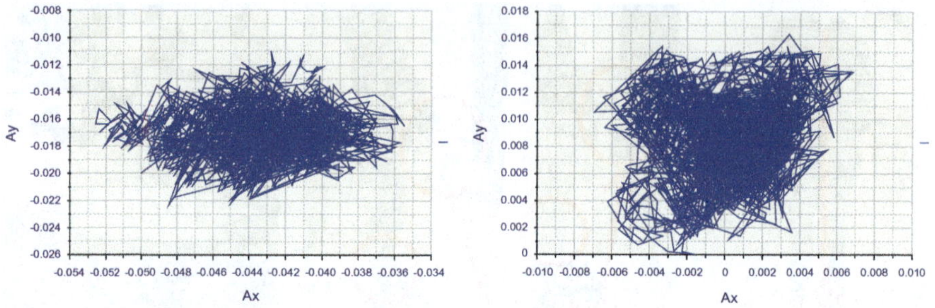

Fig. 25.6 Stability area [m × m] before/after vibrations exposure of human body subject *B*

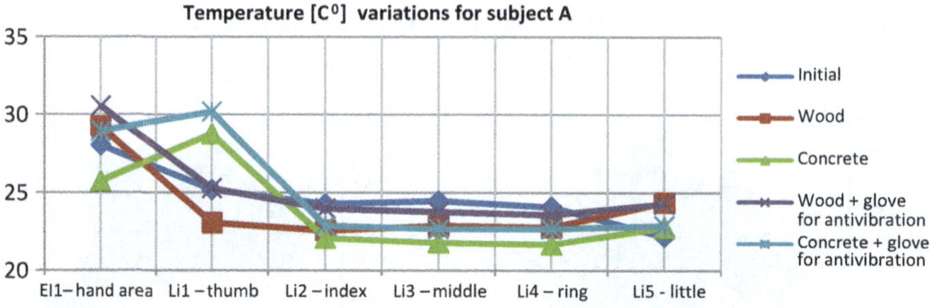

Fig. 25.7 Temperature variation for human subject A

As can be seen from Fig. 25.5, the area of stability of human subject A before exposure to vibration is extended, the subject is relaxed and the area value is 168 mm^2, unlike the situation after exposure to vibrations throughout the body when subject A presents more irregular area but with a much smaller surface, only 36 mm^2. This difference is due to the tension created in the musculature of the loco-motor system causing it to contract for bipodal support. Idem for subject B (Fig. 25.6).

In the initial phase of the investigations, the evaluations indicate, in the hand-fingers assembly, a higher temperature (about 20%) on the palmar surface and a correlated general variation, between the palm and fingers, for each human subject (Figs. 25.7 and 25.8). By comparing now the two subjects chosen in this example (A and B), in each of the analyzed situations, the data presented in the graphs of Figs. 25.9 and 25.10 are obtained.

Dexterity is assessed using MediTutor sensorial gloves, determining movement cycles and range of motion (ROM), in each case of experiments and the results are confirmed by the scores obtained in these evaluations.

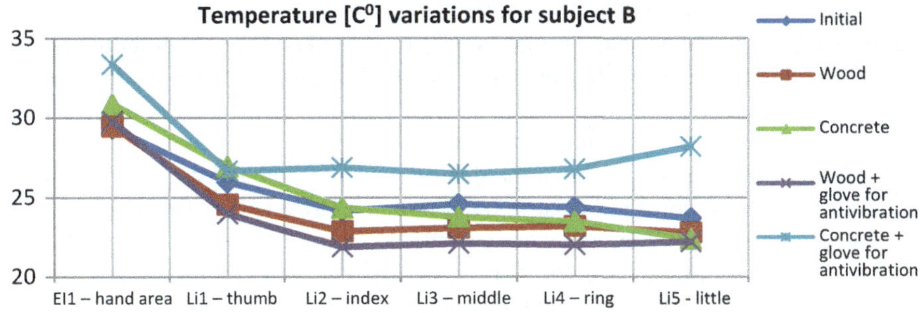

Fig. 25.8 Temperature variation for human subject B

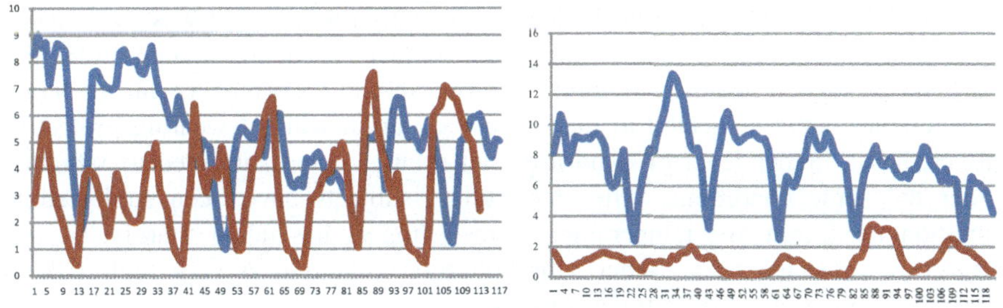

Fig. 25.9 Comparison between subject A (blue) and B (red) for vibrations (vertical [m/s^2], horizontal no. of samples) in drilling wood without (left) and with (right) anti-vibration gloves

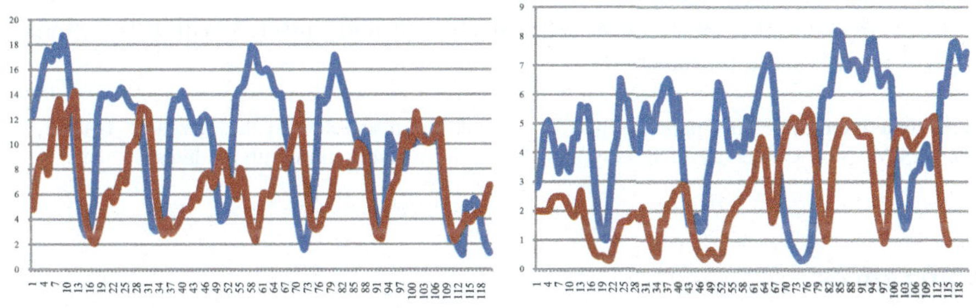

Fig. 25.10 Comparison between subject A (blue) and B (red) for vibrations (vertical [m/s^2], horizontal no. of samples) in drilling concrete without (left) and with (right) anti-vibration gloves

25.4 Conclusions

The human subject A generally shows a decrease in the maximum range of motion after inducing vibrations without antivibration gloves compared to baseline values, then increasing them with the use of antivibration gloves.

Frequency of movement increases after vibration induction. An exception is the thumb at which the values have progressively increased and the small thumb where there is a sudden increase in the frequency of wood drilling movements. For human subject B the values decrease when drilling wood and increase when drilling concrete. Frequency of movement increases after vibration induction. As a general conclusion it can be mentioned that the dexterity of a small hand decreases with the increase of the vibrations and a large hand remains close to the initial values.

In assessing the vibration exposure and the dexterity category developed by the human subjects from sample, the following aspects were found: when drilling wood, a too large glove does not cushion the vibration exposure level, even amplifies it, but a suitable protective glove dampens the level vibration approx. 4 times. When drilling the concrete: the too large glove for subject A created a higher impact force between hand and glove for subject B allowing for better contact and a slight decrease in the level of vibration to which it was subjected.

Thus, a suitable glove reduces the vibration level by approximately 40%. For optimum protection against vibration the glove should be compact with the hand.

A smaller hand, subjected to mechanical vibration without wearing protective gloves, has a higher level of exposure than a larger hand subjected to vibration under the same conditions. After the human subjects have been subjected to vibration for 2 min, the temperature of their fingers has fallen sharply, so if a subject is subjected to a longer vibration times, there is a very high risk of Raynaud's syndrome (white finger disease).

However, after being subjected to vibrations on the whole body, the human subjects recorded a remarkable rise in head temperature that shows the beneficial effects of vibrations on blood circulation and muscle toning. Vibration on the whole body, obtained from vibration plate stimulation, did not have any adverse effects on stability because they were in a reduced amount and enrolled within the physiological limits of the human body.

Acknowledgements This paper is part of the current researches in the Research Institute of TRANSIVANIA University of Brasov and due to the collaboration with students.

References

1. http://www.iso.org/iso/home/store/catalogue_ics/catalogue_detail_ics.htm?csnumber=46313. Accessed 30 Sept 2016
2. http://www.ccohs.ca/oshanswers/phys_agents/vibration/vibration_effects.html. Accessed 30 Sept 2016
3. http://www.who.int/occupational_health/pwh10_lecture_04_effects_on_human_body_vascula_system.pdf
4. https://www.ohao.org/PDF/Hand%20Arm%20Vibration%20Syndrome%20March%2020.pdf
5. Bruel and Kjaer, *Human Vibration*. http://www.bk.dk. Accessed 30 Sept 2016
6. http://clinicalgate.com/raynauds-phenomenon-3/. Accessed 30 Sept 2016

7. G. Gemne, I. Pykko, W. Taylor, P.L. Pelmear, Scand. J. Work Environ. Health **13**(4), 275–278 (1987)
8. H. Ron, *Hand-Arm Vibration Syndrome (HAVS).* Center for Research Expertises in Occupational Disease, St. Michel Hospital, Canada

Part IV
Vibration Problems in Industrial Processes

Chapter 26
An Example of Vibration and Low Frequency Noise Assessment for Railway Line

Aleksandar Gajicki, Momir Praščević and Darko Mihajlov

Abstract Reconstruction and modernization of railway section from Beograd Centar to Stara Pazova with total length of 34.7 km, provides upgrading of existing railway line with the designed speed up to 200 km/h. The paper discusses possible impacts to the environment in terms of vibration and low frequency noise during construction and exploitation of the railway line. The calculation of vibration and low frequency noise propagation and annoyance analysis were made using simulation modeling software package "VIBRA-1".

26.1 Introduction

This paper deals with vibration and low frequency noise assessment during reconstruction, modernization, construction and exploitation of the Beograd Centar—Stara Pazova section, which is a part of the following main railway lines: E 70 Beograd—Stara Pazova—Šid—State border—(Tovarnik) and E 85 (Beograd)—Stara Pazova—Novi Sad—Subotica—State border—(Kelebia). The Beograd Centar—Stara Pazova section is 34.7 km long in total [1].

Beograd Centar—Stara Pazova railway line (section) is existing source of vibration and low frequency noise which will change existing levels and disruption of the population living along the tracks after modernisation. In order to determine and minimize negative effects on the population it is necessary to calculate the future vibration and low frequency noise levels. Also, it is necessary to do annoyance analysis and propose abatement measures in accordance with obtained results.

A. Gajicki (✉) · M. Praščević · D. Mihajlov
Faculty of Occupational Safety, University of Niš, Niš, Serbia
e-mail: aleksandar@gajicki.com

M. Praščević
e-mail: momir.prascevic@znrfak.ni.ac.rs

D. Mihajlov
e-mail: darko.mihajlov@znrfak.ni.ac.rs

N. Herisanu and V. Marinca (eds.), *Acoustics and Vibration of Mechanical Structures—AVMS-2017*, Springer Proceedings in Physics 198,
https://doi.org/10.1007/978-3-319-69823-6_26

In the Beograd Centar—Stara Pazova railway corridor, which is 600 m wide, has been identified 9,109 buildings, out of them 5,385 are residential and/or vibration and low frequency noise sensitive.

The calculation of vibration propagation and annoyance analysis were made using simulation modeling software package "VIBRA-1" (ZC Ziegler Consultants AG and Swiss Rail) [2, 3].

26.2 Project Description

The existing double-track line from Beograd Centar to Batajnica will continue as it is but overhaul and modernization works will be undertaken. Only passenger traffic was planned on this section; freight traffic will be possible in exceptional circumstances.

Passenger and freight traffic will be merged and split in Batajnica station within the group of tracks belonging to Belgrade Railway Junction. In addition, Batajnica will be the terminal station for urban-suburban railway traffic in the City of Belgrade.

The existing right track on the Batajnica—Nova Pazova section will be moved to ensure 4.50 m distance between tracks (with 4.00 m wide formation) planned for passenger traffic. Additional two (freight) tracks spaced at 4.00 and 6.40 m away from the existing left track are planned after Batajnica station exit. The right freight track above double-track line for passenger traffic shall be grade-separated. After grade separation (crossing) of the right freight track over the existing two tracks, both freight tracks will run at the distance of 6.40 m from the tracks for passenger traffic.

A bed for four tracks was constructed on the Nova Pazova—Stara Pazova section; two tracks are already laid along track bed edges. The design envisages laying of two central tracks. All four tracks are designed for mixed traffic. Central tracks will be used by passenger and freight trains not planned to stop in Stara Pazova station but to run between Belgrade and Novi Sad. Peripheral tracks will be used by passenger trains planned to stop in Stara Pazova station and by trains running between Belgrade and Šid.

Planned design speeds on the Beograd Centar—Stara Pazova section:

- 100 km/h—from Beograd Centar station to Novi Beograd station and on the bridge over the Sava River;
- 120 km/h—from Novi Beograd station to Batajnica station through Bežanijska kosa tunnel;
- 200 km/h—from Batajnica station to Nova Pazova station on the tracks planned for passenger traffic;
- 120 km/h—from Batajnica station to Nova Pazova station on the tracks planned for freight traffic;

- 200 km/h—from station Nova Pazova to Stara Pazova station on the central tracks (inner two tracks), and
- 160 km/h—from station Nova Pazova to Stara Pazova station on the outer tracks.

In order to make the railway line capable for the design speeds, the following activities on subsections are proposed for modernization of Beograd Centar—Stara Pazova railway section depending on the current limitations and planned traffic organization:

1. Beograd Centar—Batajnica subsection: reconstruction of existing railway line and modernization of railway devices,
2. Batajnica—Stara Pazova subsection: construction of two new tracks, reconstruction of existing tracks and modernization of equipment and devices,
3. Beograd Centar—Stara Pazova section: reconstruction of station facilities in accordance with current and new technology tasks in the stations.

According to the design, the following stations will be retained: Beograd Centre, Novi Beograd, Zemun, Zemunsko Polje, Batajnica, Nova Pazova and Stara Pazova.

The current Tošin Bunar stopping place will be relocated towards Novi Beograd stations and two new stopping places, Altina and Kamendin will be open to traffic.

UIC GC loading gauge, which enables all modes of combined transport, was adopted. Cross section was designed in compliance with the Rulebook on design, reconstruction and construction of specific elements of railway infrastructure for particular main railway lines [4].

26.3 Vibration Suppression Measures

The laws of the Republic of Serbia do not stipulate permissible values for vibrations and low frequency noise due to railway traffic [5]. That was the reason that in order to find the impact of vibrations and to plan protective measures we had to use provisions from DIN 4150-2 [6] and Directive of the Swiss Federal Office (BEKS 1999) [7].

DIN 4150-2 standard assesses exposure of population inside houses to occasional and short time structural vibrations in the range of 1–80 Hz. The standard also specifies maximum allowed vibrations to prevent annoying people inside buildings. Maximum permitted mean values of vibrations by different use of buildings and period of day are shown in Table 26.1.

BEKS standard assessed the impact of low frequency noise from railway traffic and specified noise levels for newly built railway lines and separately for the lines subjected to modernization and repair schemes. Permissible levels of low frequency noise by zone, period of day and class of railway line are shown in Table 26.2.

Vibrations and low frequency noise due to railway traffic were calculated with the aid of “VIBRA-1” software package. Vibrations were calculated on the basis of

Table 26.1 Permitted mean values of vibrations according to DIN 4150-2

Use	Day (6^{oo}-22^{oo}) (mm/s)	Night (22^{oo}-6^{oo}) (mm/s)
Fully industrial	0.20	0.15
Predominantly industrial	0.15	0.10
Mixed use	0.10	0.07
Dwelling only	0.07	0.05
Vacation, recovery, therapy	0.05	0.05

Table 26.2 Permissible low frequency noise level according to BEKS

Zone	Newly built railway line		Modernized railway line	
	Day (6^{oo}-22^{oo}) (dB)	Night (22^{oo}-6^{oo}) (dB)	Day (6^{oo}-22^{oo}) (dB)	Night (22^{oo}-6^{oo}) (dB)
For dwelling only	35	25	40	30
Mixed use	40	30	45	35
Industrial	60	60	60	60

individual train passing while the total impact was equal to a sum of standardized procedures in DIN 4150-2.

(a) ***Vibrations during execution of works***

In the course of modernization of the Belgrade Centre—Stara Pazova railway section, vibrations and low frequency noise will increase near working sites. Vibration levels and low frequency noise will depend on the kind of works, plant or equipment and distance of housing units and buildings. Vibrations and low frequency noise due to works will be of temporary nature and will stop on completion of the works. Their level was not assessed in this design stage as knowledge lacked on the works, kind and types of machines etc. It is essential that organisation and technology are so planned that they minimise negative vibration and low frequency noise impact on population in the course of works.

(b) ***Vibrations during operations***

Data on prospective scope of railway traffic were used to analyse vibration and low frequency noise. Data for modelling and analysis were taken from the Preliminary design for reconstruction, modernization and buildings of the double track railway line Belgrade—Novi Sad—Subotica—Hungarian border, Belgrade Centre—Stara Pazova section.

Depending on operational technology international passenger trains to Novi Sad will run at maximum speed of 200 km/h and international passenger trains to Sid will run at maximum speed of 120 km/h. Maximum speed of passenger trains in regional traffic to Novi Sad will be 160 km/h and for trains to Sid will be 120 km/h. City-suburb trains that end their journey in Batajnica station run at maximum speed of 100 km/h. In Novi Sad direction passenger trains will number 9 pairs of international trains and 31 pairs of regional trains while to Sid, 8 pairs of international and 15 pairs of regional trains will run. In city-suburb transport, there will be 53 pairs of passenger trains.

Maximum permitted speed of goods trains is 100 km/h in international transport and 80 km/h in domestic transport. In the direction to Novi Sad international cargo will be hauled with 20 pairs of trains and in domestic transport with 5 train pairs. In the direction to Sid international cargo will be hauled with 6 pairs of trains and in domestic transport with 8 train pairs. It is assumed that average net train mass in domestic transport will be 500 and 900 tons in international transport with the coefficient of empty wagon running of 1.4.

Solely impacts of railway traffic on the Belgrade Centre—Stara Pazova section were considered.

The calculated mean values of vibrations and low frequency noise was done by section and period of day. The division into sections was according to changes of the parameters that condition the extent of vibrations and low frequency noise, and they are track configuration, track construction, speed, traffic volume on track, and kinds of trains running on it. Calculations was performed for following sub-section:

1. Belgrade Centre—New Belgrade section.
2. New Belgrade—TPS Zemun and TPS Zemun—Batajnica section.
3. TPS Zemun—TPS Zemun section.
4. Batajnica—km 23+762 (right side) section.
5. Batajnica—km 23+762 (left side) section.
6. km 23+762—Nova Pazova section.
7. Nova Pazova—Stara Pazova—km 36+061 section.

Example of calculation results are shown in Table 26.3. The given values refer to both the left and right side of the line unless otherwise marked. Assessed values of vibrations and low frequency noise that exceed permissible values are shaded in Table 26.3.

The running of train sets over switches increases levels of vibrations and low frequency noise. The impact initiated by switches is considered in the areas of pointwork (impact from the first switch to end of the last switch) assuming that switches which lie at mutual distances not more than 100 m will continually have an effect on an increase of vibration and low frequency noise. Calculations was performed for:

1. New Belgrade station (entry)
2. Novi Beograd (exit), Zemun, Zemun Polje and Batajnica (entry) stations
3. Batajnica (exit) (right side) station
4. Batajnica (exit) (left side) station
5. Nova Pazova (entry) station
6. Nova Pazova (exit) and Stara Pazova stations

Table 26.3 Example of mean values of vibrations and low frequency noise on the Nova Pazova—Stara Pazova—km 36+061 (end of section)

Distance from track (m)	Mean value of vibrations		Low frequency noise		Mean duration of vibrations (s)
	Day (mm/s)	Night (mm/s)	Day (dB)	Night (dB)	
5	0.072	0.062	29.1	34.2	22.1
10	0.049	0.041	24.9	29.7	22.1
15	0.039	0.032	22.4	27.0	22.1
20	0.033	0.027	20.5	25.1	22.1
25	0.029	0.023	19.0	23.5	22.1
30	0.025	0.020	17.8	22.3	22.1
35	0.023	0.018	16.7	21.1	22.1
40	0.021	0.017	15.7	20.2	22.1
45	0.019	0.015	14.9	19.3	22.1
50	0.018	0.014	14.1	18.5	22.1

Table 26.4 Example of mean values of vibrations and low frequency noise in the pointwork area in Nova Pazova (entry) station

Distance from track (m)	Mean value of vibrations		Low frequency noise		Mean duration of vibrations (s)
	Day (mm/s)	Night (mm/s)	Day (dB)	Night (dB)	
5	0.140	0.154	33.8	37.8	22.1
10	0.092	0.090	28.8	32.5	22.1
15	0.068	0.062	25.6	29.0	22.1
20	0.052	0.046	23.2	26.5	22.1
25	0.042	0.036	21.2	24.4	22.1
30	0.034	0.029	19.4	22.6	22.1
35	0.027	0.024	17.8	20.9	22.1
40	0.023	0.020	16.4	19.5	22.1
45	0.020	0.017	15.2	18.3	22.1
50	0.018	0.015	14.2	17.3	22.1

The example of the calculated mean values of vibrations and low frequency noise in stations and periods of day is given in Table 26.4. They refer to both the left and right side of the stations unless otherwise marked. Assessed values of vibrations and low frequency noise that exceed permissible values are shaded in Table 26.4.

Table 26.5 Example of mean value of vibrations and low frequency noise in the area of Bezanija tunnel

Distance from track (m)	Mean value of vibrations		Low frequency noise		Mean duration of vibrations (s)
	Day (mm/s)	Night (mm/s)	Day (dB)	Night (dB)	
5	0.085	0.036	29.8	30.2	8.9
10	0.048	0.020	23.8	24.2	8.9
15	0.034	0.015	20.2	20.6	8.9
20	0.027	0.011	17.7	18.1	8.9
25	0.022	0.009	15.7	16.1	8.9
30	0.019	0.008	14.1	14.5	8.9
35	0.017	0.007	12.7	13.1	8.9
40	0.015	0.006	11.5	11.9	8.9
45	0.013	0.006	10.4	10.9	8.9
50	0.012	0.005	9.5	9.9	8.9

On the Belgrade Centre—Stara Pazova section there are two tunnels. The Senjak tunnel is 188 m long, from km 0+443 to km 0+631, while the Bezanija tunnel is 1921 m long, from km 5+563 to km 7+485.

The example of the calculated mean values of vibrations and low frequency noise by tunnel and period of day is given in Table 26.5.

The given values refer to both the left and right side of the tunnel unless otherwise indicated. The assessed values of vibrations and low frequency noise that exceed permissible values are shaded in Table 26.5.

26.4 Conclusions

Modernized and upgraded line from Beograd Centar to Stara Pazova will contribute to a small extent in increasing existing vibration and low noise levels in areas along the railway.

Measures for suppression of vibrations and low-frequency noise are not planned for buildings located on railway land, non-residential buildings (according to town development and spatial plans) and buildings situated in the right-of-way where the construction activities are forbidden. It was estimated that other buildings will not be affected by vibrations and low-frequency noise generated by railway traffic on the Belgrade Centre—Stara Pazova section.

During the preparation of design for modernization of the existing tracks and construction of new tracks, the reduction of breaks in the running surface of a rail (rail joints, switches and crossings) should be taken into account. Rails should be welded to form continuous welded rails (CWR).

In the course of railway operation running surface of a rail shall be flat and smooth. During the repair of rails all upsweeps and downsweeps at welding points shall be removed. Maintenance plans shall include regular rail grinding.

References

1. Preliminary Design of Modernization of the Beograd—Subotica—State border (Kelebia) railway line, Beograd—Stara Pazova section, in Beograd, New Beograd, Zemun, Zemunsko Polje, Batajnica, Nova Pazova, Stara Pazova on the building plots according to the list attached to documentation, Institute of transport CIP, Belgrade (2016, In work)
2. Manual "VIBRA-1", Ziegler Consultants, Zürich (2010)
3. H. Kuppelwieser, A. Ziegler, A tool for predicting vibration and structure-borne noise emissions caused by railways. J. Sound Vib. **193**, 261–267 (1996)
4. Rulebook on design, reconstruction and construction of specific elements of railway infra-structure for particular main railway lines. Official Gazette of the RS, No. 100, Belgrade (2012)
5. P. Elias, M. Villot, Review of existing standards, regulations and guidelines, as well as laboratory and field studies concerning human exposure to vibration. RIVAS Project, UIC (2012)
6. DIN 4150-2:1999-06, Vibrations in buildings—part 2: effects on persons in buildings. DIN Deutsches Institut für Normung e.V., Berlin (1999)
7. BEKS, *Assessment of vibration and structure borne-noise from railway traffic* (Swiss Association for Standardization, Winterthur, 1999), p. 1999

Chapter 27
Vibration Analysis of the Boiler Supply Air Fan—A Case Study

Dragan Jovanović, Miomir Raos, Milena Jovanović, Milena Stanković, Ljiljana Živković and Milan Protić

Abstract The air supply fan is an important element for boiler operation, responsible for providing the necessary air flow with respect to the combustion process demands. Vibration measurements of the boiler fan were conducted on measurement points defined by standard ISO 10816-1. Vibration measurements were done with Bruel&Kjaer Vibrotest 60, vibration analyzer and offline data acquisition device. The measured parameters are related to: levels of vibrations, parameters of condition of bearing BCU (Bearing Conditions Unit) and BCS (Bearing Spectrum Conditions), analysis of the characteristic frequency of vibrations is presented on FFT Spectrum diagrams. Based on the analysis the main conclusion is: measured vibrations are high and fan is in the field of unstable work.

D. Jovanović (✉)
Faculty of Mechanical Engineering, University of Niš, Aleksandra Medvedeva 14, 18000 Niš, Serbia
e-mail: dragan.jovanovic@masfak.ni.ac.rs

M. Raos · M. Jovanović · M. Stanković · L. Živković · M. Protić
Faculty of Occupational Safety, University of Niš, Carnojevica 10a, 18000 Niš, Serbia
e-mail: miomir.raos@znrfak.ni.ac.rs

M. Jovanović
e-mail: milena.jovanovic@znrfak.ni.ac.rs

M. Stanković
e-mail: milena.stankovic1@znrfak.ni.ac.rs

L. Živković
e-mail: ljiljana.zivkovic@znrfak.ni.ac.rs

M. Protić
e-mail: milan.protic@znrfak.ni.ac.rs

N. Herisanu and V. Marinca (eds.), *Acoustics and Vibration of Mechanical Structures—AVMS-2017*, Springer Proceedings in Physics 198,
https://doi.org/10.1007/978-3-319-69823-6_27

27.1 Introduction

Boiler is, according to the definition, the flow-through system for the transformation of energy. This paper presents the measurements that were conducted on the fan for supply air of the boiler 'Minster', whose main function is the generation of low pressure steam for the production process. This kind of fan has the main role in providing the necessary amount of fresh air in order to enable the start of the process of combustion. One of the most important parts in this kind of systems is a fan whose mission is to provide fresh air circulation and its transport, with a respect to the designed parameters of flow, pressure drop through the system and with appropriate air flow and velocity [1]. In the case of malfunctioning of the boiler fan the operation of the boiler is threatened, which directly compromises the production process.

In technical practice the exact criteria are made for the assessment of the system status based on the measurement of the absolute vibration of the bearings on the outer surface of the machine. This paper presents aspects of predictive maintenance, which implied trend monitoring of vibration levels at critical points and also the status of parameters of bearing conditions, through the index of bearing condition BCU [2].

27.2 Standards and Criteria for Evaluation of Condition of the Machine

To assess the condition of all mechanical systems it is necessary to determine technical normative basis, which can be used for accomplishing verified values. The exact criteria are given for the assessment of the condition of the machine, based on the measurement of the absolute vibration of the bearings, measured on the bearing housing or bearing housings carrier. Experience in measuring the absolute vibrations are translated into a standard VDI 2056 and ISO 2372. Both of the standards are used for the criteria medium speed vibration in the frequency range from 10 to 1.000 Hz. ISO 10816-1 criteria for assessment of the condition of the machine retained a specified locations for sensors used to control vibration levels.

Machines are divided into a number of different classes, and standards are made for the most used kind of systems that are classified in four basic groups:

- Group I, which includes small machines (and fans), and electric motors with a power rating up to 15 kW;
- Group II, which includes medium machines (and fans), and electric motors with a power rating from 15 to 75 kW;
- Group III, which includes large machines (and fans), and electric motors with a power rating more than 75 kW, and large machines on rigid housing;
- Group IV, which includes turbines and large machines on flexible housing.

Although the standards provide several guidelines for the identification of machines belonging to a group, understandably not all machines are specified. Figure 27.1 presents the norms and values of the average velocity of vibration on which the assessment of the condition of the machine is based on.

Based on the listed standard ISO 10816, measurement points are selected for the determination of the vibration levels, and the machine that is presented in the paper is classified as a machine in group II according to standard ISO 10816-1.

Figure 27.1 shows that the machines in the group II, is in the field of good operation when measured vibration level is not greater than 1.12 mm/s, satisfactory to 2.8 mm/s, unsatisfactory (acceptable) within 7.1 mm/s and unacceptable when measured vibration level is greater than 7.1 mm/s.

In Figs. 27.2 and 27.3 locations of the measurement points are presented, according to standard ISO 10816-1.

Method of measurement parameters depend on the kind and type of unit subjected to the measurement process, where different points of measurement are determined for different units [3–5]. The obtained data should indicate the trend of the measured parameters of importance for the functioning of the fan unit and air conditioning installation [6].

The measured parameters are related to:

- levels of vibrations (including measurement points on the motor and on the fan) in the directions of coordinate axes, according to ISO 10816-1.
- parameters of condition of bearing through the index of bearing condition BCU (Bearing Conditions Unit) and BCS (Bearing Spectrum Conditions)
- analysis of the characteristic frequency of vibration in the spectrum, presented by FFT (Fast Fourier Transformation).

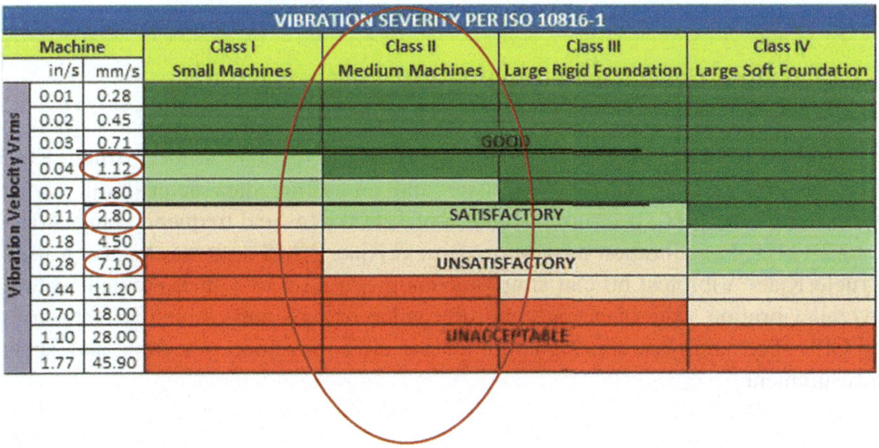

Fig. 27.1 Values of the average velocity of vibration for each group of machines

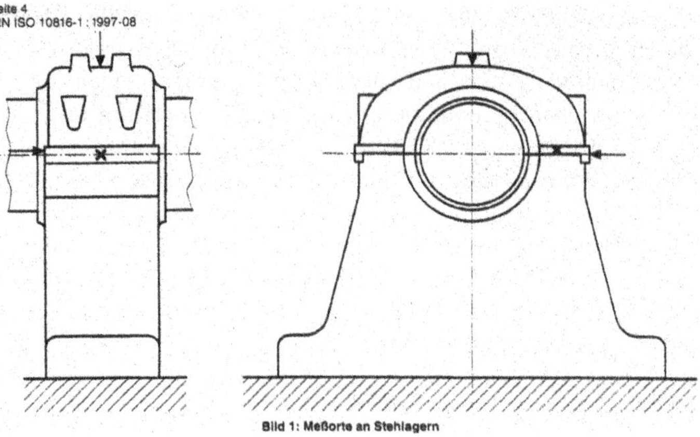

Fig. 27.2 Measurement points section a-a

Fig. 27.3 Measurement points on bearing housing section b-b

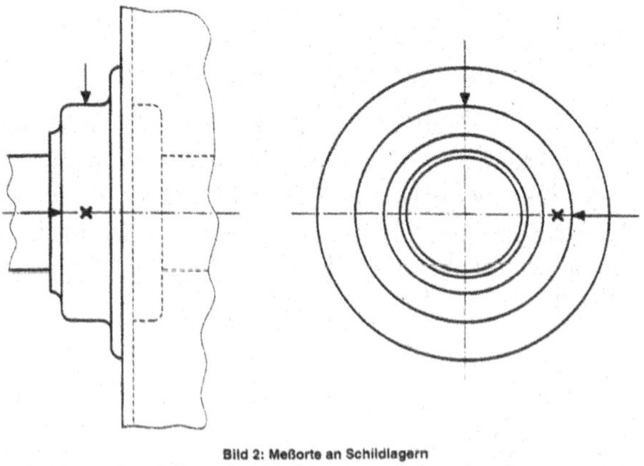

27.3 Vibration Measurement and Analysis

Vibration measurements presented in this paper were done with a Bruel&Kjaer Vibrotest 60 system, vibration analiser and an offline data acquisition device. Vibrotest 60 has FFT spectrum resolution of 12800 lines and frequency ranges from 2 to 20 kHz. For vibration analysis, a Bruel&Kjaer XMS software has been used. Bruel&Kjaer Vibrotest 60 can simultaneous process up to 5 measurement values: overall vibration value, bearing condition value or bandpass value, FFT-spectrum or CPB-spectrum, bearcon signature or selective envelope detection and speed measurement.

Technical Data of Bruel&Kjaer Vibrotest 60 are summarized as:

- measurement channels are Real Dual-channel with 4 internal measurement paths for parallel acquisition of machine vibrations and bearing condition at each channel with Plus speed channel,
- AD converter AD converter 16 Bit (96 dB) dynamic 1 Hz–20 kHz usable frequency range 1 AD converter for BCU 10 Bit (60 dB) dynamic 0–48 kHz frequency range,
- Vibration measurement types Vibration acceleration Vibration velocity Vibration displacement,
- Signal detection types: RMS value Peak value (true and calculated) Peak-peak value (true and calculated) Bearing Condition Unit (BCU),
- Broad-band Overall values High-pass: 1 Hz–10 kHz (selectable in 1/3 octave steps) Low-pass: 10 Hz–20 kHz (selectable in 1/3 octave steps),
- Bandpass measurements High-pass: 630 Hz–16 kHz (selectable in 1/3 octave steps) Low-pass: 800 Hz–20 kHz (selectable in 1/3 octave steps),
- Bearing Condition Unit (BCU) 13 kHz–48 kHz. FFT/BCS-spectrum Frequency ranges High-pass 1/2/5/10 Hz Low-pass 20/50/100/200/500 Hz 1/2/5/10/20 kHz,
- FFT/BCS/SED No. of lines 100/200/400/800/1,600/3,200/6,400/12,800,
- FFT/BCS/SED Averaging RMS, linear, exponential, peak-hold,
- FFT/BCS/SED Windowing functions: Uniform, Hanning, Flat-top. Process values inputs +/−30 V, 0/4–20 mA,
- Speed: Speed range 30–600,000 rpm Speed/Ref. Ratio (S/R) xxx/yyy $0.01 \leq S/R \leq 99$,
- Measurement accuracy Vibration measurement typically +/−2% of measured value Speed measurement typically +/−0.01% of measured value Phase angle typ. +/−1° Process values +/−0.3 V, +/−0.5 mA
- Tracking/Field Balancing Frequency range 1 Hz–10 kHz Automatic selection of bandwidth: Tracking 2–0.01 Hz Balancing 0.3–0.01 Hz Orders in tracking mode: 1st order plus selectable additional order 2–99
- CPB-spectrum Bandwidth: Steps 70%/23%/6% (selectable) High-pass: 1,1/2,2/4,5/9/18/35 Hz (selectable) Low-pass: 1,1/2,2/4,5/9/18 kHz (selectable) Averaging time: 3–999 s 2. Connections for sensors. Inputs: 2 connectors for measurement sensors 1 connector for speed/ref. sensor, each via 6-pole sockets,
- Sensor types a/v/s to max. 36 Vpp Process values +/−30 V 0/4–20 mA Reference sensor P-84/P-95.

Figure 27.4 presents the measurement points of the analyzed air supply fan of a boiler "Minster". The fan operates without a frequency regulator, within Nem = 2950 (°/min).

Measurement has been done with vibration velocity sensor type VS 080 with accessories whose operating frequency range is 1–2,000 Hz, operating temperature range is: −40···+100 °C (−40···+212 °F), transmission factor: 75 mV/mm/s.

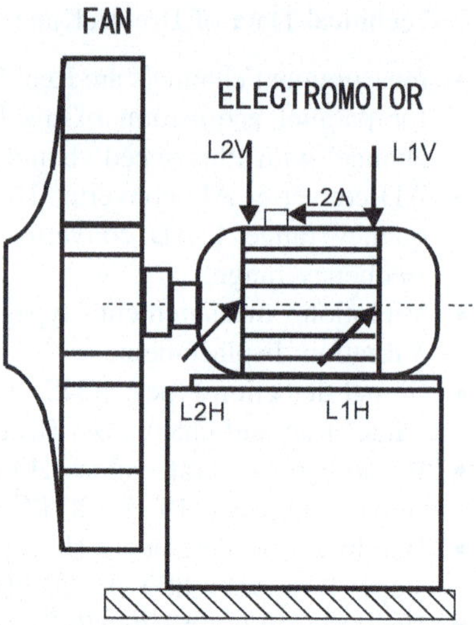

Fig. 27.4 Schematic representation for the tested fan

Table 27.1 Status display bearings and vibration levels for the displayed measurement points

Machine component	Measurement point						
Electric motor		L1H	L1V	L1A	L2H	L2V	L2A
Pem = 30 kw	v (mm/s)	6.5	11.3	–	1.0	3.0	4.6
Nem = 2950 °/min	BCU	0,11	0,16	–	0.33	0.71	0.14
L1- SKF6210	Other para	44 °C			50 °C		
L2- SKF6312	Other para						

Vibration levels during the measurement were in unacceptable area as shown in Table 27.1.

The examined values expose a procedure of estimating the state of rising trend vibration. However, the ascertainment that the level of vibration increases does not provide information about the causes that lead to an increase of vibration. The causes may be different: degeneration of bearings due to exposure to vibration, material fatigue, increased clearance due to fraying, etc., which require intervention to repair identified problem [1].

Vibration levels are in an unacceptable area which can cause malfunctioning of the boiler air supply fan. The main cause of elevated levels of vibrations is the imbalance of the rotor. Bearing parameters (BCU) indicate that the tested bearings are still in the area of satisfactory operation but damage on the bearing closer to the rotor (L2) is noticed. Initial damage on the bearing (measuring point L2) can cause the peaks on the second harmonic on FFT spectrum, and that is the case here as observed in the figures below.

The analysed boiler is the only boiler unit in the production process of the given factory, and it is used for production of the steam necessary in the production process, this point in the production process could be considered as very important i.e. a point of value. It is important to make a replacement of the bearing and to balance the rotor to avoid the possibility of unplanned outages due to a problem with the air supply fan.

27.3.1 FFT Analysis as a Diagnostic Procedure

On Figs. 27.5 and 27.6, a peak that occurs when the frequency is 49,0625 Hz is larger than other peaks for both components, but the conclusion is that for the first bearing the working field is not compromised.

FFT analysis is showing that peak is on bearing L2 rpm—1x—and it is presented trough figures from Figs. 27.7, 27.8, 27.9 and 27.10.

On Fig. 27.7 peaks that occur on frequency 99.11 and 198.21 Hz are related with rolling element.

Figure 27.8 shows the BCS spectrum for the LH2 measuring point, and a similarity with the previous figure can be noticed.

Figures 27.9 and 27.10 present the level of vibrations and BCS parameter for measuring point LV2. The peaks are noticed at same frequencies, which is related with rolling elements.

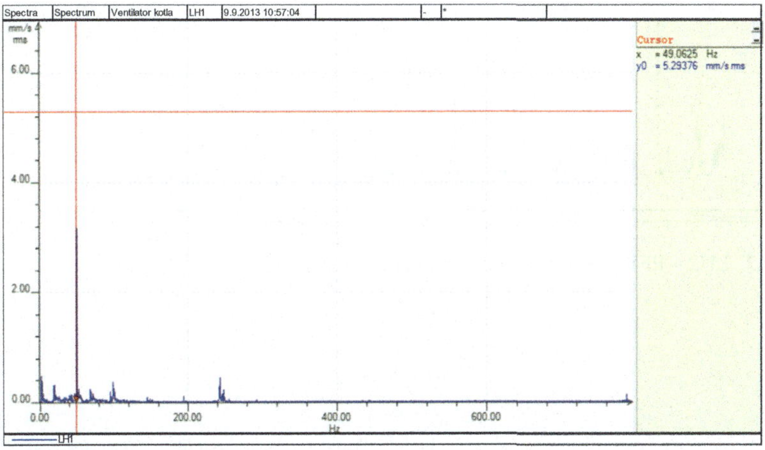

Fig. 27.5 LH1—FFT Spectrum—complete fan assembly

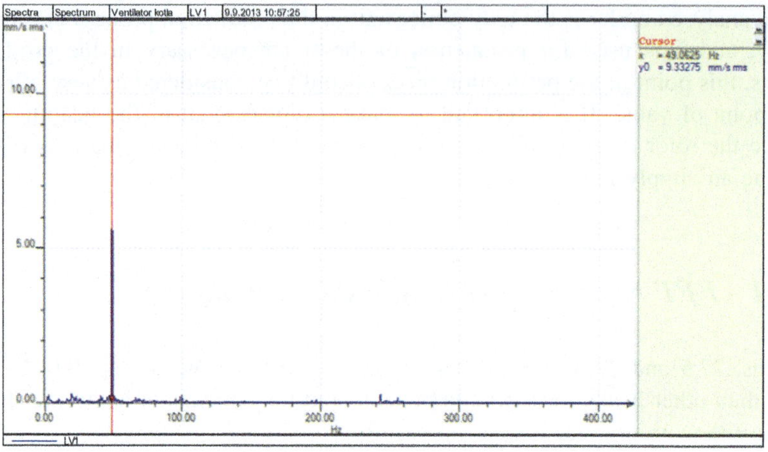

Fig. 27.6 LV1—FFT spectrum—complete fan assembly

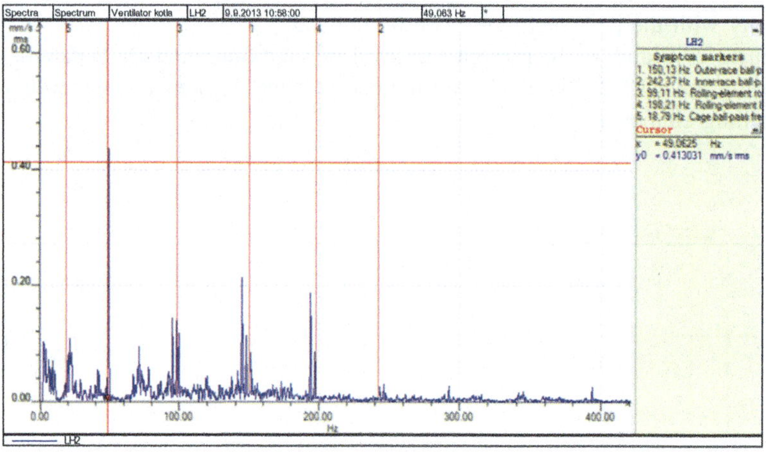

Fig. 27.7 LH2—FFT spectrum—complete fan assembly

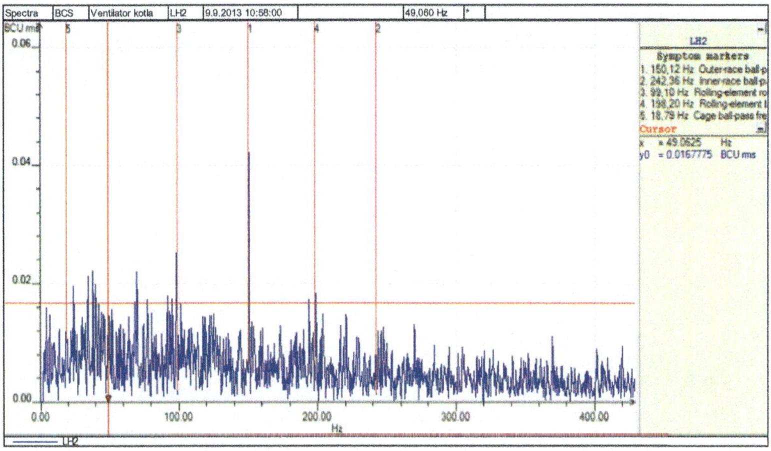

Fig. 27.8 LH2—FFT spectrum for BCS parameter

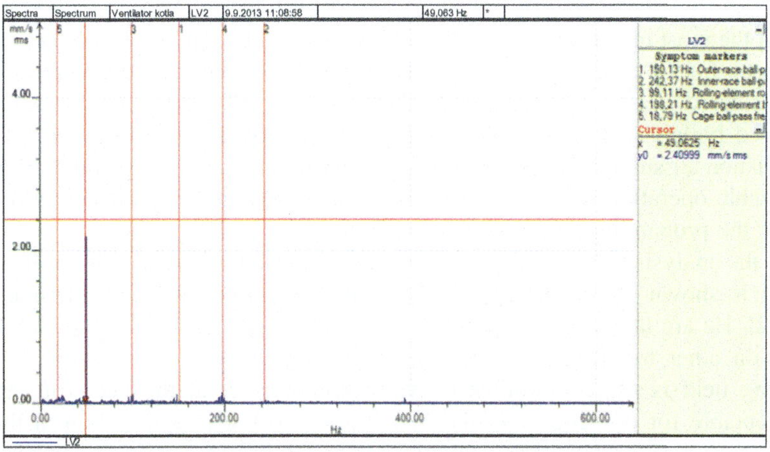

Fig. 27.9 LV2—FFT spectrum—complete fan assembly

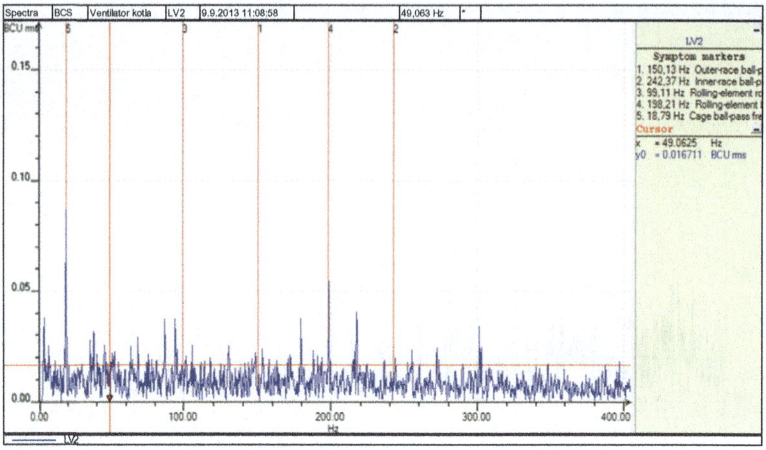

Fig. 27.10 LV2—FFT spectrum for BCS parameter

27.4 Conclusion

One of the most important part for the steam boiler is the fan, responsible to provide supply air for the combustion process. The type of production process, whose boiler is considered here, has a necessity for steam in every point of time. Since the analyzed boiler is the only boiler unit in the factory, and its operation relies on combustion air supplied by a fan, it is necessary to maintain the operation of the fan in a stable operation field and perform predictive maintenance actions in order to reduce the probability of cessation of operation.

By the analysis of the vibration measurement results performed on the air supply fan, it is shown in the paper that the peaks that occur when the frequency is 49,0625 Hz are larger than other peaks for both components (LH1, and LV1) that occur on other frequencies. Hence, it can be concluded that for this bearing the working field is not compromised. The vibration level of the fan is in the area unacceptable for operation and the first bearing to the impeller showed the first signs of weakness. However, although signs of bearing wear are present, the measured high vibration level is not the consequence of the condition of the bearing, but the unbalanced rotor state.

References

1. ISO 10816-1, Mechanical Vibration—Evaluation of Machine Vibration by Measurements on Non-Rotating Parts—Part 1: General Guidelines
2. D. Jovanović, N. Živković, M. Raos, L. Živković, M. Jovanović, M. Prašćević, Testing of level of vibration and parameters of bearings in industrial fan. Appl. Mech. Mat. **430**, 118–122 (2013). doi:10.4028/www.scientific.net/AMM.430.118

3. P. Girdhar, C. Scheffer, *Practical Machinery Vibration Analysis and Predictive Maintenance*. ISBN 0 7506 6275 1 (Elsevier)
4. S. Jovanović, D. Jovanović, Dinamičko uravnoteženje kao najvažniji postupak za poboljšanje dinamičkog ponašanja ventilatora. XXII Konferencija Buka i vibracije, Niš. ISBN 978-86-6093-019-6, pp 161–164
5. D. Jovanović, M. Jovanović, N. Živković, L. Živković, M. Raos, Belt Conveyor Drive Gearbox Problem Caused by Unpaired Gears—A Case Study
6. D. Jovaovic, M. Jovanovic, M. Raos, N. Zivkovic, M. Stankovic, M. Protic, Vibration analysis of insufficiently repaired well pump—a case study. Appl. Mech. Mater. **801**, 207–212 (2015). doi:10.4028/www.scientific.net/AMM.801.207

Chapter 28
Application of Mechanical Oscillations in Medicinal Plants Sorting Process

Augustina Pruteanu, Mihaela Nitu, Catalin Persu and Dan Cujbescu

Abstract Separators with plane sieves have an oscillating movement and are used for inhomogeneous mixture processing of medicinal plants obtained after chopping. From the analysis of the kinematics and dynamics of the separators and of the oscillating sieves type, results that the medicinal plant material movement, subjected to separation, in relation to sieve surface is influenced by the amplitude and frequency of the oscillations produced by the oscillating mechanisms and the angle of inclination to the horizontal of the separation surface. The paper presents the experimental results regarding the parameters of the sieve oscillations (amplitude and frequency) of a dimensional sorting machine, equipped with oscillating plane sieves. Also, are presented experimental results regarding the effectiveness of unbundling along sieve length of chopped medicinal plants for three oscillation frequency and three sieve inclination angles, for establishing the optimum working parameters, so that a qualitative separation is performed.

28.1 Introduction

Primary processing comprises the technical operations through which the material is prepared; respectively the harvested plant material is stored, packed, labelled or subsequently processed. Primary processing implies all the operations of conditioning, harvesting, drying, grinding by chopping (cutting), transporting, sorting,

A. Pruteanu (✉) · M. Nitu · C. Persu · D. Cujbescu
Development for Machines and Installations Designed to Agriculture and Food Industry Bucharest, National Institute of Research, 6 Ion Ionescu de la Brad, Bucharest, Romania
e-mail: apruteanu_augustina@yahoo.com

M. Nitu
e-mail: rosumihaelan@yahoo.com

C. Persu
e-mail: persu@inma.ro

D. Cujbescu
e-mail: dcujbescu@yahoo.com

N. Herisanu and V. Marinca (eds.), *Acoustics and Vibration of Mechanical Structures—AVMS-2017*, Springer Proceedings in Physics 198,
https://doi.org/10.1007/978-3-319-69823-6_28

performed by specialized equipment and through which the raw material is successively transformed, quantitatively and qualitatively, from the initial state S_0 to a state of finished product S_K [1].

The technological processing of medicinal plants for obtaining plant extracts comprises the technological operations of chopping plants at a pre-established size, separating the cut plants in dimensional fractions by the refusal method according to sieve orifices and extracting bioactive substance in different solvents, during different periods of time suitable to fraction dimensions, solvent-plant product ratio, etc. [2].

For accomplishing the operations of processing medicinal plants obtained after harvesting, special complex equipment are used, specific for this field [3]. The most used medicinal plants sorting equipment are those having included plane sieves with oscillating motion [4], which have to perform a certain sieving condition in order to obtain the desired dimensional fractions, having a high degree of separation. The plane sieve offers the possibility of sorting a mixture composed of chopped plant products in more fractions of the same or close granulations [1].

The oscillatory movement of the separation surfaces is characterized by its basic parameters. In the paper [5] have been established some correlations between the throw coefficient (the main technological parameters of the sieving process) and parameters characterizing the screen's vibratory regime (amplitude, pulsation, vibration frequency and acceleration), inclination of the sieve, the intensity of shaking, power required to produce vibrations. Based on these correlations, the authors determined the parameters of the sieves vibratory regime so that it achieves a good quality sieving.

In the paper are presented experimental data of the sieve oscillation parameters, namely amplitude and frequency, as well as experimental determinations on the efficiency of separating dried and chopped nettle fragments on length.

28.2 Materials and Methods

The species of medicinal herb used in experiments, nettle (*Urtica dioica*) were identified and harvested from the spontaneous flora according to the morphological and biological characteristics of the species [6, 7], then it was dried and chopped using the TIMATIC type chopping plants machine, set to the size of 6 mm.

Sorting machine for chopped plants (Fig. 28.1) is an equipment achieved by INMA Bucharest, used for separating plant dust, leaves, flowers of fresh or dried medicinal plants, etc., as well as for separating and removing dust and other undesired elements from the main product mass [8]. The equipment is driven by two vibrating engines mounted outside the frame, whose construction allows rotating and modifying their inclination to the horizontal in order to adjust the amplitude and frequency of oscillations. Sieves are sized according to the type of medicinal plants and production capacities. The angle of inclination of sieves can be adjusted between the limits: 12 … 15 degrees. Separation is done through the method of refusals. The standard equipment comes with a set of three sieves with orifices of different sizes that can range between 1.15 and 13.2 mm. The feed is carried out by means of an inclined conveyor belt, which discharges the plant material into the centre of the sorting feed funnel and from here on the upper sieve where the separating process takes place. The separated plant material is discharged into the collection boxes located at the end of each sieve.

The dimensional sorting machine for cut plants is provided with the possibility to adjust oscillating sieve parameters: oscillation frequency and amplitude. Oscillation frequency can be changed from the electrical engine by electric current variation and by changing the position of the actuating mechanism, in relation to the sieve (Fig. 28.1).

For determining the frequency and amplitude of vibrations of the plane sieve, an accelerometer with three-axis simultaneous measurement in the median plane of the sieves layer was placed on its surface [9] in a perpendicular position on the sieve plane, as shown in Fig. 28.2.

The measurements were made for three angles of inclination of the vibrating motors (43.7°, 67.2° and 81.5°) to the sieve plane.

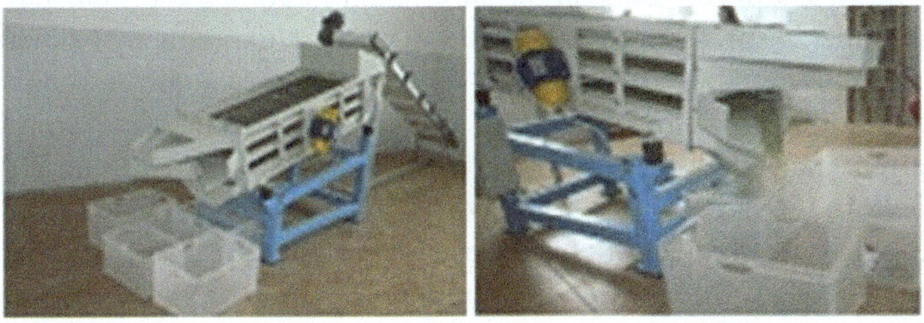

Fig. 28.1 Sorting machine for cut plants

(a) **(b)**

Fig. 28.2 Positioning the accelerometer in the median plane of the sieves, **a** overall overview; **b** detail regarding to accelerometer placement

For data acquisition, analysis, processing and display, LabVIEW software, a graphical programming language was used [10]. For conducting the measurements, the accelerometer was calibrated and the graphic recorders were created. The chosen sampling period was 0.1 ms. The measurement time was 10 s for each measurement.

Experiments on separation efficiency on sieve length were performed on a sieve with 6.3 mm square orifices, the sieve having a length of 1.395 m and a width of 0.6 m. Under the selected sieve was placed a tin collector box divided in in seven equal compartments, each compartment having 0.195 m. The working parameters varied during the course of the experiments were three sieve inclination angles (12.08°, 13.33° and 14.7°) and three vibration frequencies (1000, 950 and 900 rot/min) at a feed flow rate of 60 kg/h.

28.3 Results

The experimental researches followed on the one side the parameters of the sieve oscillations (amplitude and frequency) and on the other side the efficiency of the separation of the chopped plants on the sieve length in order to establish the optimal working parameters of separation process. The first experimental determinations aimed at the determination of the plane sieve vibration parameters under different working conditions: different oscillation frequencies regulated by variation of the electric current and different amplitudes of the oscillation of the site, regulated by three angles of inclination of the vibratory motors.

In Fig. 28.3 are rendered the measured vibrations for a frequency of 1000 rot/min, at three different inclination angles of the vibratory engines.

In Fig. 28.4 are rendered the power spectra of the measured vibrations for frequency of 1000 rot/min, at different inclination angles of the vibratory engines.

The variation of acceleration on the three axes Ox (acceleration 0), axis x with longitudinal direction, Oy (acceleration 1), axis y with transverse direction and Oz (acceleration 2) and axis z with vertical direction, were calculated at the three frequencies and the following relations were used [11, 12]:

$$\begin{aligned} \bar{a}_{x\max} &= \frac{1}{\nu} \cdot \sum_{1}^{\nu} \left|a_{x\max}\right|; \bar{a}_{y\max} = \frac{1}{\nu} \cdot \sum_{1}^{\nu} \left|a_{y\max}\right|; \\ \bar{a}_{z\max} &= \frac{1}{\nu} \cdot \sum_{1}^{\nu} \left|a_{z\max}\right| \ (\mathrm{ms}^{-2}) \end{aligned} \tag{28.1}$$

where: $\bar{a}_{x\max}$, $\bar{a}_{y\max}$, $\bar{a}_{z\max}$ represents the average of the maximum acceleration on Ox, Oy and Oz directions, in the considered interval. Maximum amplitudes, on the three directions, were calculates with the relations:

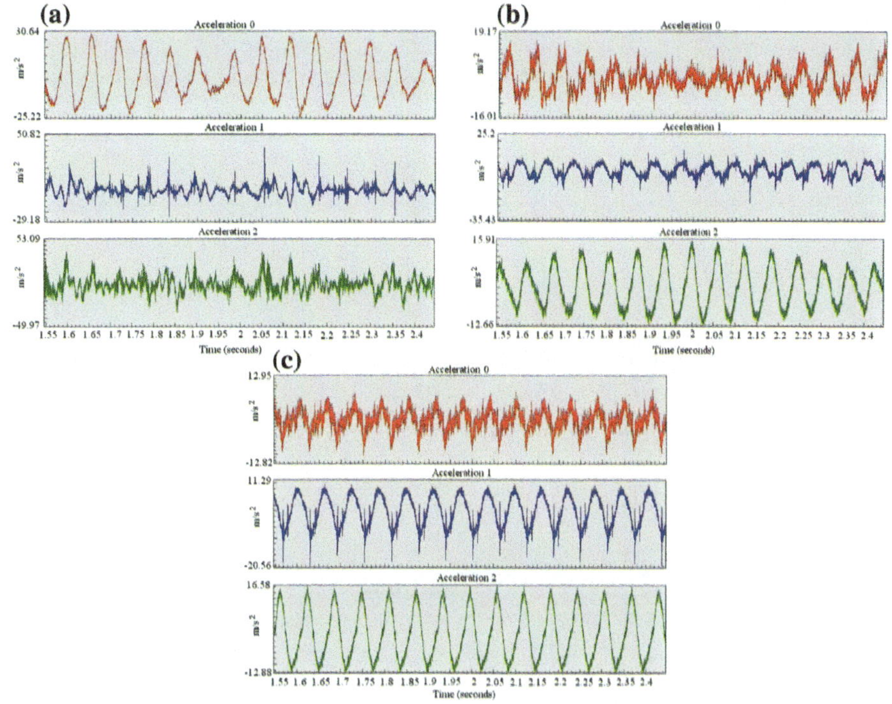

Fig. 28.3 Measured vibrations for a frequency of 1000 rot/min, at different inclination angles of the vibrating motors: **a** 43.7°, **b** 67.2°, **c** 81.5°

$$A_x = \frac{\bar{a}_{x\max}}{\omega^2} = \frac{\bar{a}_{x\max}}{4\pi^2\nu^2}; A_y = \frac{\bar{a}_{y\max}}{\omega^2} = \frac{\bar{a}_{y\max}}{4\pi^2\nu^2};$$
$$A_z = \frac{\bar{a}_{z\max}}{\omega^2} = \frac{\bar{a}_{z\max}}{4\pi^2\nu^2}\ (\text{mm}) \tag{28.2}$$

After calculation, it was found that the real amplitudes of the oscillations on Ox, Oy and Oz directions are in the limits of: 6.61 ÷ 11.57 mm.

In Fig. 28.5 are presented the calculated values of the maximum amplitudes of sieve oscillations. It was found that the working regimes corresponding to an inclination angle of vibrating motors at 67.20 are optimal, having amplitudes ranging between 6.88 and 10.55 mm.

The working conditions for the experimental determinations on cut plants were for the inclination angle of the sorting vibratory engine at 67.20, for 60 kg/h sorting feed rate and nettle herb cut at 6 mm and separated on a sieve having 6.3 mm square orifices. Separation efficiency was calculated for three inclination angles: 12.08°, 13.33°, 14.7° and three frequencies 1000, 950, 900 rpm. The experimental values obtained can be seen in Fig. 28.6.

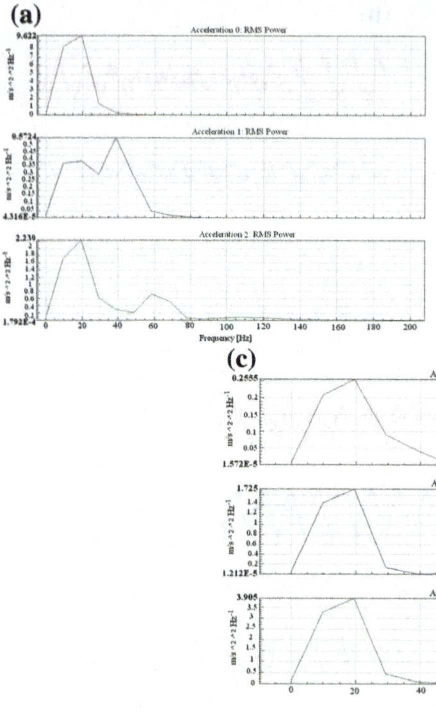

Fig. 28.4 Vibration power spectrum measured for 1000 rot/min, at different inclination angles of the vibrating motors: **a** 43.7°, **b** 67.2°, **c** 81.5°

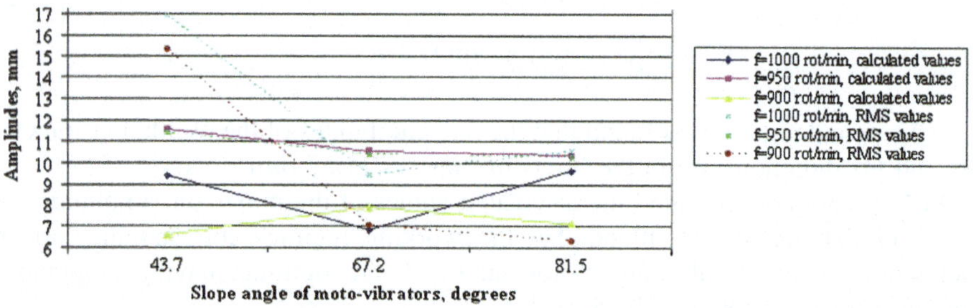

Fig. 28.5 Amplitudes for the three inclination angles of vibrating motors depending on three vibration frequencies

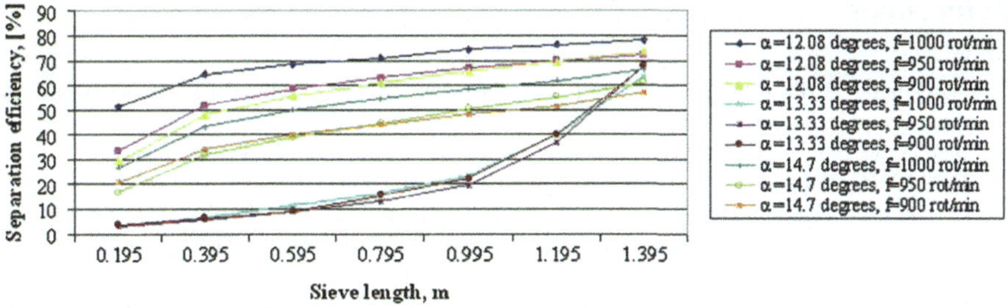

Fig. 28.6 Separation efficiency of nettle chopped fragments at different sieve inclination angles and working frequencies

28.4 Conclusions

The vibrations measured for a frequency of 1000 rot/min, at the three inclination angles of vibrating motors registered decreasing values for all three directions of the vibrating sieve. It can be seen from Fig. 28.3 that, as long as the inclination angle of the vibrating motor increases, the measured vibration values decrease.

Analyzing the graphs on vibration power spectra at 1000 rot/min for the three inclination angles of the vibrating motors with the RMS of the vibration amplitude, we consider as optimal the first case of 67.2° vibrating motors inclination angle, as the amplitude of the transverse vibration is minimal for the Oy axis, the longitudinal and vertical movements of the sieves being cursive and without beatings. From the analysis of calculated and RMS values for amplitudes (Fig. 28.5), it has been found that at 950 rot/min frequencies, the amplitudes have very close values for all three angles of inclination of the vibrating motors. At frequencies of 1000 and 900 rot/min, the amplitudes calculated in relation to those of RMS have very distant values for the small inclination angle of the vibratory engines, the 43.7° one.

From Fig. 28.6 can be seen that the efficiency of separating chopped nettle fragments is higher, having values ranging between 70 and 80% at the sieve inclination angle of 12.08°. The overall analysis of the experimental results highlights the superiority of the 1000 rot/min frequency, the sieve inclination angle of 67.2°, the amplitude between 7 ÷ 10 m and sieve inclination angle of 12.08°.

The results support the achievement of a quality separation of plant material using the plant sorting machine. From the analysis of the obtained results is found that the plant sorting machine can be successfully used for the separation of chopped medicinal plants if the working regime parameters are adequately chosen for the desired purpose.

References

1. A. Danciu, E. Postelnicu, V. Vlăduţ, I. Voicea, M. Matache, M. Ludig, M. Martinov, A. Atanasov, C. Florea, Experimenting the technology and equipment for primary processing of medicinal plants. Obtaining extractive solutions from medicinal and aromatic plants. INMATEH-Agric. Eng. J. (Bucharest) **34**(2), 57–66 (2011)
2. G. Păun, O. Gheorghe, M. Diaconu, Medicinal plants processing. MedPlaNet Project (2012)
3. S. Oztekin, M. Martinov, *Medicinal and Aromatic Crops: harvesting, Drying and Processing* (Editura Haworth Press, United States and Canada, 2007)
4. G. Ene, *Equipment for classifying and sorting the polydisperse solids materials* (Matrix Rom Publishing House, Bucharest, 2005)
5. G. Ene, T. Sima, Aspects regarding shifting materials on vibratory sieves I and II; Synthesis. J. Theor. Appl. Mech. **4**(1, 2) (2013)
6. A. Ardelean, G. Mohan, *Romanian Medicinal Flora* (All Publishing House, Bucharest, 2008)
7. O. Bojor, *Medicinal and Aromatic Guide from A to Z* (Fiat Lux Publishing House, 2003)
8. Technical book for sorting the chopped medicinal plants. INMA, Bucharest
9. G. Buzdugan, E. Mihăilescu, M. Radeş, *Vibrations Measurement* (The Academy of the Socialist Republic of Romania, Publishing House, Romania, 1979)
10. D. Selisteanu, C. Ionete, E. Petre, D. Popescu, D. Sendrescu, *LabVIEW applications for acquisition and data generating* (SITECH Publishing House, Craiova, 2004)
11. V. Vlăduţ, A. Danciu, I. Grigore, N. Herişanu, I. Dumitru, C. Sorică, I. Voicea, S. Biriş, M. Duţu, The influence of the working regime of oscillations produced by an electro-vibrator on the sorting of materials. Appl. Mech. Mater. **801**, 197–201 (2015)
12. C.C. Florea, Contributions to the technology and equipment for conditioning the medicinal plants and forest fruits before processing. Doctoral Thesis, Transilvania University, Brasov (2013)

Chapter 29
Active Vibration Control of Test Equipment Through Feedback Algorithm

Ramona Nagy, Remus Stefan Maruta and Karoly Menyhardt

Abstract Items can be exposed to complex dynamic stresses during transportation and handling. Vibration testing controls the input of these stress tests and is used to determine the item's ability to function in real world conditions. Vibration testing is used by introducing of a force function into system, usually with the use of a shaker or vibration testing machine. The basis for vibration testing is discrete closed loop control of vibratory excitation, commonly known as vibration control. The level of vibration is sensed by the control accelerometer, which is monitored in a decision module. The controller then makes the necessary adjustments to condition the drive signal so that the vibration level meets the test specifications. Throughout this paper we present the methods used for continuous closed loop, discrete closed loop control and open loop control for sine sweep and random vibration tests.

29.1 Introduction

Vibration fatigue describes a material fatigue caused by forced vibration of random nature, sine sweep or known function. Randomness implies a non-order in a sequence of numbers and the impossibility to associate a function to the test curve. The excited structure responds proportionally to its natural-dynamics modes, which results in a dynamic stress load in the material [1]. Thus the process is governed by the shape of the excitation profile and the response it produces. For a random process, the amplitude cannot be described as a function of time, because of its probabilistic nature.

R. Nagy (✉) · R. S. Maruta · K. Menyhardt
Department of Mechanics and Strength of Materials, Polytechnic University Timisoara, Timişoara, Romania
e-mail: ramona.nagy@upt.ro

R. S. Maruta
e-mail: remus.maruta@upt.ro

K. Menyhardt
e-mail: karoly.menyhardt@upt.ro

N. Herisanu and V. Marinca (eds.), *Acoustics and Vibration of Mechanical Structures—AVMS-2017*, Springer Proceedings in Physics 198,
https://doi.org/10.1007/978-3-319-69823-6_29

Random vibration tests are usually performed with the aid of a finite element method software. Commercial or noncommercial finite element method software include limited capabilities to perform probabilistic structural static or dynamic analyses with little post-processing [2]. A physical test is the closest thing to reality that can imitate and evaluate the functioning of a module. There are three types of possible fluctuating stress-time modes possible: completely reversed constant amplitude, repeated constant amplitude and random stress level amplitude and frequency.

As vibration testing evolved, methods that better represent real world data were researched, random vibration testing providing statistical confidence with random time data that has an average targeted frequency content and amplitude. By controlling frequency and amplitude, test data can be correlated to real world data sets, especially for known/desired laws of motion [3, 4].

Mainly there are two approaches to control: feedback and feedforward. The principle of feedback is represented in Fig. 29.1: the output y of the system is compared to the reference input r, and the error signal (e = r − y) is passed into a compensation unit H(s) and applied to the system G(s).

The problem consists in finding the appropriate compensator H(s) so that the closed-loop system is stable and behaves in the appropriate manner.

The control objectives are more difficult than in active damping, and we have to keep a control variable y (the displacement) to the desired value r considering external disturbances d in some frequency range. From:

$$F(s) = \frac{y(s)}{r(s)} = \frac{GH}{1+GH} \tag{29.1}$$

it is noticeable that this approach requires large values of $G{*}H$ in the frequency range where $y \simeq r$. $GH \gg 1$ infers that the closed-loop transfer function F(s) is close to 1, meaning the output y tracks the input r accurately. Generally, to achieve this, a more elaborate strategy is required involving a mathematical model of the system that can only be a low-dimensional approximation of the actual system G(s).

When a signal correlated to the disturbance is available, feedforward adaptive filtering constitutes an alternative to vibration control through feedback [5]. Its principle is explained in Fig. 29.2.

This method assumes the availability of a reference signal correlated to the primary disturbance; the signal is then passed through an adaptive filter, the output of which is applied to the system by secondary sources. The filter coefficients must

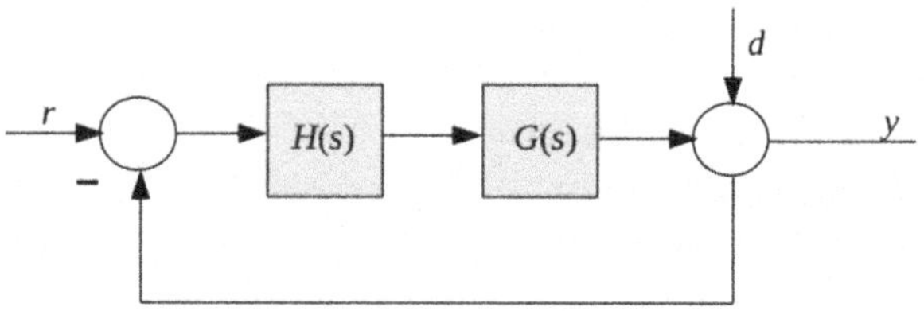

Fig. 29.1 Principle of feedback control

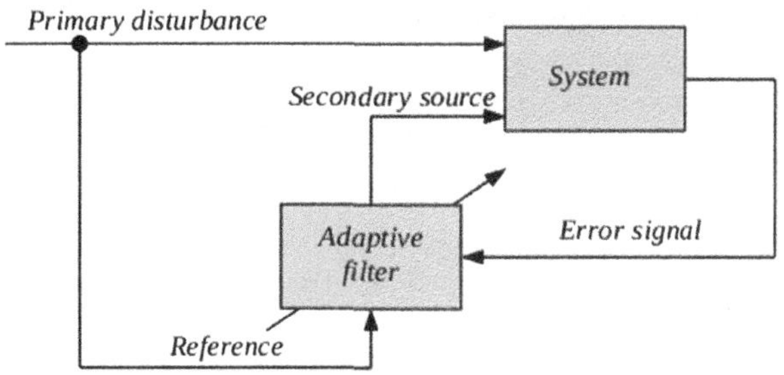

Fig. 29.2 Principle of feedforward control

be adapted in such a way that the error signal at critical points is minimized. Thus, a secondary disturbance must be produced in order to cancel the effect of the primary disturbance at the location of the sensor. The method does not need a model of the system, but the adaption procedure relies on the measured impulse response.

29.2 Continuous Closed Loop Control

One of the most ingenious way of process control is the feedback loop (Fig. 29.3).

It consists of five fundamental elements:

- A controller that reads the transmitter's signal and decides whether or not the current condition of the process is acceptable;
- An actuator functioning as the final control element that applies a corrective effort to the process per the controller's instructions;
- The process that is to be controlled;
- An instrument with a sensor that measures the condition of the process;
- A transmitter that converts the measurement into an electronic signal.

In a closed-loop control system (Fig. 29.4), information flows around a feedback loop from the process to the sensor to the transmitter to the controller to the actuator and back to the process. This type of measurement, decision and actuation sequence is known as closed-loop control and repeats as often as necessary until the desired process condition is achieved. As example, in the case of a vibration testing equipment, with the aid of accelerometers, displacement can be calculated:

$$\Delta \dot{y}_i = \dot{y}_i - \dot{y}_{i-1} = \ddot{y}_i \cdot \Delta t, \quad \text{where } \Delta t = t_i - t_{i-1} \tag{29.2}$$

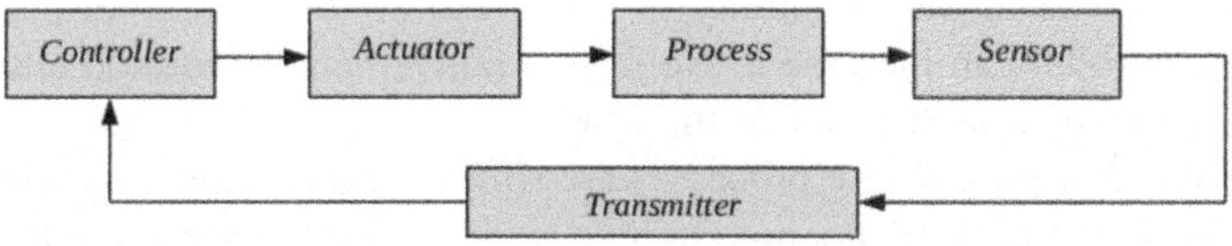

Fig. 29.3 Process control elements

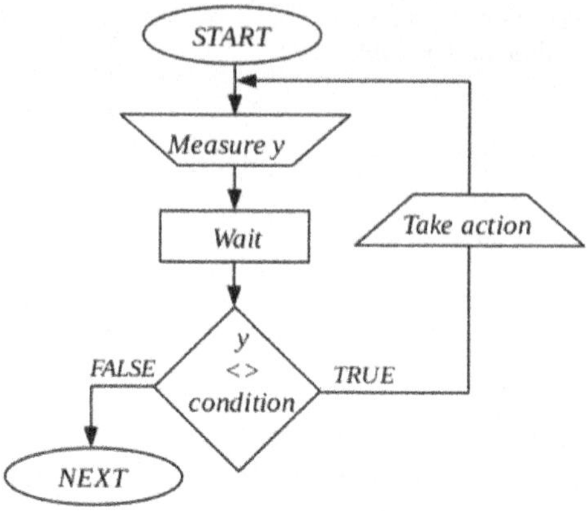

Fig. 29.4 Continous closed loop

$$\Delta y_i = \dot{y}_i \Delta t = \ddot{y}_i \Delta t \cdot \Delta t + \dot{y}_{i-1} \cdot \Delta t \Rightarrow y_i = \ddot{y} \Delta t^2 + \dot{y}_{i-1} \Delta t + y_{i-1} \qquad (29.3)$$

where $\ddot{y}$-measured acceleration, $\dot{y}$-velocity, y-displacement, t-time.

If measured/determined displacement y is outside the expected range, then some kind of correction is taken. Due to safety reasons, because there is the possibility for the oscillations to be in antiphase, limiters must be placed to avoid destruction of an electrodynamic vibrator.

If the feedback controller proves unable to maintain a stable closed-loop control (resonance), a reset or operator intervention is necessary. The easiest way to terminate an unstable oscillations is to break the loop and regain control manually.

For a continuous process, a feedback loop tries to maintain a process variable at a desired value known as the setpoint. The controller subtracts the current measurement of the process variable from the setpoint to generate an error signal. The duration and magnitude of the error signal then imposes the value of the controller's output or manipulated variable which in turn takes the corrective action applied by the actuator.

29.3 Discrete Closed Loop Control

For a discrete process (Fig. 29.5), the variable of interest is measured/determined only when a triggering event occurs, and the measure-decide-actuate sequence is typically executed just once for each event. For example, the allowed maximum displacement is set at the beginning of each test. If the current displacement is lower, the controller can increase it. No further adjustment is required until the next triggering event such as the end of the test.

Feedback loops for discrete processes are simpler than continuous control loops because discrete processes do not involve much inertia. Inertia complicates the design of a continuous control loop since a continuous controller generally needs to

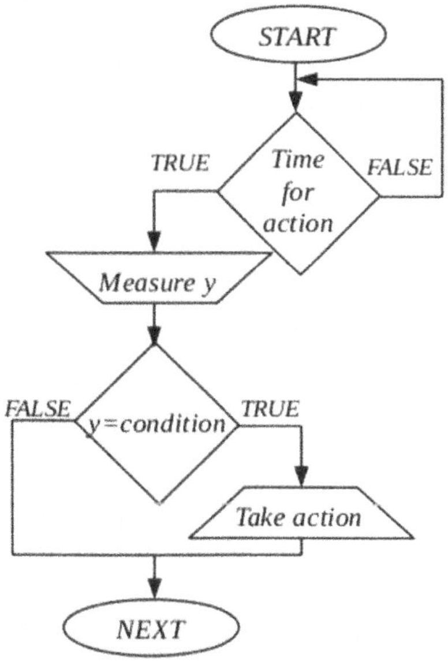

Fig. 29.5 Discreet closed loop

make a series of decisions before the results of its previous efforts are completely evident. It has to anticipate the cumulative effects of its most recent corrective efforts and plan future efforts accordingly. Waiting to see how each one turns out before trying another action simply takes too long.

29.4 Open Loop Control

Not all automatic control operations require feedback. A much larger class of control commands can be executed in an open-loop configuration without confirmation or further adjustment (Fig. 29.6). Open-loop control is sufficient for predictable operations such as starting a test cycle, or turning off the shaker.

29.5 Experimental Concept

Data acquisition systems (DAQ) provide tools and methods to measure mechanical phenomena as voltage or current. The most notorious systems from Bruel & Kjaer or National Instruments provide mature means to do this but, at a hefty price. Our affordable solution, even though not out of the box, was the combination of the Raspberry Pi, a single-board computer that runs the Linux kernel-based operating system with a USB DAQ. One Linux compatible DAQ (as in driver ready) is the USB-1208FS device that provides analog measurement option to the digital ones

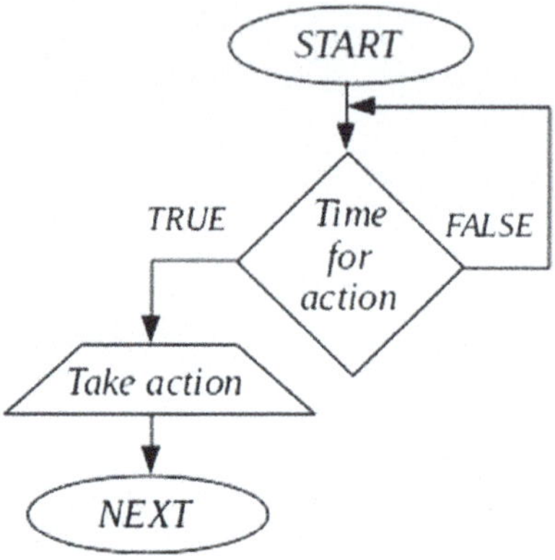

Fig. 29.6 Open looped process

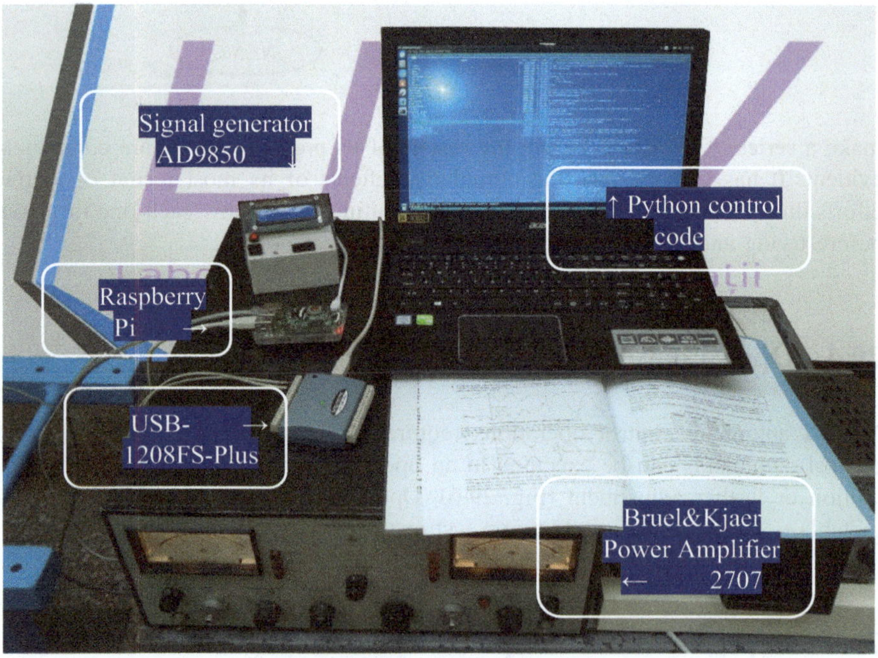

Fig. 29.7 Data acquisition system

available on the Raspberry Pi. Unfortunately, it lacks the visual programming language that lets the user create programs by manipulating program elements graphically, so all the coding was done in Python (Fig. 29.7).

The aim of our study was to test polyurethane foams at fatigue in a controllable environment. The test signal was generated from the Raspberry Pi through an

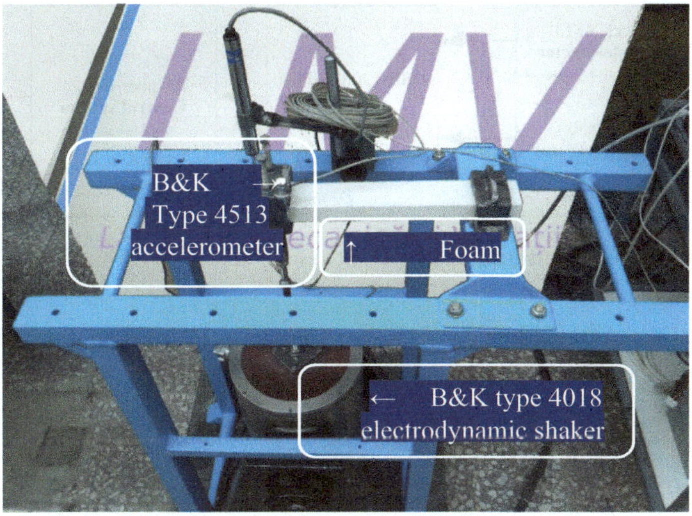

Fig. 29.8 Polyurethane foam fatigue test rig

AD9850 Direct Digital Synthesizer module. The signal was then amplified by Bruel & Kjaer Power Amplifier Type 2707 and fed to a B&K type 4018 electrodynamic shaker (Fig. 29.8).

Due to system inertia, the output from the signal generator was correlated with the output from the signal amplifier and a double integrated response signal from a B&K type 4513 monoaxial accelerometer. Table 29.1 shows the measured peak voltages at the electrodynamic shaker's coil for three given generator signal amplitudes (U_{EX}) over various tested frequencies.

With lower frequencies, keeping a constant generator signal amplitude, the shaker coil voltage was smaller due to the power amplifier's nonlinearity.

Continuous close loop control method used in polling data from the signal amplifier and accelerometer peaked out around 90 kHz. Processing the data, double integration of the accelerometer to obtain the displacement amplitude and frequency, comparing it with the desired/generated signal and correction value generation was observable for about 10,000 cycles due to the processor-memory limitation and writing latency of the system. The block diagram for the developed control method is show in Fig. 29.9, where the analog data is captured by the

Table 29.1 Shaker coil peak voltage based on input signal amplitude

Frequency (Hz)	U_{EX} = 1.83 (V_p)	U_{EX} = 3.66 (V_p)	U_{EX} = 5.5 (V_p)
10	1.70	3.82	5.75
20	2.10	4.58	6.68
40	2.19	4.80	7.20
60	2.20	4.90	7.25

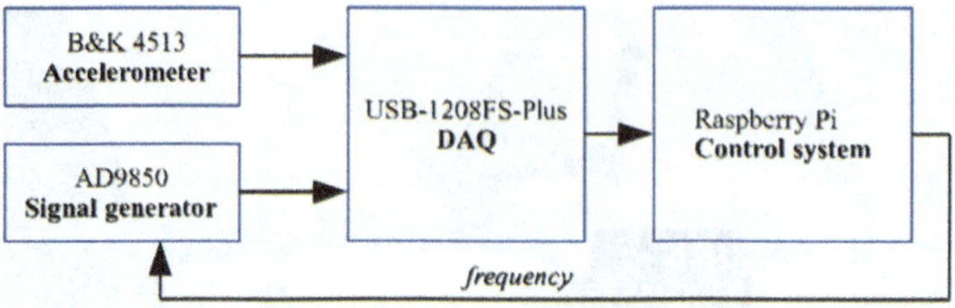

Fig. 29.9 Continuous/discrete closed loop control using Raspberry Pi

USB-1208FS DAQ and transferred to the Raspberry Pi for digital processing and decision making.

29.6 Conclusions

Most of the times there is little difference between continuous closed-loop control, discrete closed-loop control, and even open-loop control. For the first two cases the time for action expects a particular step to be reached or condition to be met in the control sequence. At that point, an open-loop controller (unsupervised, uncontrolled) would just execute an action and then proceed to the next step in the algorithm. A continuous closed-loop controller is always ready to make corrections. It takes actions, measures conditions, waits if needed and repeats the loop unless a specified condition has been reached, but as processing power goes, this is not feasible, and within 0.1 s received data (10,000 readings), there is no observable correction, but rather and oscillating response and control loop.

A discrete closed-loop controller on the other hand, will measure or determine a displacement in the process at given steps in time to perceive if action needs to be executed or not. We successfully obtained Discrete Closed Loop Control polling frequencies lower than 1 kHz. In this case, both read and control signals are steadily converging and depending on the amplitude and frequency, corrections can be done in under 1 s.

References

1. K. Menyhardt, R. Nagy, R.S. Maruta, Design of modular vibration testing equipment. Appl. Mech. Mater. **801**, 333–337 (2015). doi:10.4028/www.scientific.net/AMM.801.333
2. T.L. Paez, Random vibration—a brief history. Sound Vib. The Noise and Vibration Control Magazine, Jan 2012
3. D.S. Steinberg, Vibration Analysis for Electronic Equipment, 2nd edn (Wiley-Interscience, New York, 2000). ISBN 978-0471376859

4. L.D. Lutes, S. Sarkani, Random Vibrations: Analysis of Structural and Mechanical Systems (2003). ISBN 978-0750677653
5. A. Preumont, Vibration Control of Active Structures, An Introduction 3rd edn. (Springer, Berlin, 2011). ISBN 978-94-007-2033-6

Chapter 30
Shaking Table for the Analysis of Pillars with Top Mass

Vasile Iancu, Iosif Galuska and Tiberiu Manescu

Abstract In this paper, we present a shaking table designed by the authors, used in laboratory experiments to simulate earthquakes. This equipment makes possible studying the dynamic behavior of civil engineering structures subjected to seismic motion. Among these structures, bridges won a special attention, because they are easily harmed during earthquakes. In addition, bridges have importance in assuring a facile communication in the case of hazards. The most sensitive bridge element is the pillar, which supports important bending moments if horizontal forces act on his top. The shaking table developed by the authors is designed to analyze the dynamical behavior of pillars with a tip mass that represents the superstructure. The soil displacement is simulated by the main plate, actuated by an electric engine which ensures different excitation frequencies. The analyzed pillar is mounted on the main plate via a reinforcement element, in order to increase its stability. Tests were performed with different frequencies, in order to find out the actuation precision. During tests, the accelerations at the base and the top of the pillar were acquired with the LabView program of National Instruments, and the results shows that the shaking table can be controlled with high accuracy.

30.1 Introduction

There are several different experimental techniques that can be used to test the response of structures to verify their seismic performance, one of this is the use of an earthquake shaking table. This is a device for shaking structural models or

V. Iancu (✉) · I. Galuska · T. Manescu
Universitatea "Eftimie Murgu" Reşiţa, P-ta Traian Vuia 1-4, 320085 Reşiţa, Romania
e-mail: v.iancu@uem.ro

I. Galuska
e-mail: iosif.galuska@gmail.com

T. Manescu
e-mail: manescu.tiberiu@gmail.com

N. Herisanu and V. Marinca (eds.), *Acoustics and Vibration of Mechanical Structures—AVMS-2017*, Springer Proceedings in Physics 198,
https://doi.org/10.1007/978-3-319-69823-6_30

building components with a wide range of simulated ground motions, including reproductions of recorded earthquakes time-histories.

The modern tables consist of a rectangular platform that is driven in up to six degrees of freedom using servo-hydraulic or other types of actuators. The earliest shaking table was made up in 1893, at the University of Tokyo to categorize types of building construction [1], this shaking table was set in motion by a simple wheel mechanism. Test specimens are fixed to the platform and shaken, often to the point of failure.

Shaking tables are also used in other fields of engineering to test and qualify vehicles and components that must respect heavy vibration requirements and standards i.e. road and railway vehicles, aerospace, military standards, or tried and de- sired properties of the soil [2], for civil and industrial structures [3], etc.

Meanwhile, bigger and more powerful shaking tables have been put in operation allowing for the adoption of lower scaling factors and therefore involving very important dynamic forces [4].

30.2 Shaking Table Design

The shaking table is built of a metal frame build from rectangular pipes having a plate made from wood and metal with a thickness of (40 mm wood and 10 mm steel plate), the height of the table is 850 mm. Due to this height, the gravity center of the shaking table is at 450 mm above the ground.

The dimensions for the sliding table are 600 × 300 × 40 mm, this is made of a 40 × 20 × 2 mm rectangular metal frame and a 40 mm thick wooden board. The height of the shaking table is 850 mm. Due to this height, the center of gravity of the stand is 450 mm above the bottom. The sliding table in its uniform and rectilinear motion tends to require the structural resistance of the metal frame of the shaking table in length. For the attenuation of this movement, two legs were related to the main frame of the shaking table for increase here stability Fig. 30.1a and in Fig. 30.1b is presented the pillar and same earthquake isolators.

For the connection between the sliding table and the metallic case of the shaking table, we use a metallic ball-shaped guide Fig. 30.2, which take the translation movements at high speeds at a low noise level.

The movements requirement of the sliding plate of the shaking table being from 0 to 400 rpm, for a fine translation motion it was necessary to mount a control panel with a frequency converter, that and the wiring diagram are presented in Fig. 30.3. For the frequency converter, we choose FRENIC-Mini series of general use, made by Fuji Electric C. Ltd.

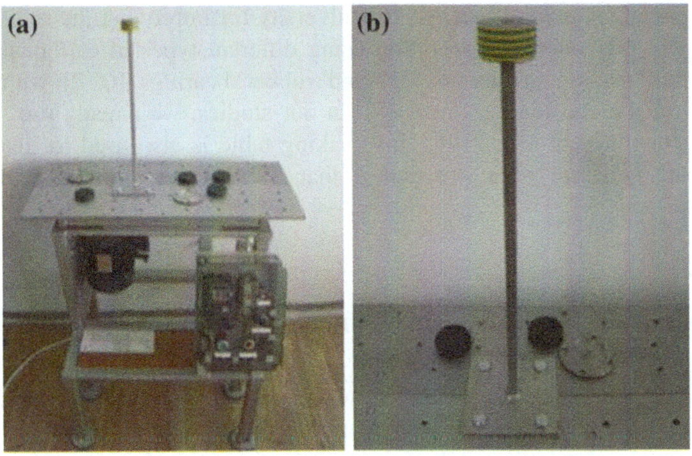

Fig. 30.1 The shaking table (**a**), the pillar and same earthquake isolators (**b**)

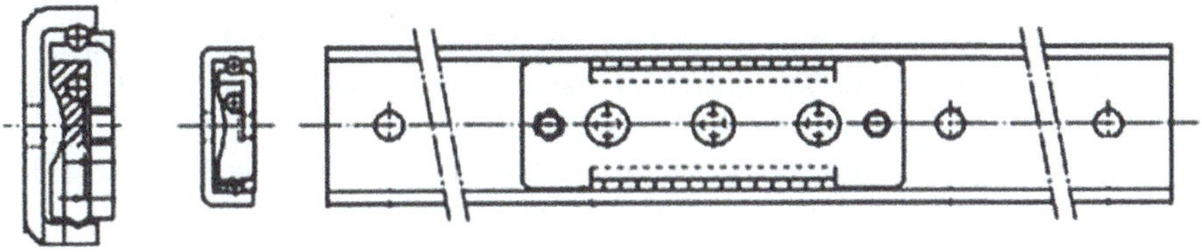

Fig. 30.2 The metallic ball-shaped guide of the shaking table

(a) (b)

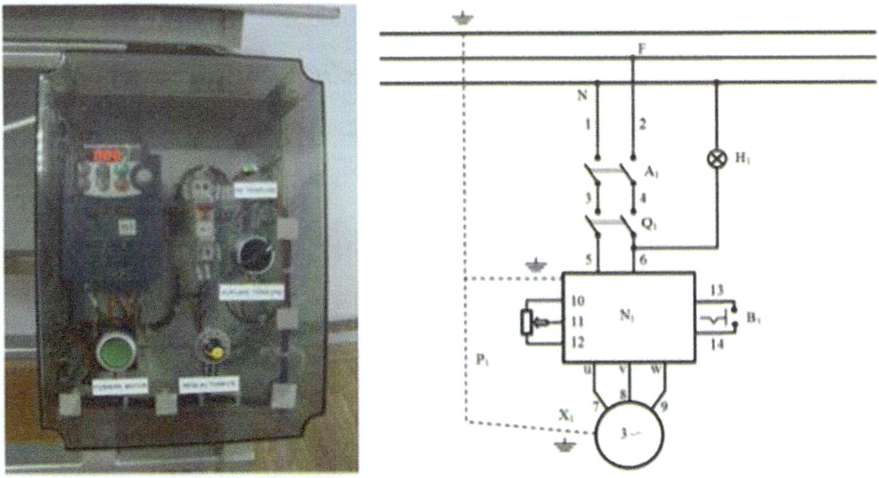

Fig. 30.3 The control panel (**a**) and the wiring diagram (**b**)

The shaking table was made in our University for improving the stability of the structure and their mode of operation using different types of earthquake sealing devices, like friction pendulums [5] and rubber bearings [6, 7] with different properties of the elastomeric layers [8]. In our studies, we already use such insulators [9] for different tests [10]. This shaking table is also used for teaching the students the behavior of a structure if that is subject to seismic waves of an earthquake.

30.3 Shaking Table Calibration

For the table calibration, we use a pillar with a head mass Fig. 30.1b. After starting the engine of the shaking table we increase the level of his power using a switch with nine levels, one level is 1% of the power of the electric motor of the shaking table who has 0.750 kW AC.

For calibrating the shaking table we use the LabView program, the calibration was done on 9 levels engine rotation from 1 to 9. The experiment was stopped at level 9 because we achieve the 3 Hz frequency, rarely encountered in earthquakes. One level means 1% of the power of the electrical motor. The results of the experiment are shown in Table 30.1.

With the experimental results of the shaking table shown in Table 30.1, we plotted a chart of the frequencies and amplitudes Fig. 30.4.

Table 30.1 The working parameters of the shaking table

Level	Frequency (Hz)	Amplitude (g)
1	0.307	0.00004
2	0.644	0.00100
3	0.973	0.00720
4	1.312	0.00745
5	1.641	0.00760
6	1.978	0.13000
7	2.306	0.30600
8	2.651	0.63000
9	2.989	1.15100

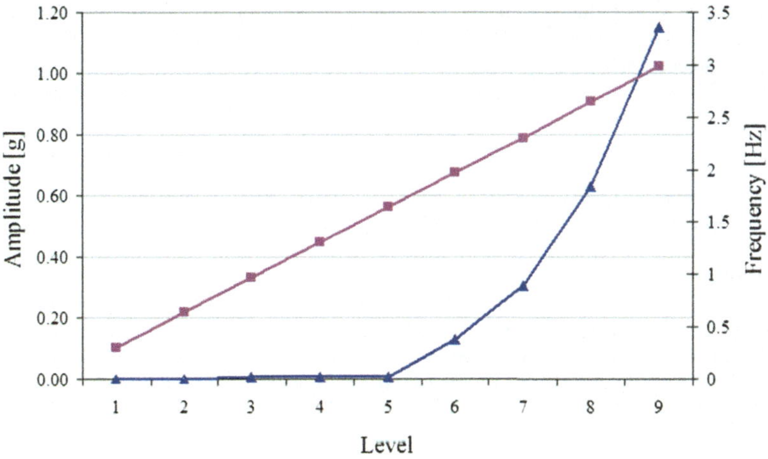

Fig. 30.4 The chart of the amplitude (magenta) and frequency (blue)

30.4 Conclusions

As a result of the tests, we have concluded that for a safe operation at high levels of the shaking table, that must be reinforced by lowering the center of gravity (a metal plate will be mounted at the bottom of the table).

The measurements made with the LabView program reveal some parasitic frequencies aspects of the shaking table do to the cinematic chain and the metallic ball-shaped guide. These parasitic frequencies and others will be mitigated by reinforcing the shaking table and using some elastomeric materials (like rubber).

Acknowledgements The work has been funded by the Sectoral Operational Programme Human Resources Development 2007–2013 of the Ministry of European Funds through the Financial Agreement POSDRU/159/1.5/S/132395.

References

1. R. Reitherman, Earthquakes and Engineers: An International History (ASCE Press, Reston, 2012), pp. 126–127
2. S.K. Prasad, I. Towhata, G.P. Chandradhara, P. Nanjundaswamy, Shaking table tests in earthquake geotechnical engineering. Curr. Sci. **87**(10), 1398–1404 (Special Section: Geotechnics and Earthquake Hazards, 2004)
3. X. Weixiao, S. Jingjiang, Y. Weisong, D. Ke, Shaking table comparative test and associated study of a stepped wall-frame structure. Earthq. Eng. Eng. Vib. **13**(3), 471–485 (2014)
4. R. Bairrao, C.T. Zav, Shaking table testing of civil engineering structures—the LNEC 3D simulation experience, in *12th World Conference on Earthquake Engineering*, Auckland, New Zeeland (2000)

5. C.Ş. Jurcău, G.R. Gillich, V. Iancu, D. Amariei, Evaluation and control of forces acting on isolated friction pendulum, in the *3rd WSEAS International Conference (EMESEG'10)*, Corfu Island, Greece (2010), pp. 220–225
6. G.R. Gillich, D. Amariei, V. Iancu, C.Ş. Jurcău, Aspects behavior of bridges which use different vibration isolating systems, in *10th WSEAS International Conference on Automation & Information (ICAI'09)*, Praga (2009), pp. 140–145
7. V. Iancu, G.R. Gillich, C.M. Iavornic, N. Gillich, Some models of elastomeric seismic isolation devices, *12 International Symposium AVMS' 2013*, Timişoara (2013)
8. G.R. Gillich, P. Bratu, D. Frunzăverde, D. Amariei, V. Iancu, *Identifying Mechanical Characteristics of Materials with Non-Linear Behavior Using Statistical Methods* (Harvard University, Cambridge, 2010)
9. V. Iancu, O. Vasile, G.R. Gillich, Modelling and characterization of hybrid rubber-based earthquake isolation systems. Revista Materiale Plastice **49**(4), 237–241 (2012)
10. I. Negru, G.R. Gillich, Z.I. Praisach, M. Tufoi, E.W. Gillich, Nondestructive evaluation of piers, in *Proceedings SPIE 9438, Health Monitoring of Structural and Biological Systems 2015*, San Diego, USA, Mar 23, 2015, p. 943817

Chapter 31
Dynamic Analysis of the Reaction Chamber for the ELIADE Array

Sorin Vlase, Paul Nicolae Borza, Gabriel Suliman, Cristian Petcu, Maria Luminita Scutaru, Marius Ghitescu and Cristi Nastac

Abstract The ELI-NP Array of DEtectors (ELIADE) is one of the experimental setups being built at ELI-NP. The reaction chamber for the ELIADE array is, together with the CCD camera the goal of research team. The precision of the experiment make important the small deformation and the vibration of the equipment. To study this, a model using Finite Element Method is used. The device consists of elastic elements to a smaller or greater extent. Usually, if the velocities and the occurring loads are low then the rigid elements hypothesis can lead to an excellent model. But in the experiment that will be conduct in the laboratory with ELIADE, precision is so important that it require extremely small deformation of the device. For this a model that takes into account the elasticity of the body must be used in order to study the movement of the part of the device during the experiment.

S. Vlase (✉) · P. N. Borza · M. L. Scutaru · M. Ghitescu · C. Nastac
TRANSILVANIA University of Braşov, Braşov, Romania
e-mail: svlase@unitbv.ro

P. N. Borza
e-mail: borzapn@gmail.com

M. L. Scutaru
e-mail: lscutaru@unitbv.ro

M. Ghitescu
e-mail: marius.ghitescu@unitbv.ro

C. Nastac
e-mail: ndcproiect@gmail.com

G. Suliman · C. Petcu
Institut of Nuclear Physics-IFIN-HH, Măgurele, Bucharest, Romania
e-mail: gabriel.suliman@eli-np.ro

C. Petcu
e-mail: cristian.petcu@eli-np.ro

N. Herisanu and V. Marinca (eds.), *Acoustics and Vibration of Mechanical Structures—AVMS-2017*, Springer Proceedings in Physics 198,
https://doi.org/10.1007/978-3-319-69823-6_31

31.1 Introduction

The Gamma Beam System (GBS) is based on the creation of high energy photons after the collision between a visible light laser (2.3 eV, 515 nm) and a high energy electron beam. The collision between the free electrons (not bound "inside" an atom) and photons is described by a process named "Compton scattering". In nature, this kind of collisions usually takes place between an electron at rest and a high energy photon, and results in the electron gaining energy at the expense of the photon which loses some of its energy. Because in the GBS, the roles are reversed (the electron has much more energy than the photon), the photon is the one gaining energy, and as a result, we say that we are dealing with "an inverse Compton scattering" [1].

The ELI-NP Array of DEtectors (ELIADE) is one of the experimental setups being built at ELI-NP which could benefit from the present project. From the physics point of view the array is made of 8 (up to 12) HyperPure Germanium (HPGe) segmented clover detectors. The experiments envisaged for ELI-NP require this detectors to be placed on two rings, relative to the forward direction of the gamma beam. The detector's axis are all converging to a "theoretical" point which we call "the center of the array". Ideally, this is point where the photon beam intersects the target inside the reaction chamber (Fig. 31.1).

An important distinction that needs to be done is the difference between the "center of array" (described above) and the "desired center of the target". The surface of the gamma beam spot size on the target depends on a number of factors, but will typically range between 0.5 and 2 mm in diameter. Ideally, the two are identical, but deviations from the ideal case have very different impacts on the experiments. The problem becomes the precision obtains in the experiment. To achieve this is necessary to make a suitable model in order to obtain a dynamical response of the system.

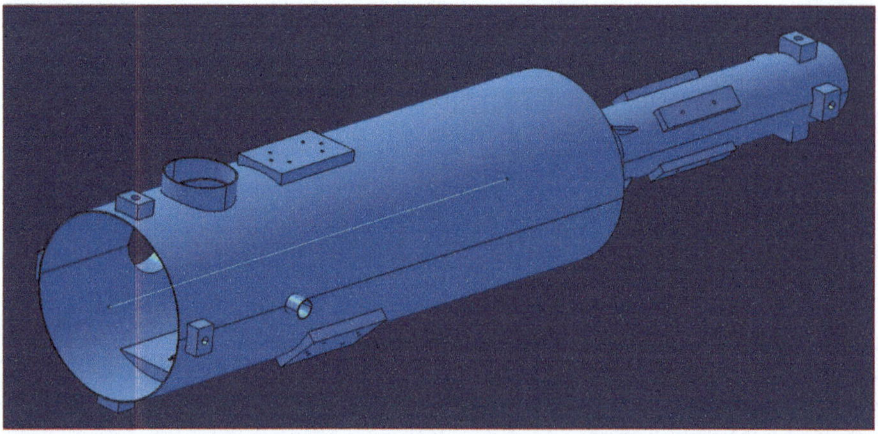

Fig. 31.1 The sketch of the interaction chamber

31.2 Current State and Research Context

The hypothesis of rigid body, frequent used in the dynamical analysis of multibody systems, may be satisfactory in most applications. But the different part of a device consists of elastic elements to a smaller or greater extent. If the precision of the experiment imposes to know the deformations, that must be small, the elasticity becomes a significant element. In our case the deformations and the possible vibration may have, generally, an unfavorable influence on the operation of the device.

Continuous mathematical models can be applied, from theoretical point of view, but are not useful in practical applications. The best way of approaching the problem is to apply the finite element method. The advantages of this approach result from [2–6].

The papers approaching this field have performed an analysis of a single deformable element, having a plane motion and then the study extended to the mechanisms with plane-parallel motion [7, 8] with all the deformable elements. In [9] the results obtained in this field are being synthesized and some theoretical assumption are presented in [10–14].

31.3 FEM Model

The type of finite element used shall determine the chosen shape functions and the equations of motion shall be determined in a general case without considering certain shape functions. In what follows it is considered that deformations are small not influencing the general rigid motion of the whole system. It is considered that a chosen arbitrary finite element together with the solid a component of which it actually is, participates to the entire rigid motion of the mechanical system. The velocities and accelerations of the points of this finite element shall be entirely determined if the velocity and acceleration of the origin of coordinate system the finite element is related to, the angular velocity and angular acceleration are known. In order to write the equations of motion for the studied finite element the equations of Lagrange shall be used. The kinetic energy and the strain energy for the finite element and the work of the distributed and concentrated forces shall be also determined. If these values have been determined we can write the Lagrange function and apply the equations of Lagrange in order to be able to determine the equations of motion of the nodal points. The bodies forming the mechanical system will be considered as being linear elastic.

Let's consider a finite element belonging to an elastic element of the mechanical system studied. The finite element considered shall be related to the local mobile coordinate system, which participates to the general rigid motion (Fig. 31.2). The

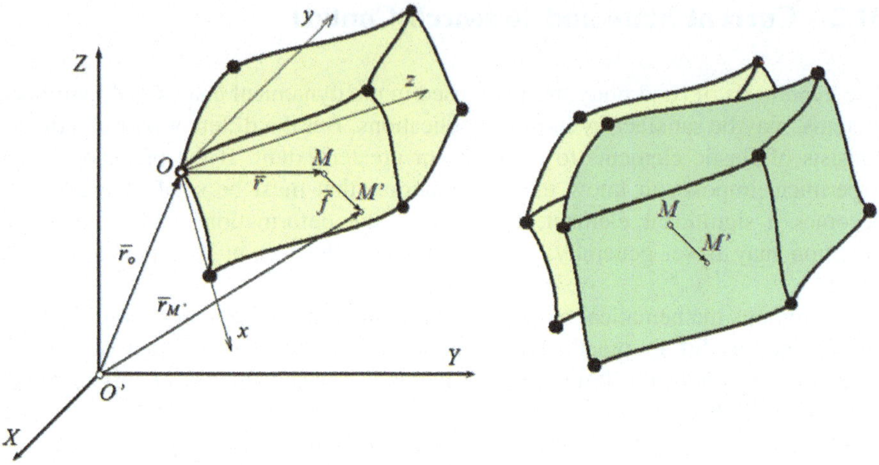

Fig. 31.2 A representative finite element

multi-body system consisting of several solids, these vectors will be different for each solid composing the system. The transformation of a vector from the local system of coordinates into the global system of coordinates occurs by means of a matrix $\boldsymbol{R}$.

If we note the position vector of point M with $\boldsymbol{r}_{M,G}$ we may write:

$$\boldsymbol{r}_{M,G} = \boldsymbol{r}_{O,G} + \boldsymbol{r}_G = \boldsymbol{r}_{O,G} + \boldsymbol{R} \cdot \boldsymbol{r}_L, \tag{31.1}$$

where index G indicates a vector with the components expressed in the global coordinate system and index L a vector with the components expressed in the local coordinate system.

If point M has a displacement $\boldsymbol{f}_L$ changing into M′, we may write:

$$\boldsymbol{r}_{M',G} = \boldsymbol{r}_{O,G} + \boldsymbol{R} \cdot (\boldsymbol{r}_L + \boldsymbol{f}_L) \tag{31.2}$$

where $\boldsymbol{r}_{M',G}$ is the position vector of point M' with components expressed in the global coordinate system. The continuous displacement field $\boldsymbol{f}(x, y, z)_L$ is approximated in the finite element method, depending on the nodal coordinates, by relation:

$$\boldsymbol{f}_L = \boldsymbol{N}(x, y, z)\boldsymbol{\delta}_e(t)_L \tag{31.3}$$

where the matrix elements N—the shape functions—, will depend on the type of the finite element chosen.

31.4 Equation of Motion for a Three-Dimensional Finite Element

The equations of motion shall be obtained in the local coordinate system. For this purpose the equations of Lagrange shall be used [15]. The kinetic energy of the considered finite element is given by the expression:

$$E_c = \frac{1}{2}\int_V \rho v^2 dV = \frac{1}{2}\int_V \rho \boldsymbol{v}^T_{M',G}\boldsymbol{v}_{M',G} dV. \tag{31.4}$$

The potential energy (internal work) is:

$$E_p = \frac{1}{2}\int_V \boldsymbol{\sigma}^T \boldsymbol{\varepsilon} dV. \tag{31.5}$$

We remind you of the Hooke law which we write as follows:

$$\boldsymbol{\sigma} = \boldsymbol{D}\boldsymbol{\varepsilon} \tag{31.6}$$

for an homogeneous, isotropic material. The differential relations which link the strains to the finite deformations that can be expressed in a concise form:

$$\boldsymbol{\varepsilon} = \boldsymbol{a}\boldsymbol{f} \tag{31.7}$$

where $\boldsymbol{a}$ represents the differentiation operator (see [6]). If using the relations (31.6) and (31.7) the strains energy results:

$$E_p = \frac{1}{2}\int_V \boldsymbol{\delta}^T_{e,L}\boldsymbol{k}_e\boldsymbol{\delta}_{e,L} dV, \tag{31.8}$$

where $\boldsymbol{k}_e$ is the stiffness matrix:

$$\boldsymbol{k}_e = \int_V \boldsymbol{N}^T\boldsymbol{a}^T\boldsymbol{D}^T\boldsymbol{a}\boldsymbol{N} dV. \tag{31.9}$$

If the distributed forces vector is noted with $\boldsymbol{p} = \boldsymbol{p}(x, y, z)$, then the external work is:

$$W = \int_V \boldsymbol{p}^T_L \boldsymbol{f}_L dV = \left(\int_V \boldsymbol{p}^T_L \boldsymbol{N} dV\right)\boldsymbol{\delta}_{e,L}. \tag{31.10}$$

The nodal forces $\boldsymbol{q}_e^T$ give an external work:

$$\boldsymbol{W}^c = \boldsymbol{q}_{e,L}^T \boldsymbol{\delta}_{e,L}. \tag{31.11}$$

The Lagrangean for the considered element will be:

$$L = E_c - E_p + W + W^c. \tag{31.12}$$

The equations of motion are obtained by applying the equations of Lagrange. After a series of elementary calculations and rearranging of terms we get the equations of motion for the finite element considered:

$$\begin{aligned} &\left(\int_V \boldsymbol{N}^T \boldsymbol{N} \rho dV\right) \ddot{\boldsymbol{\delta}}_{e,L} + 2\left(\int_V \boldsymbol{N}^T \boldsymbol{R}^T \dot{\boldsymbol{R}} \boldsymbol{N} \rho dV\right) \\ &\dot{\boldsymbol{\delta}}_{e,L} + \left(\boldsymbol{k}_e + \int_V \boldsymbol{N}^T \boldsymbol{R}^T \ddot{\boldsymbol{R}} \boldsymbol{N} \rho dV\right) \boldsymbol{\delta}_{e,L} \\ &= \boldsymbol{q}_e + \int_V \boldsymbol{N}^T \boldsymbol{p}_L dV - \left(\int_V \boldsymbol{N}^T \rho dV\right) \boldsymbol{R}^T \ddot{\boldsymbol{r}}_O - \int_V \boldsymbol{N}^T \boldsymbol{R}^T \ddot{\boldsymbol{R}} \mathbf{r} \rho dV \end{aligned} \tag{31.13}$$

The equations of motion can be written in a condensed form:

$$\begin{aligned} &\boldsymbol{m}_e \ddot{\boldsymbol{\delta}}_{e,L} + 2\boldsymbol{c}_e \dot{\boldsymbol{\delta}}_{e,L} + \left[\boldsymbol{k}_e + \boldsymbol{k}_e(\varepsilon) + \boldsymbol{k}_e(\omega^2)\right] \boldsymbol{\delta}_{e,L} \\ &= \boldsymbol{q}_e + \boldsymbol{q}_{e,L}^* - \boldsymbol{q}_{e,L}^*(\varepsilon) - \boldsymbol{q}_{e,L}^i(\omega^2) - \boldsymbol{m}_{Oe}^i \boldsymbol{R}^T \ddot{\boldsymbol{r}}_O. \end{aligned} \tag{31.14}$$

The equations of motion are related to the local system of coordinates and can be linearized the system of reference being considered as "frozen" in that particular position, in which the field of velocities and accelerations is known. The matrix coefficients can be calculated after choosing the shape functions and the nodal coordinates for expressing the displacement of a point.

The assembling operation leads to the elimination of the liaison forces (see [14–16]). The involved matrices can be computed by choosing the nodal coordinates and the shape functions for the chosen finite element.

31.5 Conclusions

The obtained equations in the study of the dynamic response have additional terms. The first additional term is $2\boldsymbol{c}_e \dot{\boldsymbol{\delta}}_{e,L}$ and is due to the relative motion of nodal coordinates relative to the mobile coordinate systems attached to the moving bodies—Coriolis effects. The second additional term is $\boldsymbol{k}_e(\varepsilon) = \boldsymbol{k}_e(\omega^2)$ and represents a change in stiffness determined by the accelerations field of relative motion.

These two terms can become significant in the dynamic response of the multibody system and can change not only quantitative but also qualitative this response. In order to obtain a very high accuracy in the motion and control of the interaction chamber the presented model was used in the project of the reaction chamber for ELIADE array.

References

1. C. Sun, Characterisation and Diagnostics of Compton Light Source, Ph.D. thesis (2009)
2. T. Heitz, H. Teodorescu-Draghicescu, S. Lache, A. Chiru, S. Vlase, M.R. Calin, Advanced T700/XB3585 UD carbon fibers-reinforced composites. J. Optoelectron. Adv. Mat. **16**(5–6), 568–573 (2014)
3. M.N. Velea, C. Schneider, S. Lache, Second order hierarchical sandwich structure made of self-reinforced polymers by means of a continuos folding process. Mat. Design **102**, 313–320 (2016)
4. J. Gerstmayr, J. Schöberl, A 3D finite element method for flexible multibody systems. Multibody Syst. Dyn. **15**(4), 305–320 (2006)
5. Y.A. Khulief, On the finite element dynamic analysis of flexible mechanisms. Comput. Meth. Appl. Mech. Eng. **97**(1), 23–32 (1992)
6. S. Vlase, Contributions to the Elastodynamic Analysis of the Mechanisms with the Finite Element Method (In Romanian). Ph.D., Transilvania University (1989)
7. A.G. Erdman, G.N. Sandor, A. Oakberg, A general method for kineto-elastodynamic analysis and synthesis of mechanisms. J. Eng. Ind. ASME Trans. 1193 (1972)
8. S. Vlase, Finite element analysis of the planar mechanisms: numerical aspects. In *Applied Mechanics*, vol. 4. (Elsevier, 1992), pp. 90–100
9. A. Ibrahimbegović, S. Mamouri, R.L. Taylor, A.J. Chen, Finite Element Method in Dynamics of Flexible Multibody Systems, Multibody System Dynamics, vol. 4, No. 2–3, pp. 195–223 (2000)
10. M. Marin, S. Vlase, M. Paun, Considerations on double porosity structure for micropolar bodies. AIP Adv. **5**(3), art. Number 037113 (2015)
11. M. Marin, D. Baleanu, S. Vlase, Effect of microtemperatures for micropolar thermoelastic bodies. Struct. Eng. Mech. **61**(3), 381–387 (2017)
12. M. Marin, Harmonic vibrations in thermoelasticity of microstretch materials, J. Vibr. Acoust. ASME **132**(4), 044501–044501-6 (2010)
13. M. Marin, Weak solutions in elasticity of dipolar porous materials. Math. Probl. Eng. **20-018**, Art. ID 158908 (2015)
14. W. Blajer, K. Kołodziejczyk, Improved DAE formulation for inverse dynamics simulation of cranes. Multibody Syst. Dyn. **25**, 131–143 (2011)
15. S. Vlase, A method of eliminating Lagrangean multipliers from the equations of motion of interconnected mechanical systems. J. Appl. Mech. ASME Trans. **54**(1) (1987a)
16. S. Vlase, Elimination of Lagrangean multipliers. Mech. Res. Commun. **14**, 17–22 (1987)

Chapter 32
Considerations Regarding Vibrating Platforms Electromagnetically Driven

Carmen Brăcăcescu, Valentin Vlăduţ, Elena Sorică, Simion Popescu and Cristian Sorică

Abstract The paper presents the diagram and the operational principle of electromagnet driven vibrating platforms and the constructive and functional factors influencing the working parameters of this type of equipment. Diagrams are presented of automatic amplitude and frequency control systems of the vibrations generated by the studied electromagnetic device, in view of its successfully utilization for the drive of vibrating platforms used in the interphase transport of grain seeds and products resulting from their processing. Also there are presented the advantages of using these types of vibration systems as resulting from experimental research.

32.1 Introduction

In order to perfect the primary technology of cereal seeds at agricultural producers, to reduce wastes along processing stages links (cleaning, sorting, storing, transport, processing, packing, pre-commercialisation storing, distribution and commercialising) as well as to assure the food security, it is necessary to correlate the specific high performing technical equipment, with low energy and materials consumption.

C. Brăcăcescu (✉) · V. Vlăduţ · E. Sorică · C. Sorică
National Institute of Research-Development for Machines and Installations Designed to Agriculture and Food Industry—INMA, Bucharest, Romania
e-mail: carmenbraca@yahoo.com

V. Vlăduţ
e-mail: valentin_vladut@yahoo.com

E. Sorică
e-mail: postelnicu.elena@yahoo.com

C. Sorică
e-mail: cri_sor2002@yahoo.com

S. Popescu
Transilvania University of Brasov, Braşov, Romania
e-mail: simipop38@yahoo.com

N. Herisanu and V. Marinca (eds.), *Acoustics and Vibration of Mechanical Structures—AVMS-2017*, Springer Proceedings in Physics 198,
https://doi.org/10.1007/978-3-319-69823-6_32

Also, continuous progresses on improving primary processing as prim operations of conditioning and of technical-material base, periodically determine the demands substantially increasing to foreign bodies missing, seeds uniformity, sanitary conditions and batch homogeneity [1–3].

Operating the vibrating conveyors destined to agricultural products transport is based on, micro-throwing" effect of the particles within material mass [4].

In turn, the equipment for impurities separating, accomplished in form of oscillating mass, provided with sieves, have an operating principle similar to vibrating platforms conveyors and stirring the bulk materials by vibrations is regularly induced, through a vibrating platform inclined from horizontal [5–7].

Modern systems equipment, generating electromagnetic vibrations type, gives technical equipment a high yield and an improved efficiency. To establish the optimum functioning regime of the vibrating platforms is necessary to know the properties of the transported/processed material (particles size, distribution and their shape) and to establish the parameters which influence the technological process.

32.2 Materials and Methods

In Fig. 32.1 are presented base schemes of vibrating platforms, connected by antennas of different mechanisms systems, which generate transport movements [4, 7].

In turn, the equipment for impurities separating, achieved in form of sieves oscillating masses, receive oscillating movements generated by the crank mechanism (Fig. 32.1a), by eccentric masses vibrating devices (Fig. 32.1b) or electromagnets vibrating devices (Fig. 32.1c).

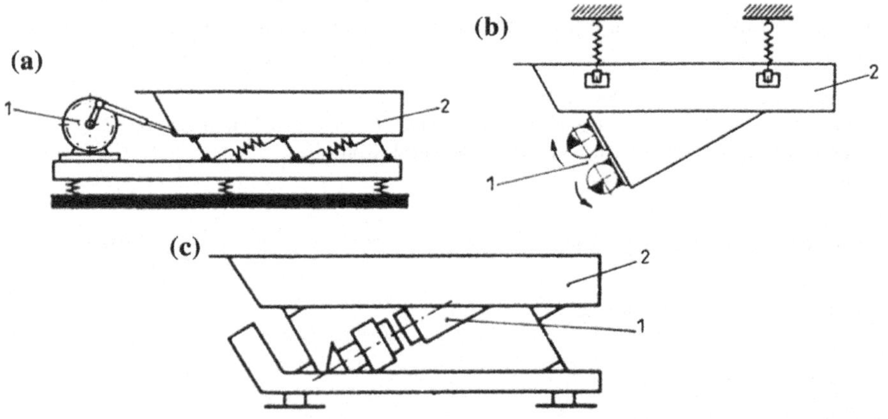

Fig. 32.1 Types of vibrating platforms: **a** rod-crank driven; **b** eccentric masses vibrating devices (unbalanced); **c** electromagnetically vibrating driven: 1—mechanism (device) for vibrations generating; 2—transport area

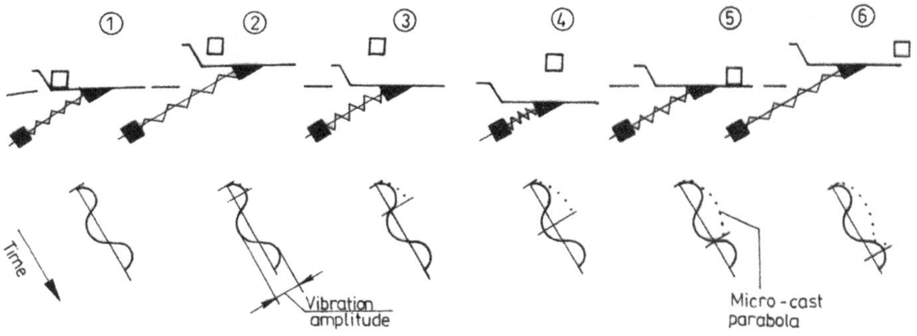

Fig. 32.2 Transport process stages of the material particles on vibrating platforms

Materials transport on platform is performed by its vibrations, process in which is fulfilled throwing materials on oscillation line. Thus, in transport stage the material fall of the platform and move to a throwing trajectory (Fig. 32.2).

Particles are accelerated starting from an established frequency/amplitude ratio of the vibrations; this acceleration regime is used to achieve a parabolic upward movement of the particles, which after a specific movement, hit again the surface of the transport body. The required condition for excitation of the micro-throwing effect to occur is that the vertical upward acceleration of the particles, induced by the vibrations of the platform, is greater than the downward gravitational acceleration. The most important parameters which influence this process of particle dislocation are vibrations, frequency and amplitude, angle between axis of excitation and platform, solid material properties such as: particles size, particles distribution, particles shape and friction. Electromagnetic vibrator driven conveyors consist of a mechanical transport device by means of which the solid material layer moves, as well as the operation unit/drive unit for vibration excitement supported by a helical spring suspension (Fig. 32.3) [7–9].

From a dynamic point of view, the vibrating systems of the conveyors or separators represent dynamic spring—mass systems [8, 9], which take into account two masses connected by springs:

- the first mass m_a represents the mass of the product feed subassembly, consisting of the vibratory platform 1, the mass of transported material 10 and the magnetic vibrator (made up of elements 2–4);
- the second mass m_f represents the mass of the free part, consisting of the moving bodies 5 and 6.

Magnetic operating/drive system used as free vibrators or along with steering suspensions operates in resonance conditions with safer excitement energies, thus allowing the creation of very compact units. Under the action of the vibrations generated by the electromagnetic vibrator, the vibration frame 1, elastically suspended on the support by the springs 9, is subjected to vibrations with the elongation S, oriented in a given direction α. The effective amplitude S_a of the

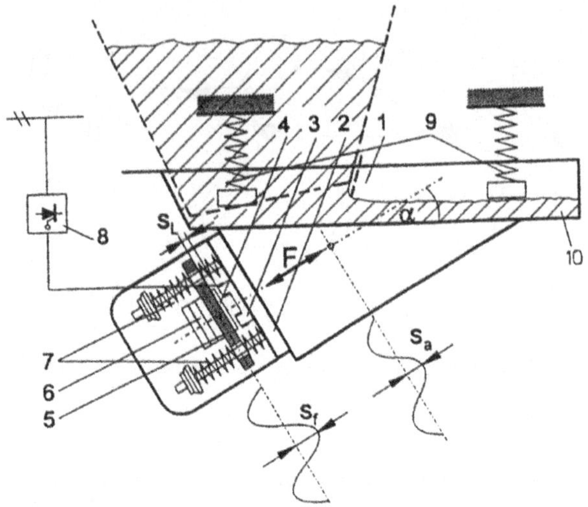

Fig. 32.3 The schematic diagram of an electromagnetically driven vibratory platform: 1—vibration frame; 2—vibrator housing and other components; 3—electromagnet; 4—mobile magnetic core plate; 5, 6—additional masses; 7—spring; 8—control device; 9—suspension springs; 10 - transported material, S_f—vibrations amplitude of the electromagnet plate; S_a—effective amplitude of the gutter

platform depends on the constructive, functional and control parameters, as well as the kind of material transported. The movement of the platform vibratory mass can be considered, with sufficient approximation, as a sinusoidal harmonic oscillation motion and elongation S of the oscillatory motion is given by the relation:

$$S = S_a \sin(\omega t) \tag{32.1}$$

In which: S_a is oscillation amplitude; ω—oscillation pulsation;
Speed of the oscillatory motion is given by relation:

$$v = S_a \omega \cos(\omega t) \tag{32.2}$$

In which $v_0 = S_a \omega$
Oscillatory motion acceleration is expressed by the relation:

$$a = -S_a \omega^2 \sin(\omega t) \tag{32.3}$$

In which $a_0 = -S_a \omega^2$ is acceleration amplitude;
If the value of the acceleration amplitude a_0 of the platform is related to the gravitational acceleration g, the transport characteristic K_M is obtained by the relation:

$$K_M = \frac{a_0}{g} = -\left|\frac{S_a \omega^2}{g}\right| \tag{32.4}$$

As the vibratory conveyor platforms on movement direction are inclined with angle α in relation to the transport surface (Fig. 32.4), the components in the

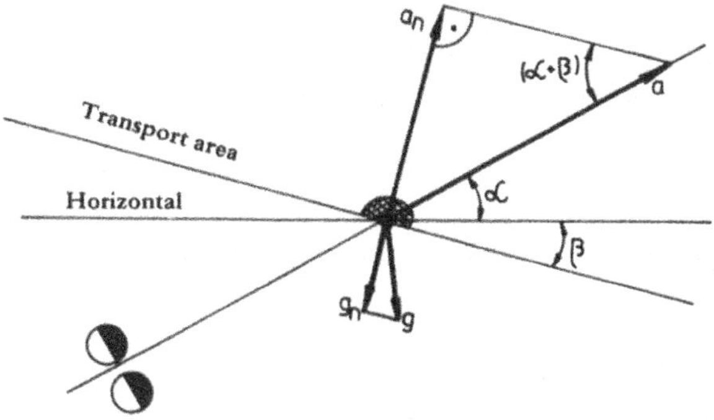

Fig. 32.4 Scheme of accelerations that act on particles in the vibration

horizontal direction of the elongation and the acceleration are obtained with the relations:

$$S_h = S\cos\alpha\,,\quad a_h = a\cos\alpha \tag{32.5}$$

and the components in the vertical direction are given by the relations:

$$S_v = S\sin\alpha \tag{32.6}$$

$$a_v = a\sin\alpha \tag{32.7}$$

Taking in account transport area inclination angle β in relation to horizontal, equipment presents K_W throwing characteristic, which is expressed in function of K_M machine characteristic, by relation:

$$K_w = K_M \frac{\sin(\alpha+\beta)}{\cos\beta} \tag{32.8}$$

From relation (32.8) is noticed that throwing characteristic value $K_W = a_n/g$ represents, at a certain scale, the value of normal acceleration value at a_n transport surface.

If consider a material particle to be located on the transport surface, depending on the value of the K_W throwing characteristic, three distinct domains of the process kinematics are distinguished. If the $K_W < 1$, the particle remain stationary on transport surface. Due to the inertia force, of course there is a certain sliding of the particle on the supporting surface. This represents the working domain of sock sliding.

If $1 < K_W < 3, 3$, then, at the A moment the particle detaches from the support surface in the upward direction (Fig. 32.5). At the time of throwing Δt_F the material particle will describe a parabolic trajectory and at point B encounters the transport surface again and in the event of an ideal collision, it remains in the time interval Δt_B in contact with the transport surface, after which the process is repeated. This mode of operation corresponds to vibratory platforms. The layers of material, with a good approximation, are considered to have a behaviour corresponding to a plastic collision. In the case of the two-mass vibration system shown in Fig. 32.3, at which the electromagnet is fed with 50 Hz industrial frequency, the material carrier (trough) oscillates at a double frequency (6000 rpm), which in most of cases is inadmissible. To reduce the platform frequency to 3000 rpm, a rectifier of a half-period 8, is introduced, but the rectified voltage modifies the action of the magnetic flux and the movement of the gutter. This movement is described by another equation with a disturbing force represented by a Fourier trigonometric series.

The f_e, natural not amortized frequency of the spring-mass system in Fig. 32.3 can be determined based on the masses m_a and m_f (see previous explanations in correlation with Fig. 32.3) and the spring constant C:

$$f_c = \frac{1}{2\pi} \cdot \sqrt{\frac{C}{m_r}} \tag{32.9}$$

where m_r is the reduced system mass in Fig. 32.3, calculated with relation known from vibration theory:

$$m_r = \frac{m_a \cdot m_f}{m_a + m_f} \tag{32.10}$$

This vibrational 2-mass system responds to a frequency of excitation f_a, the amplitude increasing as it approaches the natural frequency f_e (for $f_a/f_e = 1$ the resonance appears). The natural frequency decreases as the damping increases (Fig. 32.5), with the V—factor being heavily influenced by damping. For vibrating conveyors to operate at resonance values, the system must be calibrated accordingly (additional masses 6 in Fig. 32.3). Effective depreciation is friction-induced; for example friction from inside the solid materials, and external rubbing on surfaces.

Figure 32.6 highlights the fact that the operation of the vibrating system at a slightly subcritical excitation value is the most recommended, because an increased damping and mass of the solid material on the platform surface—which usually coincide—have a stabilizing effect (see Fig. 32.6, DE, EF).

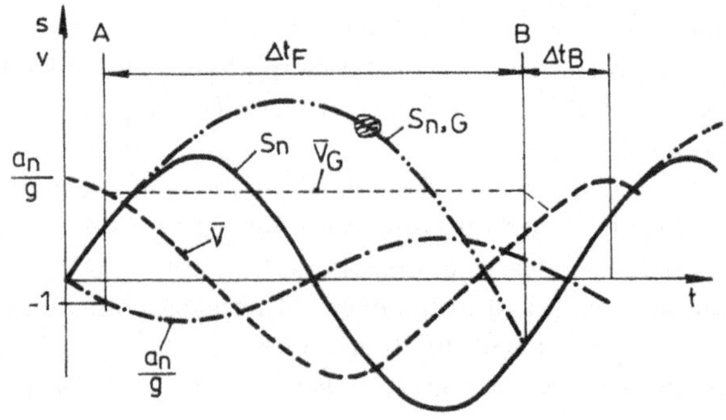

Fig. 32.5 The trajectory of the transported material and the conditions of movement at the vibratory platforms

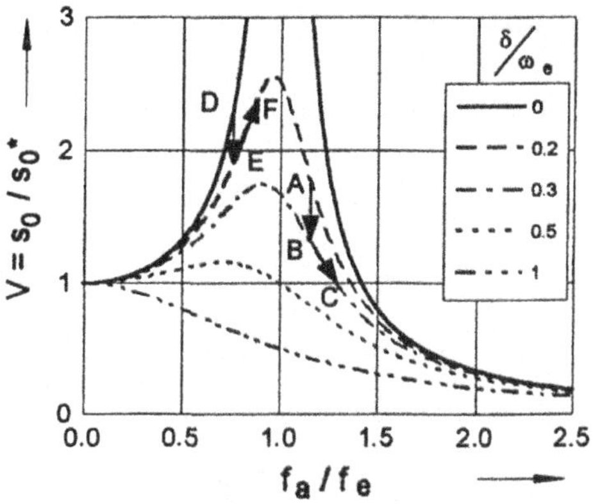

Fig. 32.6 The vibration amplitude amplification factor (V) depending on the excitation frequency (f_a/f_e) ratio

To achieve magnetic excitation and electromagnetic vibrator amplitude control, we use electronic modules for the optimal excitation frequency (25, 50, 100 Hz) and adjust to a preset vibration amplitude value via voltage. It is strongly recommended to automatically control the vibration amplitude to maintain the jamming potential within tight limits.

32.3 Conclusions

The theoretical study of vibrating platforms electromagnetically driven can be achieved by replacing the real systems with equivalent vibrating dynamic systems with external excitation, comprising one or two vibrating masses.

By solving the differential movement equations of vibrating systems and their applicability for concrete situations, we can study by computer simulation the kinematic and dynamic behaviour of systems endowed with existing or in course of designing vibrating platforms or vibrating sieves.

The results of theoretical researches performed on basis of those presented in the paper allow the optimization of constructive and functional parameters of vibrating platforms, in view of manufacturing state-of the art installation.

Taking into account the technical and economic advantages of the vibrating systems with electromagnets they are used not only at vibrating conveyors for grain matters (including seeds) but also at driving the vibrating sieves of technological primary processing installations for grain seeds.

References

1. C. Brăcăcescu, C. Sorică, G. Bunduchi, Experimental researches regarding the optimization of working process of technical equipment for cereals primary processing, Tractors and power machines, UDK 631.372, Godina **19**(1), 77–84 (2014)
2. T. Căsăndroiu, Primary processing and agricultural products storing, lithographed course Politehnica, University of Bucharest (1993)
3. C. Sorică, C. Brăcăcescu, D. Manea, E. Sorică, M. Epure, C. Covaliu, Considerations regarding the optimization of the mechanical conditioning process of grain into the indented cylinder separators. Ann. Fac. Eng. Hunedoara—Int. J. Eng. Tome XII—Fasc. **3** (2014)
4. H. Krampe, *Transport, Umschlag* (Fachverlag, Leipzig, Lagerung. VEB, 1990)
5. L. Brîndeu, in *Vibration and vibropercussion*, Basic of Vibration and Vibropercussion Mechanics, (Publishing House Politehnica, Timişoara, 2001)
6. J. Yan, Liu C., Zhao L., Dynamic characteristics of vibrating screen with determinate structure and statistically indeterminate structure. Appl. Mech. Mater. 34–35 (2010)
7. S. Popescu, Influence of functional parameters of the gravimetric dosing process of granular agro-food material. Bul. Transilvania Univ. Braşov, Serie A **11**(47), 169–176 (2005)
8. M.H. Pahl, *Lagern, fodern und Dosieren* (Verlag TÜV Rheinland, Köln, 1989)
9. G. Vetter, in *The Dosing Handbook*, (Vulkan—Verlag, Essen, 1994)

Chapter 33
A MEMS Electro-Mechanical Device for Scanning and Energy-Harvesting Based on Moving Magnets Matrix

Constantin Daniel Comeaga and Viorel Gheorghe

Abstract The article presents the improved design, modeling and simulation of a new type of MEMS electromagnetic scanning system, based on a moving matrix of permanent magnets. The previous design solution, using a mobile part as mirror and actuator, has some limitations explained through modal analysis. The second design, whose functioning is explained also through modal analysis, allow better performances and reduce disadvantages observed in the first case.

33.1 Introduction

Laser micro scanning devices are actually widely used in a large scale of applications, from virtual reality devices to miniature medical equipment. Other domains which integrates MEMS laser scanner are co-focal microscopy, finger print sensing, optical coherence spectroscopy, printing, light detection and ranging systems for automotive industry or even miniature portable micro projectors.

Different actuating solutions exist, ranging from electrostatic, piezoelectric, electromagnetic principles to the use of shape memory materials. In practice only few of them imposed as a feasible solution. The electro-dynamic or electromagnetic (EM) actuators [1, 2] have complex structures but develop high forces at any scale, and can be used at resonance or far from it. This paper is continuing the work regarding a new type of electromagnetic actuator (EA) that may replace conventional scanners [3, 4]. The goal of this work is to combine, using modal analysis, the electromagnetic actuation with an improved design, usable for scanning and energy harvesting.

C. D. Comeaga (✉) · V. Gheorghe
University Politehnica Bucharest, Bucharest, Romania
e-mail: comeaga@hotmail.com

V. Gheorghe
e-mail: viorel_gheorghe@yahoo.com

N. Herisanu and V. Marinca (eds.), *Acoustics and Vibration of Mechanical Structures—AVMS-2017*, Springer Proceedings in Physics 198,
https://doi.org/10.1007/978-3-319-69823-6_33

33.2 Design of the 1-Stage MEMS

The initial design is based on the tendency of physical system to attain the lowest energy, including the magnetic systems. The system contain an array of magnets, having the same magnetization orientation, placed at the free end of an elastic hinge. The magnets are placed in the magnetic field obtained with a underneath coil, connected to a signal source. A force appear on the magnets, the value and orientation of force is so as to minimize the magnetic energy (Fig. 33.1a). Due to this force the elastic hinge bend, resulting a displacement but also a rotation of the end (on the elastic hinge, on the opposite side to the magnets array, is placed a mirror). This design have the advantage that allows to use the system for energy harvesting also, when the scanner is not used. The vibration of the base (on the base is fixed the hinge and also are placed the coils) conduct to a vibration of magnets, producing in turn an induced voltage in coils. With a proper electric circuit, this electric energy can be collected. The dimensions of the mobile part are included in Fig. 33.1b.

The scanner can be used in static and dynamic conditions. In static conditions is necessary a high electric current, with negative thermal effects, but the angular speed and position are completely controllable [3, 4]. The control is difficult due to the strong non-linearity. The displacements are low and the efficiency poor.

A better working regime is the resonant one, when the actuator is driven at the mechanical resonance. In this case the consumed electric energy is very low and the efficiency very high. For the case of internal stress equal to the yield stress the force acting on magnets drop form 93 mN for the static case, producing a rotation of 18° at the free end, to 1.48 mN for resonant functioning, producing 16° longitudinal rotation. The main disadvantage is that the system is working at a fixed frequency, without any possibility to have a controlled variation of displacement and speed. Currently, as many systems involving MEMS scanners can work easily at a fixed frequency (scanners used in virtual reality devices as in smart glasses, scanners used for Lidar systems, medical appliances, etc.) the resonance functioning is preferred (with piezoelectric, capacitive or electrodynamic actuators). The commonly

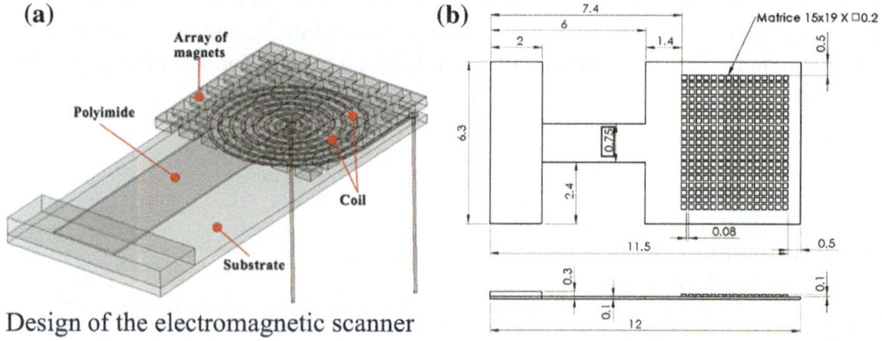

Fig. 33.1 Placement of magnets array and the overall dimensions

mechanical design used for scanners, having capacitive or electrodynamic actuators, has a mirror connected to torsional hinges. It is not possible to use torsional hinges for electromagnetic actuators because requires a change of magnetization for magnets, from one side to the opposite side of the array, and it is difficult to use for energy harvesting because the base displacement is usually translational and not rotational.

33.3 The Dynamic Behavior of the System

The dynamic response was obtained through simulation in Comsol [3, 4], the control values being selected so as the mechanical stress is limited to the yield limit for the elastic hinge material in the useful frequency range (2…2400 Hz).

The displacement of the magnets array center and also the longitudinal rotation (in respect with an axis perpendicular to the hinge length) and the transversal rotation (in respect with an axis on the direction of the hinge length) show variations specific to systems having multiple resonances (Fig. 33.2).

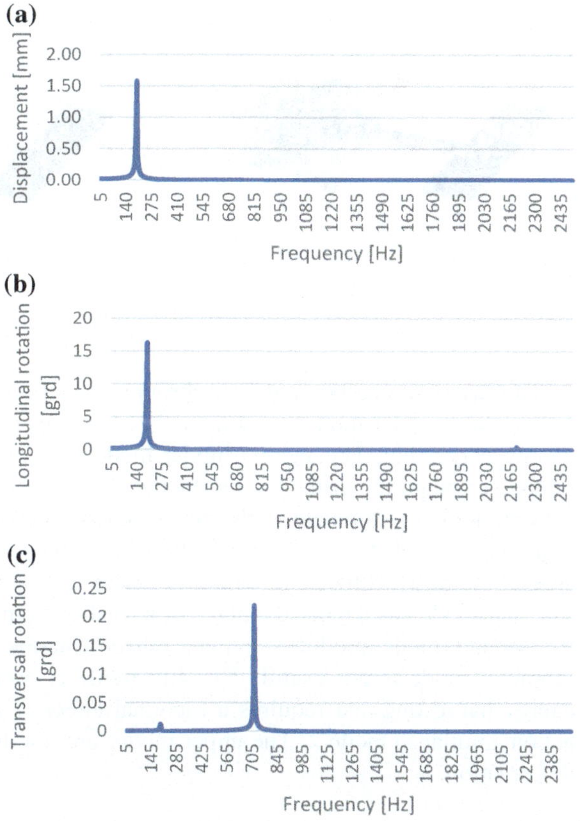

Fig. 33.2 Mirror movements: **a** Displacement of central magnet, **b** longitudinal rotation, **c** transversal rotation

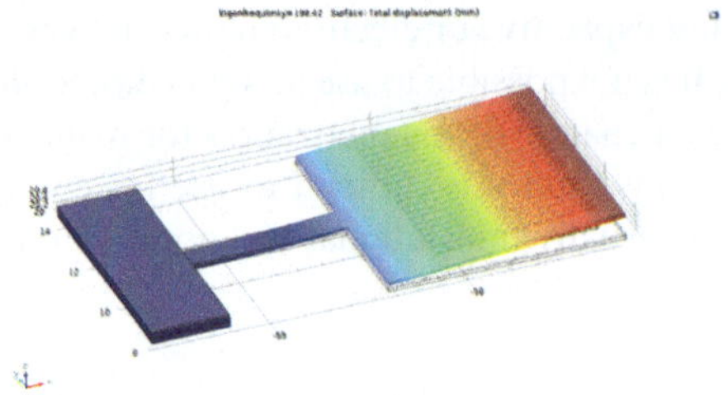

(a) The first resonance at 198.4Hz.

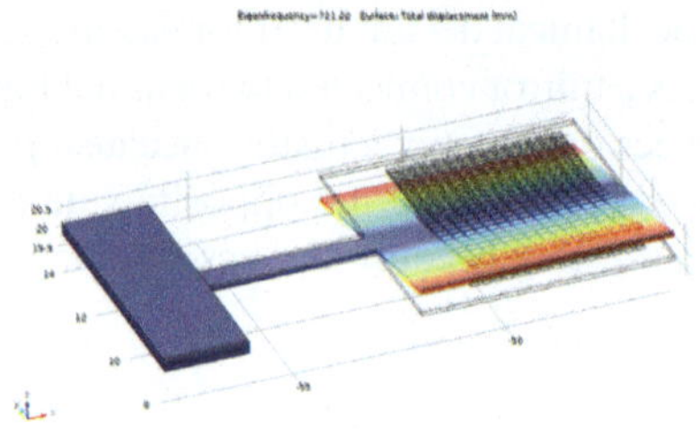

(b) The second resonance at 721.2Hz.

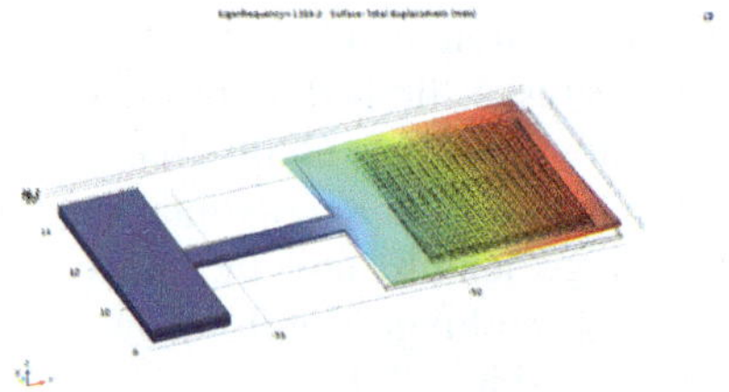

(c) The 3-rd resonance at 1389.2Hz.

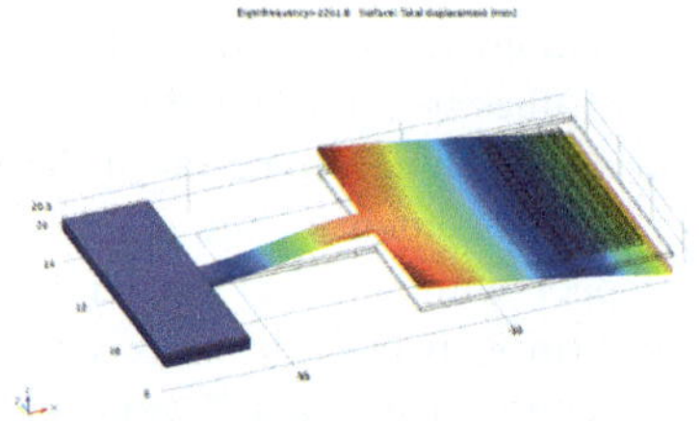

(d) The 4-th resonance at 2201.8Hz

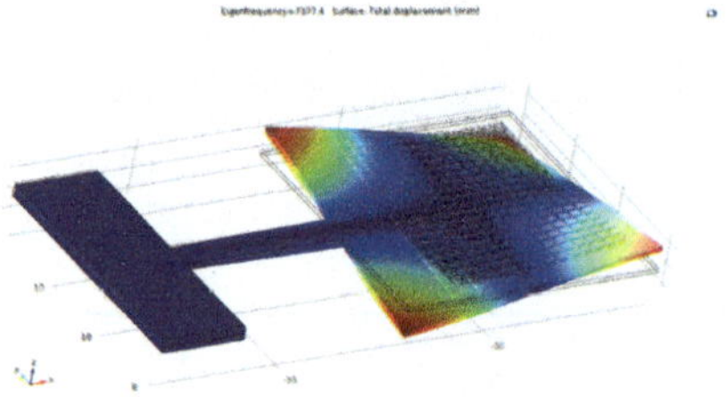

(e) The 5-th resonance at 7377.4Hz.

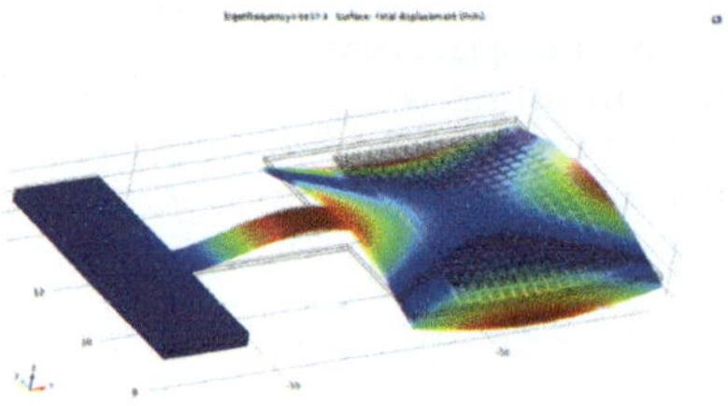

(f) The 6-th resonance at 9837.4Hz

Fig. 33.3 Modal shapes and frequencies of 1-stage design

The response can be explained through modal analysis (Fig. 33.3).

The first modal mode is a transversal bending mode, useful for scanning and energy harvesting. The second modal mode is a torsional mode, more useful for scanning comparing with the first mode, but not useful for energy harvesting and very difficult to stimulate with the designed solution. Stimulation of this mode requires a fast change of magnetization in magnets from one side to the opposite side of the array, difficult to obtain in practice, or a very complex magnetic field. The first mode has the disadvantage of a rotation combined with a translation while the second mode produces a pure mirror rotation. The 3-rd in-plane bending vibration mode is not useful. The 4-th mode is useful for scanning and less for energy harvesting and requires a torsional effect of magnets-magnetic field interaction, similar to mode 2. The upper modes are not useful because creates a mirror deformation.

Table 33.1 Displacement and rotations for a maximum stress of 750 MPa

Modal frequency [Hz]	Displacement of central magnet [mm]	Longitudinal rotation [grd]	Transversal rotation [grd]
198.42	1.584264	15.526	0.019
721.22	5.045036E-02	0.077	0.180
1389.20	1.470825E-01	1.405	0.008
2201.84	1.574122E-02	0.055	0.0016
18294.51	4.931575E-03	0.097	0.0227

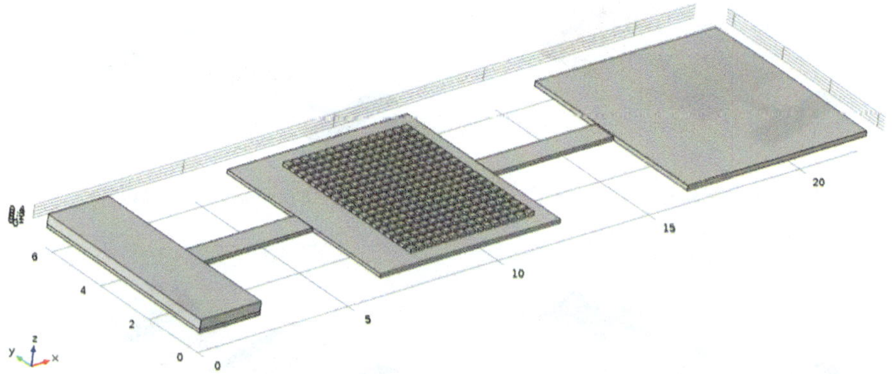

Fig. 33.4 Improved design with the mirror placed on an additional resonator

When selecting the vibration mode used is mandatory to consider the variation of stress with frequency (for upper frequency modes, with high bending or torsion, the stress rise fast for a given displacement). The displacement and rotations computed for each frequency mode for attendance of maximum stress of 750 MPa indicates a drop of allowed rotations with frequency (Table 33.1).

The maximum rotation is around 16°, associated to mode 1, acceptable for a scanning device but the displacement is two big, considering the variations of magnetic field (the field is dropping faster when moving from the coil surface).

33.4 The Dynamic Behavior of the 2-Stage System

An improved design use a two stage structure (Fig. 33.4).

The modal shapes and frequencies indicate that some vibration modes contain a strong rotation of the mirror with small displacement of actuator (Fig. 33.5).

The useful frequencies could be identified in Table 33.2, showing the displacement and rotations computed for each frequency mode for attendance of yield stress. The best modes are 454.7 and 1666.7 Hz, with a low displacement of

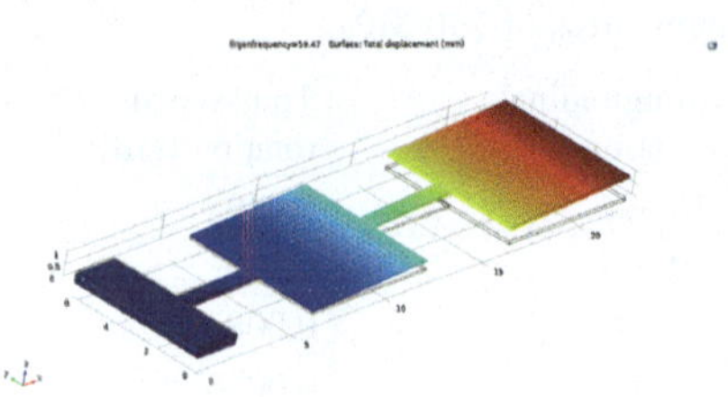

(a) The first resonance at 59.47Hz.

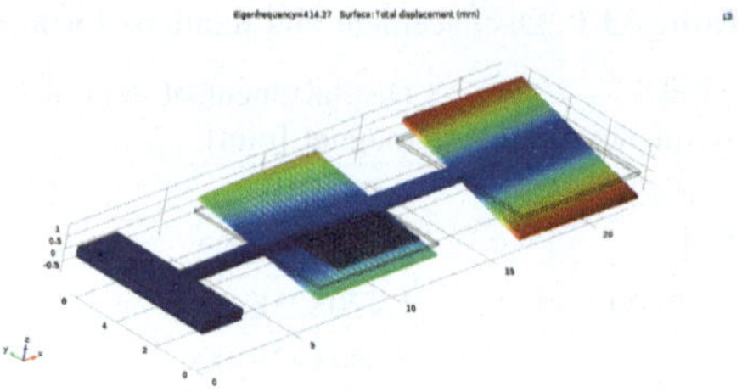

(b) The second resonance at 414.27Hz.

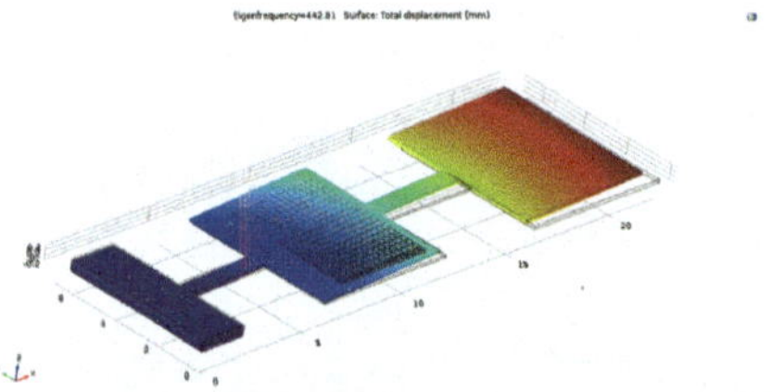

(c) The 3-rd resonance at 442.81Hz.

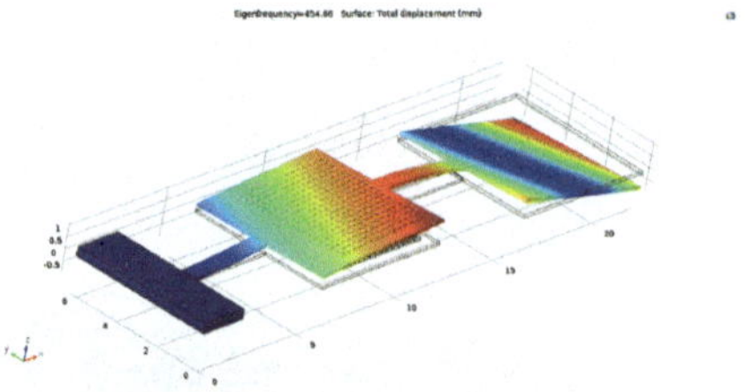

(d) The 4-th resonance at 454.66Hz

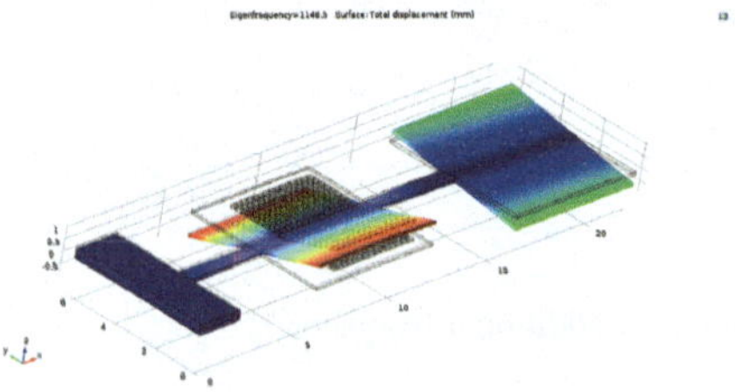

(e) The 5-th resonance at 1148.51Hz.

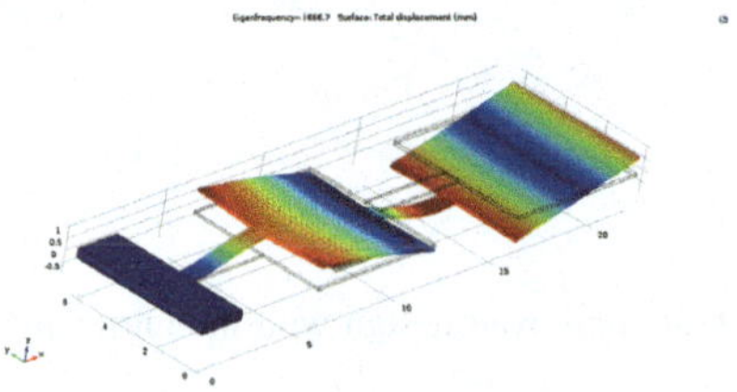

(f) The 6-th resonance at 1666.7Hz

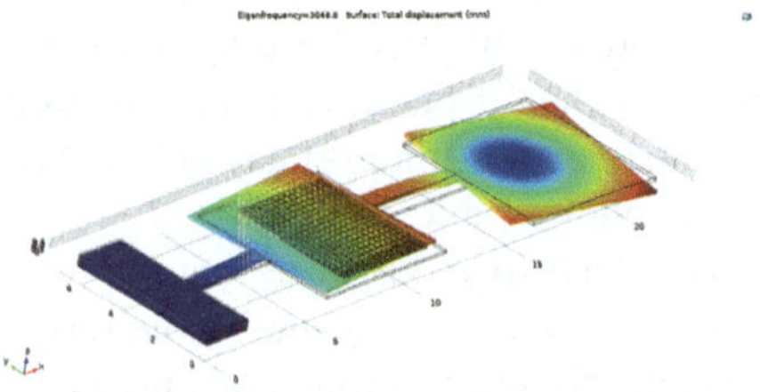

(g) The 7-th resonance at 3048.6 Hz.

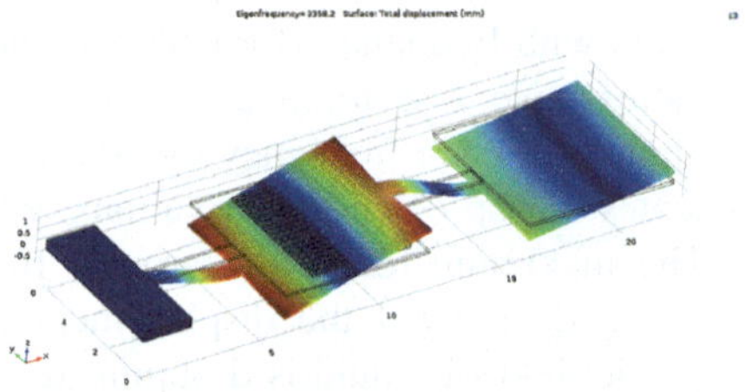

(h) The 6-th resonance at 3358.2Hz.

Fig. 33.5 Modal shapes (not at real deformations) and associated frequencies of 2-stage design

actuator assuring a linear control, but is easier to produce the excitation for 454 Hz as at 1666.7 Hz is necessary to obtain a magnetic torque acting on magnets. The 6-th mode is apparently useful but shown a mirror deformation higher than the limit imposed through design, and similar for the upper frequencies (radius higher than 2 m in any section). The system response to the base vibration, important for energy harvesting, also is expected to show an improvement as the system with an

Table 33.2 Displacement and rotations for a maximum stress of 750 MPa

Modal frequency [Hz]	Displacement of central magnet [mm]	Longitudinal rotation actuator [grd]	Transversal rotation actuator [grd]	Longitudinal rotation mirror [grd]	Transversal rotation mirror[grd]
59.47	1.75476689	18.2041	0.0017	25.4872	0.0025
414.27	0.01126328	0.0411	15.056	0.6302	143.7129
442.81	−0.00252433	0.0142	1.3585	0.4037	21.8306
454.66	0.54291531	3.6507	0.1022	59.2586	0.9710
1148.51	0.00360442	0.0143	9.9814	0.2764	56.7453
1666.70	0.05607754	1.7656	0.0100	92.9751	0.0779

additional mirror behave like a 1DOF mechanical system with an additional dynamic absorber.

Using the 1 degree-of-freedom system with dynamic absorber produces only a narrowband filter with two frequency bands. Tuning the electric circuit allows to obtain a system with a large frequency band comparing with the 1DOF system which behave like a filter with a very narrow frequency band [5].

33.5 Conclusions

The study proved that limitations specific to 1DOF electromagnetic actuators can be reduced using a 2-stage design, with a first actuating stage having limited displacements and a second stage having large amplitudes of rotation. The simulations showed a good improvement of performances for this design comparing with the 1-stage design. Also this design allow use of the system for actuating, acting as a scanner, and also energy recovery, acting as an energy harvesting device. The system will be improved, focusing on the simultaneously optimal design of the mechanical and the electrical parameters of the supply/recovery circuit.

Acknowledgements The work was conducted in the framework of AEEL PNCD II PT PCCA 249/2014 research grant, code PN-II-PT-PCCA-2013-4-1557, financed by the Romanian National Authority for Research, Development and Innovation - UEFISCDI. The authors thank to professor Alexandru Morega from the Laboratory for Multiphysics Modeling at the University POLITEHNICA of Bucharest for the permanent support.

References

1. J. Yan, S. Luanava, V. Casasanta, Magnetic actuation for MEMS scanners for retinal scanning displays. Proc. SPIE **4985**, 115–120 (2003)
2. M.A. Stafne et al., Positional calibration of galvanometric scanners used in laser doppler vibrometers. Measurement **28**, 47–59 (2000)

3. A.M. Morega, N. Tanase, N.M. Popa, M. Morega, J.B. Dumitru, Numerical simulation of an electromagnetic bending-mode cantilever microactuator, in *Proceedings of the 8th International Symposium on Advanced Topics in Electrical Engineering (ATEE)*, 23–25 May 2013. doi:10.1109/ATEE.2013.6563478
4. A.M. Morega, N. Tanase, M. Morega, C.D. Comeaga, C. Ilie, *in Bending Mode Cantilever Actuators For Micro-Electromechanical Systems*, Proceedings of the 9th International Symposium on Advanced Topics in Electrical Engineering (ATEE), 07–09 May 2015. doi:10.1109/ATEE.2015.7133869
5. M. Mitiu, C.D. Comeaga, Dynamic response control of a mass-spring-viscous damper system by use of an additional electro-mechanical impedance. U.P.B. Sci. Bull., series D **75**(4), 149–162 (2013)

Chapter 34
Optimization of Vibro-Compaction Technological Process Considering Rheological Properties

Cornelia-Florentina Dobrescu and Eugeniu Brăguţă

Abstract The present paper is aimed to improve the compaction technology using rheological models for describing soil response parameters under impact loading of vibrating rollers. The technological parameters during the compaction are influenced by rheological characteristics of compacted soils. In contrast, the soil parameters depend on the parameters and operation procedures of compaction equipments. Due to insufficient or inappropriate soil compaction demonstrated in practice, that can cause severe failures of road layers, a particular attention of vibro-compaction process and interaction of surfaces with compactors is given to rheological properties of deformed strata closely connected to the nature of the force action applied. Based on experimental data, the Voigt-Kelvin and Maxwell rheological models were applied to predict the effect of dynamic action under known conditions on granular soils behavior during compaction process. The dynamic ground response is expressed in terms of displacement amplitude of vibrating roller and force amplitude transmitted and for each rheological model the influence of increasing ground stiffness on the parametric values is shown. The end results can be used to optimize the compaction technology, in order to increase the structural capacity of road layers and quality of engineering works.

34.1 Introduction

The issue of dynamic compaction procedures based on vibrations applied to granular soils (sand, gravel, and crushed stone), with reduced fines content, is considered one of the most recent research topics in the practice of ground

C.-F. Dobrescu (✉)
National Institute for Research and Development in Construction, Urban Planning and Sustainable Spatial Development, Bucharest, Romania
e-mail: corneliadobrescu@yahoo.com

E. Brăguţă
„Dunărea de Jos", University of Galaţi, Galaţi, Romania
e-mail: bragutae@gmail.com

N. Herisanu and V. Marinca (eds.), *Acoustics and Vibration of Mechanical Structures—AVMS-2017*, Springer Proceedings in Physics 198,
https://doi.org/10.1007/978-3-319-69823-6_34

improvement methods [1]. Therefore, the road structures and the external platforms of buildings under medium loadings can be constructed by using mineral aggregates in admixture with clay particles as stabilization binder.

The vibration parameters are significant to achieve the desired compaction effect. Therefore, the amplitude and frequency parameters become important factors in the correction of technological procedures. Consequently, the vibro-compaction equipments need special accessories and devices that adjust the vibration regime in order to obtain the required parameters [2] resulting from specific rheological modelling. The experimental results obtained in laboratory and in situ conditions reveal that the rheological behaviour of mixtures prepared in optimized gravimetric doses based on mineral aggregates (sand, gravel, and crushed stone) with clay can be modelled by using Voigt-Klein or Maxwell rheological models. For an accurate assessment of the compaction effect, a parametric correlation of vibratory compaction requirements with physical and mechanical parameters of the soil is necessary [3, 4]. Compound rheological models, which are able to describe the dynamic behaviour of different soil categories, were developed in the past by several researchers and equipment companies. The viscoelastic-plastic models [5, 6], dependent on soil type, indicate a good agreement between experimental and theoretical assumptions. Among them, the Bathlet, Ephremides, Dvorak-Peter, Hartmann, Voigt-Kelvin and Maxwell [7] compound models can be mentioned. In this context, two models were selected as relevant for analytical modelling of simple viscoelastic behaviour during the real-time compaction process. Thus, the parametric variation of mechanical soil properties for Voigt-Kelvin and Maxwell viscoelastic models will be analyzed in the present study, due to the significant information that may be obtained on the soil response during vibro-dynamic compaction [8].

In the paper, the effect of dynamic on ground properties is analyzed for each rheological model, considering that the vibration regime is stationary and for a specified frequency of the perturbing force. It is considered that the compaction process of aggregate mineral mixtures is carried out using the dynamic rolling-compressing procedures. A comparative analysis of dynamic responses for two rheological models is made, using the model for the dynamic compaction of viscoelastic materials presented in [9].

34.2 Analytical Modelling of Soil Response Parameters During the Compaction Process

The main viscoelastic effects of the mineral aggregates in admixture with clay particles as a stabilization binder, arranged in a layered structure, are put in evidence by using the two considered models. For the present study, the Voigt-Kelvin and Maxwell rheological models were used in order to reflect the process of interaction between compaction and intrinsic parameters of granular soils.

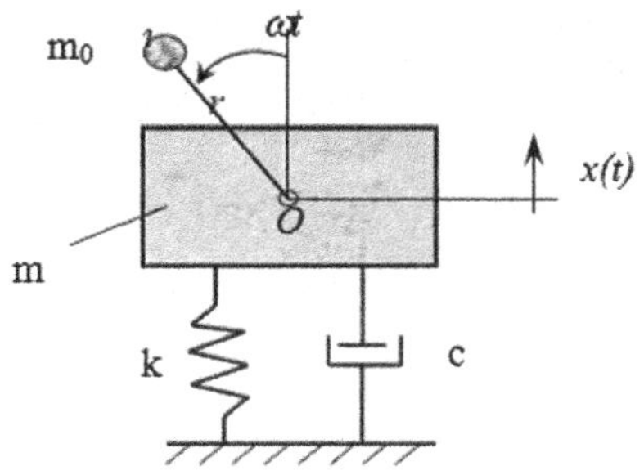

Fig. 34.1 Schematization of the Voigt-Kelvin model

The model Voigt-Kelvin can be adopted for clayey soils mixed with mineral aggregates of 0–4 and 4–8 mm grain size and moisture content ranging from 12% up to 13%. The weight of mineral aggregates represents 60–70% of the mass of mixtures, while 30–40% is clay content. According to the rheological model, the response parameters were estimated (Fig. 34.1) in terms of dynamic force transmitted to the soil, Q, and amplitude of technological vibrations, A.

The perturbing force has the forms presented in (34.1):

$$F_0 = \frac{m_0 r}{m} k\Omega^2 \tag{34.1}$$

where $\Omega = \omega\sqrt{m/k}$.

The dynamic force amplitude transmitted to the soil, $\tilde{Q}$, can be expressed by (34.2):

$$\tilde{Q} = \frac{1 + 2\zeta\Omega_j}{\left(1 - \Omega^2\right) + 2\zeta\Omega_j}\Omega^2 F_0^{st} e^{j\omega t} \tag{34.2}$$

The displacement amplitude is given by the following equation:

$$A = \frac{\Omega^2 A_{st}}{\sqrt{\left(1 - \Omega^2\right)^2 + 4\zeta^2\Omega^2}} \tag{34.3}$$

The Maxwell viscoelastic model, schematized in Fig. 34.2, can be used in the dynamic analysis of cohesionless soils with high content of mineral aggregates, with particle size range from 0–4, 4–8 to 8–16 mm, mixed with 10–15% clay content and moisture content of 15%.

The force amplitude transmitted to the soil has the form in (34.4):

$$Q_0 = F_0^{st}\Omega^2 \frac{2\zeta\Omega}{\sqrt{\Omega^4 + 4\zeta^2\Omega^2\left(1 - \Omega^2\right)^2}} \tag{34.4}$$

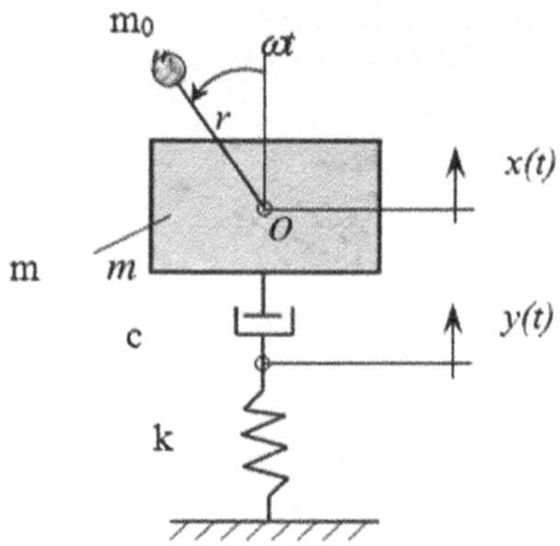

Fig. 34.2 Schematization of Maxwell model

The displacement amplitude A is given by (34.5):

$$A = A_{st}\Omega^2\sqrt{\frac{1+4\zeta^2\Omega^2}{\Omega^4+4\zeta^2\Omega^2(1-\Omega^2)^2}} \tag{34.5}$$

The behaviour of optimized mixtures during vibro-compaction, prepared in gravimetric doses by using mineral aggregates (sand, gravel, and crushed stone) with clay, is modelled using the Voigt-Klein and Maxwell rheological models. The parametric families of soil response curves are developed in order to identify its capacity to work under vibration regime. The evolution of granular soils response in compaction process is determined by assessing the changes of different stiffness coefficient for the compacted layer after each passing of the vibrating roller. For the analysis, the following experimental input data were considered as follows: soil stiffness, k, with values ranging from 1×10^8 N/m up to 4×10^8 N/m for different successive passes of vibratory system and the pulsation excitation, ω, ranging from 0 to 500 rad/s. For each new value of k, a value of damping ratio, ζ, is set, with the limit $\zeta \leq 0,7$, meaning that the relation $c = 2\zeta\sqrt{km}$ is valid. The families of curves developed based on Voigt-Kelvin (Figs. 34.3 and 34.4) and Maxwell (Figs. 34.5 and 34.6) rheological models are illustrated below, considering the displacement amplitude and the force amplitude as a function of pulsation and stiffness.

34.3 Discussion of the Results

The correlation between the technical parameters of vibratory roller and the specific properties of compacted layer is achieved by varying m, k, Ω, ζ factors, according to soil category and vibration regime. The viscoelastic rheological modelling of soils can be made by using the Voigt-Kelvin and Maxwell models, dependent on stiffness and damping ratio soil characteristics and vibration field in anti-resonance

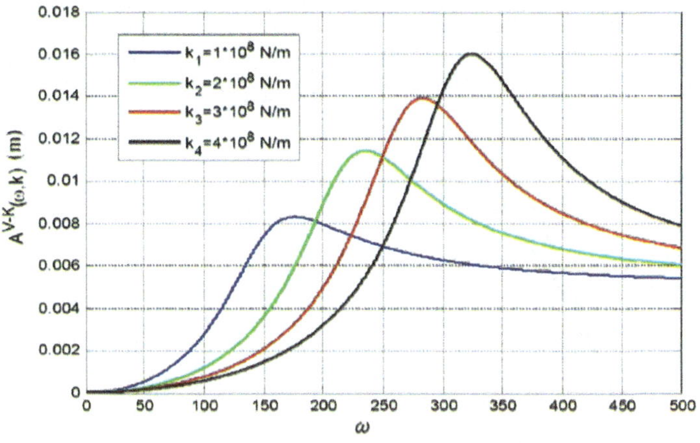

Fig. 34.3 Displacement amplitude dependent on pulsation and stiffness (Voigt-Kelvin model)

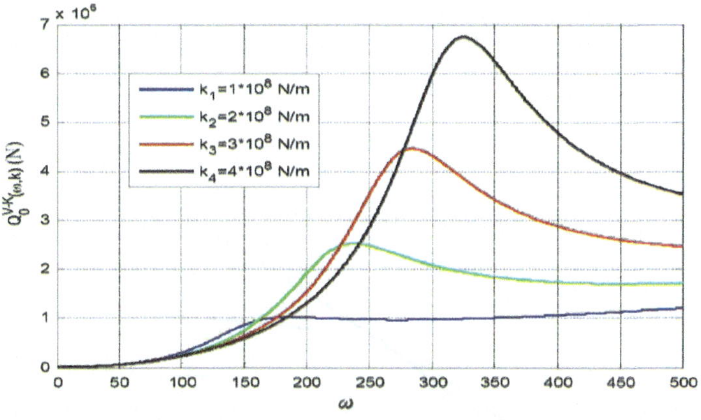

Fig. 34.4 Force amplitude dependent on pulsation and stiffness (Voigt-Kelvin model)

or post-resonance regime, in parametric stability. Based on the diagrams modelled by using Voigt-Kelvin and Maxwell model, the high influence of soil characteristics on compaction parametric values was pointed out. By using the Voigt-Kelvin model, the compacted layer (clayey sand class) is mainly characterized by a significant elastic component, linked to the self-pulsation of value system for ensuring the post-resonant regime in the conditions of stationary forced vibrations with frequencies of 40–50 Hz. Also, the elastic character of the Voigt-Kelvin model is evidenced by the significant elastic ground characteristic compared to the viscous one. For granular soils, by using the Maxwell model, the viscous component was outlined, so that for the compaction process, the post-resonant regime can be ensured at 25–35 Hz, with the optimal amplitude to provide the force transmitted to the compacted layer.

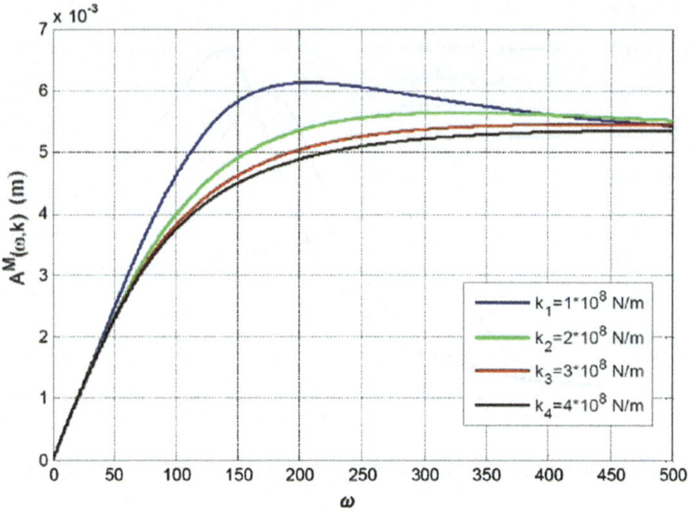

Fig. 34.5 Displacement amplitude dependent on pulsation and stiffness (Maxwell model)

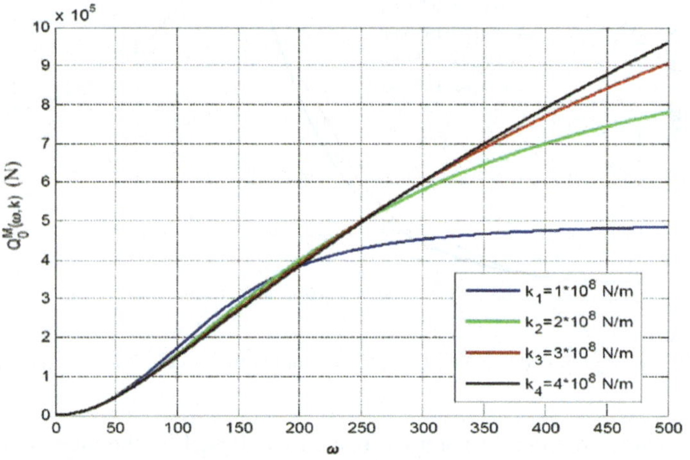

Fig. 34.6 Force amplitude dependent on pulsation and stiffness (Maxwell model)

34.4 Conclusions

The dynamic compaction effect can be optimized based on analytical models, by developing families of curves taking into account the viscoelastic soil properties, as well as the dynamic excitation characteristics and the technological specifications of vibrating rollers. Based on the use of rheological models for the assessment of the dynamic response during vibro-compaction process for different soil categories, the mainly elastic behaviour of the "Voigt-Kelvin" system and the viscous behaviour of the "Maxwell" system were revealed.

To optimize the dynamic compaction process, it is necessary to carry out "in situ" experimental testing section, designed to ensure the physical and mechanical soil conditions correlated with the dynamic parameters of equipment. The changes in vibration frequency, forces, local variability of moisture content and compositions, depth and stiffness of layer influences the vibration parameters. The variability of vibratory system technological parameters can be controlled and adjusted according to existing ground conditions and specific properties. It is recommended that for each soil category a geotechnical study is performed in order to establish the viscoelastic characteristics during the compaction process. Numerical simulation performed in the study by considering Voigt-Kelvin and Maxwell rheological models has led to a realistic assessment of soil dynamic response during the vibro-compaction process. This can be considered as an important tool for ensuring a high quality of engineering compaction works and improving design techniques.

Acknowledgements A part of the present work was made possible by financial support of Ministry of Research and Innovation in the framework of the National Research Programme (CONSUS) and some of the data were collected from several experimental testing performed during applied research work.

References

1. K.R. Massarsch, B.H. Fellenius, Vibratory compaction of coarse-grained soils. Can. Geotech. J. **39**(3), 25 (2001)
2. M.A. Mooney, P.B. Gorman, J.N. Gonzalez, Vibration based health monitoring during earthwork construction. Struct. Health Monit. **2**(4), 137–152 (2005)
3. Z.G. Ter-Martirsyan, A.Z. Ter-Martirsyan, E.S. Sobolev, Vibration of embedded foundation at multi-layered base taking into account non-linear and rheological properties of soils. Proc. Eng. **153**, 747–753 (2016)
4. S. Mihailescu, P. Bratu, in *Technologies and equipments for road structures* (*in Romanian*) (Impuls Publishing House, Bucharest, Romania, 2008)
5. P. Zafiu, *Technologies and equipments for works mechanization (in Romanian)* (Technical University of Civil Engineering, Bucharest, Romania, 2005)
6. P. Bratu, *Vibration of elastic systems* (Technical Publish. House, Bucharest, Romania, 2000)
7. P. Bratu, *Elastic bearing systems for machines and equipment* (Technical Publishing House, Bucharest, Romania, 1990)
8. C.F. Dobrescu, Highlighting the change of the dynamic response to discrete variation of soil stiffness in the process of dynamic compaction with roller compactors based on linear rheological modelling. Appl. Mech. Mater. **801**, 242–248 (2015)
9. C.f. Dobrescu, in *The rheological behaviour of stabilized bioactive soils during the vibration compaction process for road structures*, Book of abstracts of the 22nd International Congress on Sound and Vibration, Italy, 2015; Technical University of Civil Engineering, Bucharest, Romania (2005)

Chapter 35
Determination of Flexural Critical Speed of a Rotor by Vibroacustical Signal Analysis

Ion Crastiu, Dorin Simoiu, Eva Nyaguly and Liviu Bereteu

Abstract The paper aims is to present a vibroacoustic method in order to determine the critical bending speeds of a rotor assimilated to a shaft with a disc. Experimental determinations are based on signal analysis recorded with a microphone. The eigenfrequencies and the shape of their eigenmodes are determined using the Finite Element Method (FEM). Signals acquired under boundary conditions for the considered sample, with free-free ends or simple supported ends, are processed using the Fast Fourier Transform (FFT). To confirm the results the eigenfrequencies obtained using the modal analysis, with their eigenfrequencies experimentally determined are compared.

35.1 Introduction

The vibrations of a bar or shaft with concentrated masses and the determination of the frequencies of this mechanical system have been the problem of several works. In the paper [1], Dirac's function is used to determine these parameters. The analysis of the dynamic behavior of a single concentrate mass was made in the paper [2]. The paper [3] analyzes the free vibrations of a Timoshenko bar with concentrated masses connected to it by springs. This paper explores the dynamic behavior of a cylindrical shaft with a centrally located disk. Analytical solving was

---

I. Crastiu (✉) · D. Simoiu · E. Nyaguly · L. Bereteu
Mechanics and Materials Strength Department, Politehnica University of Timişoara, Bd. Mihai Viteazul, nr. 1, 300222 Timişoara, Romania
e-mail: ion_crastiu@yahoo.com

D. Simoiu
e-mail: dorin.simoiu@yahoo.com

E. Nyaguly
e-mail: pintiutza@gmail.com

L. Bereteu
e-mail: liviu.bereteu@upt.ro

N. Herisanu and V. Marinca (eds.), *Acoustics and Vibration of Mechanical Structures—AVMS-2017*, Springer Proceedings in Physics 198,
https://doi.org/10.1007/978-3-319-69823-6_35

done using the Rayleigh-Ritz method, based on which the critical bending speed of the rotor is obtained.

35.2 Materials and Methods

In order to experimentally determine the bending critical speeds of a rotor from the vibroacoustic analysis of a vibration signal, a steel sample was made of a shaft with the length L = 294 mm and the diameter D = 20 mm and a disk-shaped rotor. The rotor was made of steel, having the thickness h = 10 mm, the external diameter D_{ex} = 60 mm and the inner diameter equal to D. To validate the analytical results by analyzing the free vibrations of a shaft with a rotor, in which the shaft is considered a bar but also of the results obtained from the modal analysis, experimental measurements were done. The four-dimensional differential equation describing the motion of a shaft with a rotor, as shown in Fig. 35.1, while respecting the Euler-Bernoulli assumptions, is given below

$$EI\frac{\partial^4 V(x,t)}{\partial x^4} + \rho A\frac{\partial^2 V(x,t)}{\partial t^2} + m\delta(x - L/2)\frac{\partial^2 V(x,t)}{\partial t^2} = 0, \qquad (35.1)$$

having the following conditions at the ends:

$$V(0) = V(L) = 0, \text{and} \quad \frac{\partial^2 V(0)}{\partial x^2} = \frac{\partial^2 V(L)}{\partial x^2} = 0. \qquad (35.2)$$

In (35.1) the following notations are used: E is the longitudinal elastic modulus, I is the moment of inertia, V(x, t) is bending deformation, ρ is the density, A is the area of the cross section, m is the mass of the rotor and δ is the Dirac function.

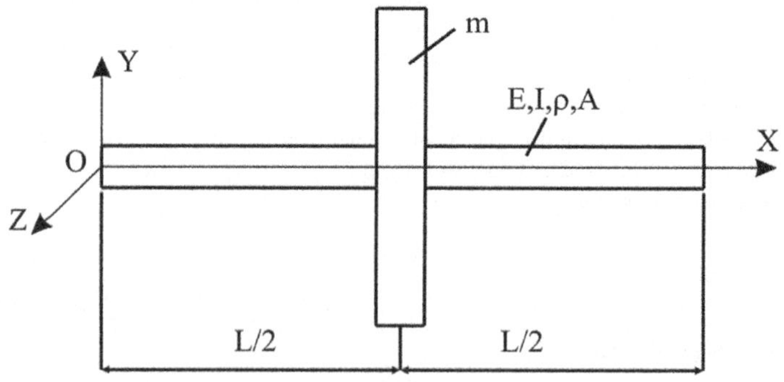

Fig. 35.1 The shaft with a centrally placed rotor

According to the Rayleigh-Ritz approximate analytical method, the solution of this differential equation can be taken as a linear combination of functions of the form:

$$V(x,t) = \sum_{i=1}^{n} \phi_i(x) q_i(t) = \{\phi\}^T \{q\} \tag{35.3}$$

where $\phi_i(x)$ are eigenmodal functions that must satisfy the boundary conditions imposed by the type of joint to which the two ends of the bar are subjected. Customized (35.3) for the rotor mass center, this speed can be:

$$\frac{\partial V_G(x,t)}{\partial t} = \{\phi\}^T \{\dot{q}\}, \tag{35.4}$$

and kinetic energy of the rotor shaft assembly:

$$\begin{aligned} E_c &= \frac{1}{2} \int_0^L \rho A \left(\frac{\partial V(x,t)}{\partial t} \right)^2 dx + \frac{1}{2} m \left(\frac{\partial V_G(x,t)}{\partial t} \right)^2 \\ &= \frac{1}{2} \{\dot{q}\}^T [M_1] \{\dot{q}\} + \frac{1}{2} \{\dot{q}\}^T [M_2] \{\dot{q}\} \end{aligned} \tag{35.5}$$

where

$$[M_1] = \rho A \int_0^L \{\phi\}\{\phi\}^T dx, \quad [M_2] = m \left[\{\phi\}\{\phi\}^T \right]_{x=L/2}. \tag{35.6}$$

Assuming the rigid disk, the potential energy will be given by the shaft deformation energy, thus:

$$E_p = \frac{1}{2} \int_0^L EI \left(\frac{\partial^2 V(x,t)}{\partial x^2} \right)^2 dx = \frac{1}{2} \{q\}^T [K] \{q\}, \tag{35.7}$$

where

$$[K] = EI \int_0^L \frac{\partial^2}{\partial x^2} \{\phi\} \frac{\partial^2}{\partial x^2} \{\phi\}^T dx \tag{35.8}$$

is the stiffness matrix.

The transverse displacement V(x, t) can be considered as a result of the adding together of two motions: one in the plane OXY denoted by Y(x, t) hereinafter called in plane motion, respectively the other in the OXZ plane denoted Z(x, t), and hereinafter called out of plane motion. The two displacement are given by

$$Y(x,t) = \sum_{i=1}^{n} Y_i(x)q(t)_i \text{ and } Z(x,t) = \sum_{i=1}^{n} Z_i(x)q(t)_i \tag{35.9}$$

The number of acceptable functions is chosen taking into account the number of critical speeds to be determined. It is intended to determine the first critical speed so choose a single acceptable function. As a result the two movements, in plane and out of plane, will be approximated by the same function

$$Y(x,t) = Z(x,t) = \sin\frac{x\pi}{L} \cdot \cos(pt - \varphi) \tag{35.10}$$

where p represents the first natural circular frequency.

Using the relations given by (35.5) and (35.7) in the mechanical energy conservation theorem, the square of the first circular frequency corresponding to the movement in the OXY plane or in OXZ plane is determined

$$p^2 = \frac{EI\left(\frac{\pi}{L}\right)^2 \frac{L}{2}}{\frac{\rho AL}{2} + m}. \tag{35.11}$$

For this approximate analytical method, the critical speed will have the expression

$$n = \frac{30}{L}\sqrt{\frac{EIL}{\rho AL + 2m}} \tag{35.12}$$

Consequently, in the determination of the critical bending speed of the rotor, it is necessary to measure the first natural frequency both in the OXY plane and in the OXZ plane.

In order to confirm the shape of the vibration modes in connection with the experimental determinations of the natural frequencies, the method of modal analysis is applied. Following the modeling of the sample using FEM, a matrix system of differential equations of the second order can be obtained which can be written

$$[M]\{\ddot{q}\} + [K]\{q\} = \{Q(t)\} \tag{35.13}$$

where $\{q\}$ is the generalized coordinate vector, $\{Q(t)\}$ is the vector of the generalized forces and [M], [K] are the inertia and stiffness matrices. The characteristic equation attached to the differential equation given by (35.13) is:

$$\left|-p^2[M] + [K]\right| = 0 \tag{35.14}$$

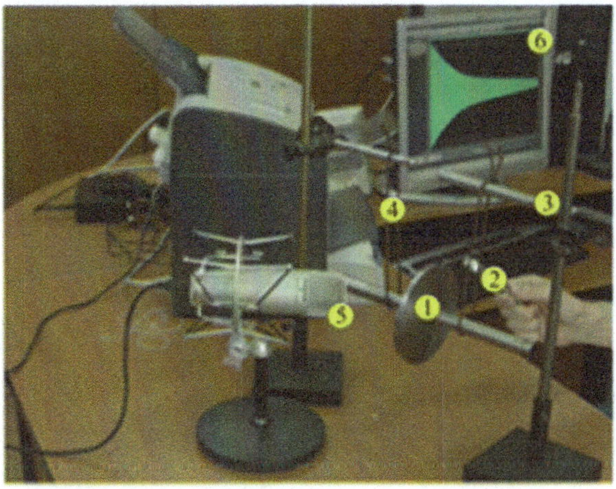

Fig. 35.2 Photo of experimental setup

To determine the eigenmodes of the problem given by (35.14) the block Lanczos method was used. It is recommended because it can give a large number of eigenmodes and because they can be in a wide range of frequencies.

The experimental setup, used to record the acoustic signals obtained at the excited samples by a mechanical impulse, is shown in Fig. 35.2. It has the following composition: the sample 1, which is a mechanical structure, made of shaft and rotor, a hammer 2 for producing mechanical impulses, a frame 3 for sample support, elastic threads 4 for achieving the border conditions, acoustic transducer 5 is a microphone, and computer 6 having a signal acquisition board. This stand was designed and used for the papers: [4, 5].

35.3 Results and Discussions

The modal analysis is carried out on the steel simple supported shaft with single rotor and on the free-free shaft. The simple supported shaft is designed in the graphical environment of the ANSYS [6]. Mesh of the shaft and rotor is generated automatically by ANSYS, while is used the spatial element SOLID187. The element is defined by 10 nodes while each node has three degrees of freedom. The SOLID187 has a quadratic shifting behavior and is suitable for modeling of the finite element irregular mesh. The maximum size of the element is 5 mm. The mesh is created of 3200 elements and of 22,459 nodes.

The rotor and the shaft were weighed with a digital balance with the precision of one hundred grams. The mean values after six weights were m = 800.333 g for the rotor and m_s = 700.231 g for the shaft. Both the rotor and the shaft were made of stainless steel having the modulus of elasticity $E_1 = 210$ GPa, and density$\rho_1 = 7850\ \mathrm{kg/m^3}$. The first natural frequencies of the sample obtained by modal analysis are given in Table 35.1 and in Figs. 35.3 and 35.4 for F-F ends, and in Figs. 35.5 and 35.6 for S-S ends.

Table 35.1 The first natural frequencies of the shaft in XOY and XOZ planes

No.	Boundary conditions	First natural frequency in XOY plane (Hz)	Critical speed in XOY plane (rpm)	First natural frequency in XOZ plane (Hz)	Critical speed in XOZ plane (rpm)
1	F-F	828.88	49,732.8	829.30	49,758
2	S-S	596.23	35,733.8	596.66	35,799.6

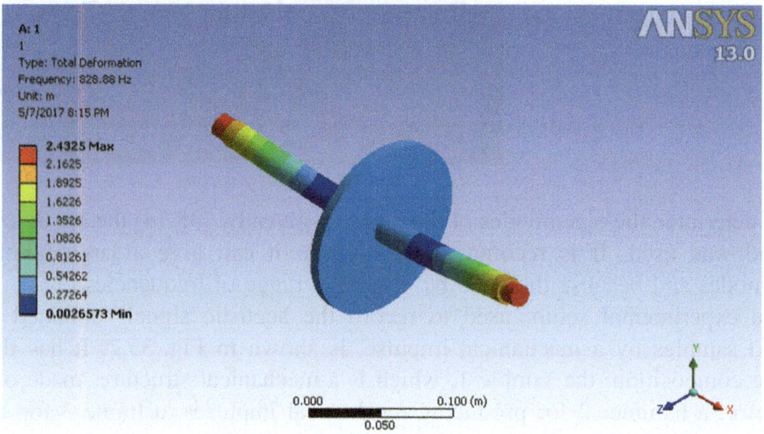

Fig. 35.3 The first modal shape in XOY plane for free-free shaft

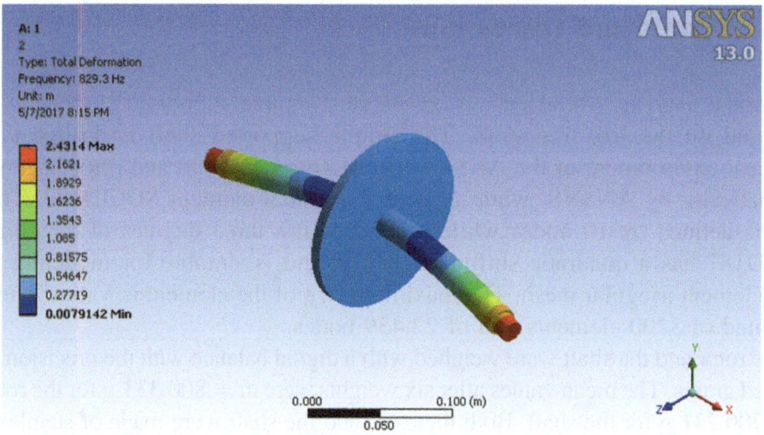

Fig. 35.4 The first modal shape in XOZ plane for free-free shaft

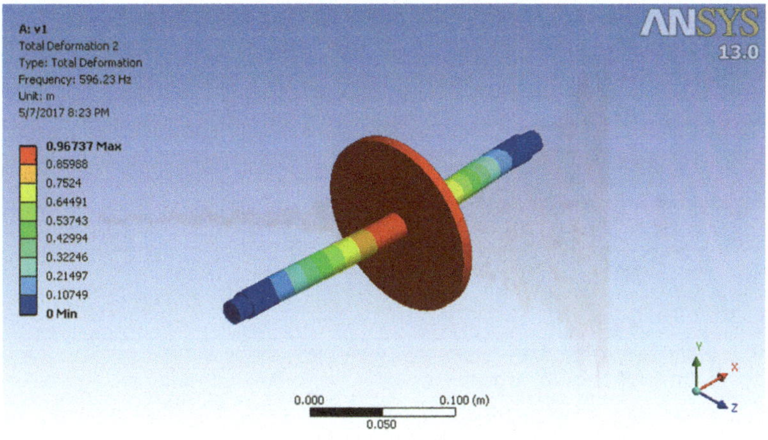

Fig. 35.5 The first modal shape in XOY plane for simple supported shaft

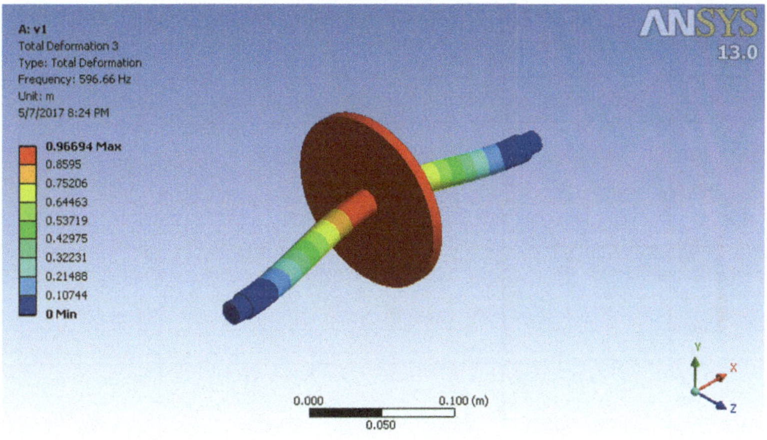

Fig. 35.6 The first modal shape in XOZ plane for simple supported disk

To determine the critical speed of the rotor, FFT is required for four signals: two for F-F boundary and of the rotor and others two for S-S boundary of the rotor. The recorded signal is shown in the Fig. 35.7 for S-S disc and the first natural frequency from the Fourier spectrum is shown in Fig. 35.8.

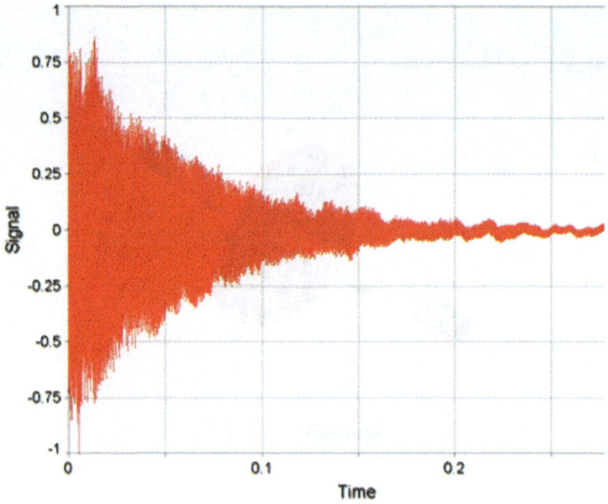

Fig. 35.7 Vibroacustical signal in plane

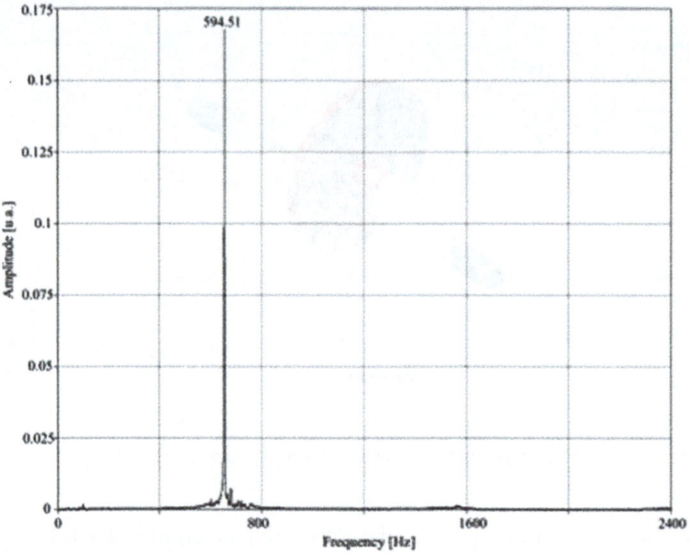

Fig. 35.8 First natural frequency from spectrum

The experimental results obtained for the determination of the first natural frequency after the analysis of six recorded signals and their mediation are presented in Table 35.2.

Table 35.2 The first experimental frequencies of the rotor in XOY and XOZ planes

No.	Boundary conditions	First natural frequency in XOY plane (Hz)	Critical speed in XOY plane (rpm)	First natural frequency in XOZ plane (Hz)	Critical speed in XOZ plane (rpm)
1	F-F	822.73	49,363.8	825.32	49,519.2
2	S-S	594.51	35,670.6	595.42	35,725.2

35.4 Conclusions

Very small differences are found between experimental results for the determination of the first natural frequency and the corresponding frequency determined by numeric method using modal analysis. The close results of the critical speed obtained from the numerical analysis of those obtained from the experimental determinations validate this method. From a practical point of view, it is interesting only the case for the shaft with simple support ends. The other case where the shaft ends are free-free was done only to strengthen the experimental results. If the rotor shaft were considered to be a Timoshenko bar, taking into account the effect of the shear forces and the effect of the rotation inertia, it is hoped to achieve much closer results.

References

1. Y. Chen, On the vibration of beams or rods carrying a concentrated mass. J. Appl. Mech. **30**(2), 310–311 (1963)
2. M.A. De Rosa, C. Franciosi, M.J. Maurizi, On the dynamic behaviour of slender beams with elastic ends carrying a concentrated mass. Comput. Struct. **58**(6), 1145–1159 (1996)
3. Y. Yesilce, O. Demirdag, S. Catal, Free vibrations of a multi-span Timoshenko beam carrying multiple spring-mass systems. Sadhana **33**(4), 385–401 (2008)
4. L. Bereteu, M. Vodă, D.R. Pascu, E. Nyaguly, Influence of some oxide coating layers on the mechanical properties of steel plates. Adv. Mater. Res. **1111**, 91–96 (2015)
5. A. Perescu, L. Bereteu, D. Simoiu, E. Nyaguly, Nondestructive method for the determination of the elastic properties of welded aluminum plates. Adv. Mater. Res. **1111**, 73–78 (2015)
6. http://www.ansys.com, Simulation driven product development ANSYS (2017)

Part V
Structural Vibration

Chapter 36
Vortex Induced Vibration and Wind Flow Around Bridge Cables

Irena Gołębiowska and Maciej Dutkiewicz

Abstract In the paper two models are analyzed. First model is the wind flow around the ellipse cylinder that can correspond to the cable of the suspended and cable-stayed bridge. Computations are performed for subcritical regime from the range of Re = 6.5×10^4 to Re = 1.4×10^5. In the analyzed case, the RNG $k - \varepsilon$ method, belonging to the Reynolds-Averaged Navier-Stokes models, is used. The numerical simulations are performed for aspect ratio from the range of B/D = 0.60 to B/D = 2.2. The analyzed ellipse cylinder reflects the possible change of the cable's section due to the ice. The behavior of lift coefficient in time domain is presented. The value of lift coefficient was calculated for the purpose of the analysis in the model of vortex induced vibrations. The spectra density analysis for lift force is presented and the velocity contour around ellipse cylinder is drawn. On the basis of the calculated lift coefficient from the first model, the vortex induced vibration model is formulated in second model. In the study the dimensionless amplitudes of motion A/D for different damping parameter η_c is presented. Also the phase angle between displacement $y(t)$ and the force $F(t)$ is shown. The dependence of dimensionless air velocity $S_t v_r$ and response frequency $\frac{f_s}{f_s^0}$ is performed. The analysis of influence of Reynolds number on dimensionless displacement amplitude and aspect ratio on the r.m.s. lift coefficient are investigated.

I. Gołębiowska · M. Dutkiewicz (✉)
Department of Building Construction, Faculty of Civil, Architecture and Environmental Engineering, University of Science and Technology in Bydgoszcz, Al. Prof. S. Kaliskiego 7, 85-796 Bydgoszcz, Poland
e-mail: macdut@utp.edu.pl

I. Gołębiowska
e-mail: irena.golebiowska@utp.edu.pl

N. Herisanu and V. Marinca (eds.), *Acoustics and Vibration of Mechanical Structures—AVMS-2017*, Springer Proceedings in Physics 198,
https://doi.org/10.1007/978-3-319-69823-6_36

36.1 Introduction

Vortex induced vibrations (VIV) are caused by vortex shedding, located behind the bluff bodies. VIV are especially important in the design and service of land constructions such as overhead transmission lines, suspended and stayed cable bridges and offshore structures such as pipelines, risers, mooring lines. On the basis of formulation of the fluid forces applied to the cylinder in the direction of lift [1, 2], one can classify the VIV models as forced model where force is independent of vertical displacements and depends on time only, the second model is fluid elastic model where force depends on vertical displacement and time, and the third model is coupled model where force is resulting from the coupling of cylinder and wake. Bishop and Hassan [3] found that the lift force is increased with the amplitude of cylinder motion. They also concluded from the tests that the fluctuating lift force acting on the cylinder is a self-excited oscillatory mechanism in the flow field. Hartlen and Currie [4] assumed that the oscillating lift force on the cylinder could be represented by an equation in the form of the van der Pol equation. This model was extended by Skop and Griffin [1]. The nature of flow-induced oscillations of cylinders in the streamwise direction was extended by Sainsbury and King [5]. Flow-induced streamwise oscillations of flexible cantilevered cylinders was investigated by King et al. [6]. In-line oscillations were also studied by Martin, Curie and Naudascher and they gave some explanations to them [7]. Other methods rely on wake oscillators. Significant contributions to the development of these models were performed in [8].

The article is divided into four main parts, introduction, formulation of the flow and vortex induced vibrations models, results, discussion and conclusions and concerns the forced model described above.

36.2 Formulation of the Flow and Vortex Induced Models

Two separates models were studied: flow and vortex induced analysis. First model concerns the air flow around the stationary ellipse cylinder. On the basis of this model the suitable lift coefficient C_{Le} is calculated. The difference in pressure on the upper and lower part of the analyzed body provides the lift force which makes the movement of the body in the vertical direction. The lift coefficient is calculated from pressure distribution around the body, i.e. its upper and lower part [9]. The second model concerns the released cylinder under the lift force with value calculated in the flow model. The numerical simulations are performed for fixed aspect ratio from the range of B/D = 0.60 to B/D = 2.2 for the purpose of presenting the general trend of relations between B/D and C_{LRMS} and to compare with other results [10, 11].

36.2.1 Air Flow Model

In the analyzed problem, the equations of motion of the air flow around the cylinder take the following form [12]:

$$\frac{D\bar{v}_i}{Dt} = \bar{f}_i - \frac{1}{\rho}\frac{\partial \bar{p}}{\partial x_i} + \frac{1}{\rho}\frac{\partial}{\partial x_j}\left(\bar{\tau}_{ij} + \tau^*_{ij}\right) \tag{36.1}$$

where $\bar{\tau}_{ij} = \mu\left(\frac{\partial \bar{v}_i}{\partial x_j} + \frac{\partial \bar{v}_j}{\partial x_i}\right)$

Additional continuity equations describing mean and fluctuating components are:

$$\frac{\partial \bar{v}_i}{\partial x_i} = 0, \quad \frac{\partial v'_i}{\partial x_i} = 0 \tag{36.2}$$

To solve the above equations, the additional stress Reynolds tensor is adopted in the following form:

$$\tau^*_{ij} = \mu_t\left(\bar{v}_{i,j} + \bar{v}_{j,i}\right) - \frac{2}{3}\bar{\rho}k\delta_{ij}, \quad \mu_t = \frac{\rho c_\mu k^2}{\varepsilon} \tag{36.3}$$

The $k - \varepsilon$ dependences are formulated in additional transport equations:

$$\frac{\partial}{\partial t}(\bar{\rho}k) + (\bar{\rho}k\bar{v}_i)_{,i} = \left(\bar{\tau}_{ij}\bar{v}_j\right)_{,i} - \bar{\rho}\varepsilon + \left(\mu_k k_{,i}\right)_{,i} \tag{36.4}$$

$$\frac{\partial}{\partial t}(\bar{\rho}\varepsilon) + (\bar{\rho}\varepsilon\bar{v}_i)_{,i} = c_{\varepsilon 1}\left(\bar{\tau}_{ij}\bar{v}_j\right)_{,i} - c_{\varepsilon 2}\bar{\rho}\frac{\varepsilon^2}{k} + \left(\mu_\varepsilon \varepsilon_{,i}\right)_{,i} \tag{36.5}$$

where: k—turbulence kinetic energy; ε—turbulence kinetic energy dissipation; $\mu_{k(\varepsilon)} = \mu + \frac{\mu_t}{\sigma_{k(\varepsilon)}}$, with the following assumption: $c_\mu = 0.09$, $c_{\varepsilon 1} = 1.45$, $c_{\varepsilon 2} = 1.92$, $\sigma_k = 1.00$, $\sigma_\varepsilon = 1.30$.

The above formulated method belongs to the Reynolds-Averaged Navier-Stokes models. The cylinder with aspect ratio of the range from B/D = 0.60 to B/D = 2.2 was solved that reflect the section of the bridge cable with ice on it (Fig. 36.1). Cylinder is located in the same distance from the flow inlet and outlet, that is 20B. The initial conditions are: uniform flow inlet with velocities $v_x = V_\infty$, $v_y = 0$. No slip condition is applied on the cylinder surface and $v_x = v_y = 0$. The flow for stream simulation around cylinders with fixed Reynolds number from the range of $Re = 6.5 \times 10^4$ to $Re = 1.4 \times 10^5$ is used. For improving accuracy of results, 21,630 elements were applied. Navier Stokes equations were solved numerically with use of Semi Implicit Method for Pressure Linked Equations (SIMPLE) with sequenced calculations of velocity and pressure's components. The coefficients of sub relaxations were additionally used for stabilizing of the calculations process. Additionally, second order upwind method for momentum equations was adopted.

36.2.2 Vortex Induced Vibration Model

The rigid cylinder of mass m_c, aspect ratio B/D, and length L is assumed to be exposed to a flow of velocity v_x. The cylinder is assumed to be mounted at both ends with two elastic springs of total stiffness k_c, and with viscous dampers of total damping coefficient c_c. The coordinates axis is located at the center of the cylinder at its stationary position. The displacement perpendicular to the stream direction is denoted by (t). The external lift force acting on the cylinder is denoted by $F(t)$ (Fig. 36.1b). The motion of the cylinder in the lift direction is defined as:

$$m_c\ddot{y}(t) + c_c\dot{y}(t) + k_c y = F(t) \tag{36.6}$$

Excitation force is described as [4]:

$$F(t) = \frac{1}{2}\rho\, v_x^2 D\, C_{Le}\sin(\nu t + \delta) \tag{36.7}$$

where ρ is air density, v_x is air velocity, C_{Le} is the lift coefficient that is calculated in the air-flow model, $C_{Le} = \sqrt{2}C_{Le}^{RMS}$, $\nu = 2\pi f_s$, $f_s = S_t\frac{v_x}{D}$. In the above S_t is the Strouhal number and f_s is frequency of motion. The solution of (36.6) is assumed as:

$$y(t) = A\sin(\nu t + \gamma) \tag{36.8}$$

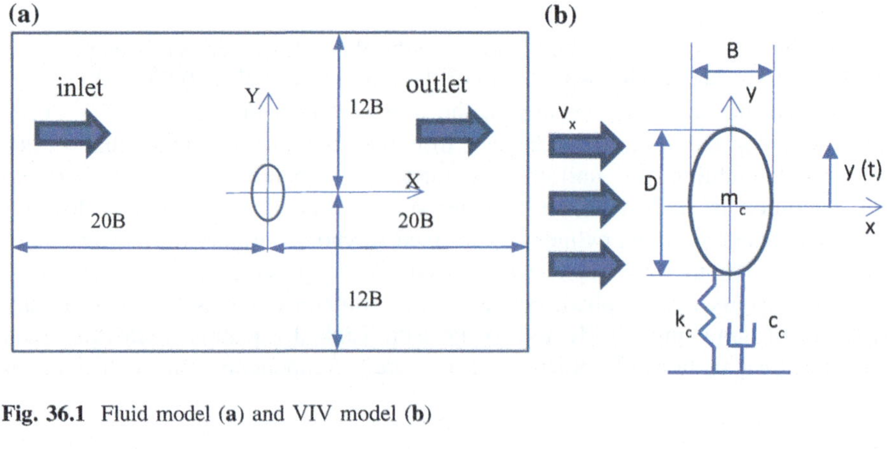

Fig. 36.1 Fluid model (**a**) and VIV model (**b**)

After rearrangements one can receive the solution of (36.6) in the following form:

$$\frac{A}{D} = \frac{8\pi D v_r^2 C_{Le} v_x}{m^* \left[v_x^2 - 8\pi^2 v_r^2 v_x^2 S_t^2 + 16\,\pi^4 v_r^4 v_x^2 S_t^4 + 4\,\eta_c^2 S_t^2 \pi^4 D^2 v_r^4\right]^{\frac{1}{2}}} \tag{36.9}$$

where: $\eta_c = \frac{c_c}{2\,m_c}$, $m^* = \frac{m_c}{\rho\pi D^2/4}$, $v_r = \frac{v_x}{f_s^0 D}$, f_s^0 is the frequency of oscillation of the cylinder in still air. The phase angle has the following form

$$\tan(\delta - \gamma) = \frac{8\,\eta_c \pi^2 D\,S_t}{v_x\left(1 - 4\,S_t^2\,\pi^2 v_r^2\right)} \tag{36.10}$$

A resonance occurs at $S_t\,v_r = 1$, corresponding to the frequency of shedding $f_{sLe} = f_s$ being equal to the frequency of the structure f_s^0, i.e. $S_t\,v_r = S_t \frac{v_x}{f_s^0 D} = \frac{f_{sLe}}{f_s^0}$.

36.3 Results and Discussion

The results of numerical analysis of flow for cylinder for Re = 6.5×10^4, B/D = 0.6 are shown in Figs. 36.2a and 36.3a. On Fig. 36.2a the behavior of lift coefficient in time domain are presented. In the analyzed time domain the lift coefficient is characterized by root mean square (R.M.S.) value $C_{Le}^{RMS} = 1.17$. On Fig. 36.2b. the spectra density analysis for lift force is presented. The dominant frequency of lift force is f_{sLe} = 0.174 Hz. The velocity contour around ellipse cylinder and in the

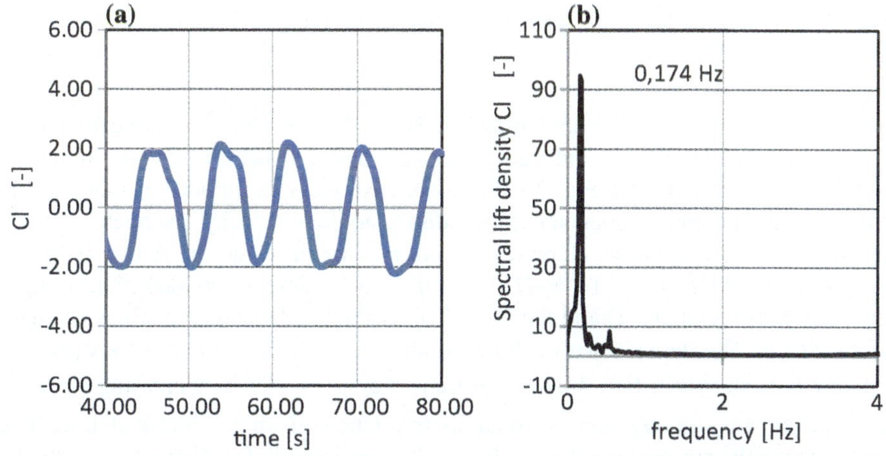

Fig. 36.2 Variation of lift coefficient with time (**a**) and spectral lift density for ellipse cylinder (**b**) for Re = 6.5×10^4

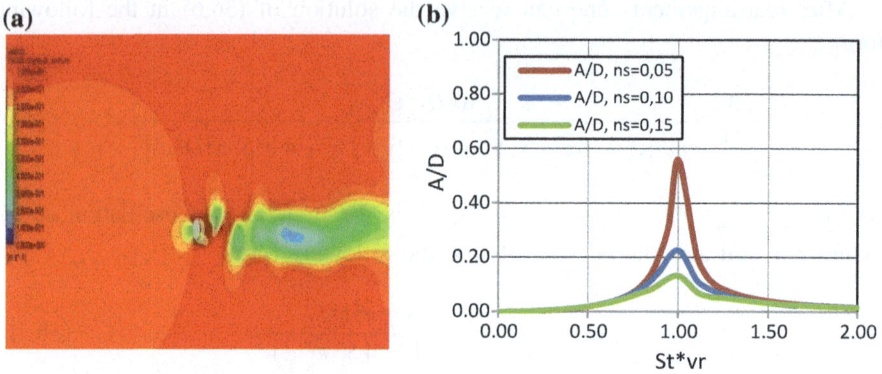

Fig. 36.3 Velocity contour for ellipse cylinder and Re = 6.5 × 10^4 (**a**) and amplitude of motion for air velocity variation and different damping coefficients (**b**)

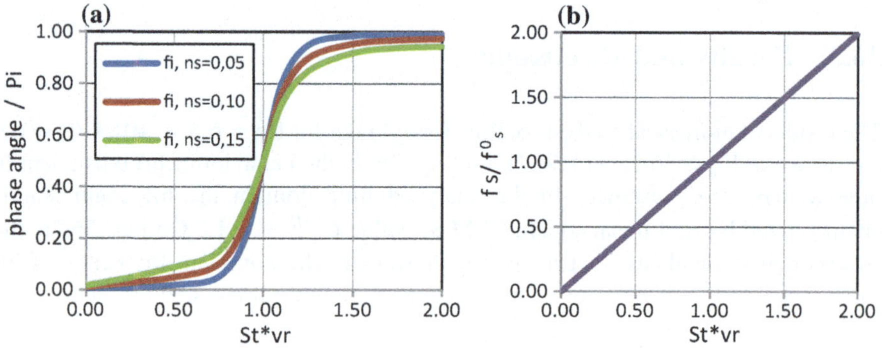

Fig. 36.4 Phase angle between displacement and force (**a**) and dependence of dimensionless air velocity and response frequency (**b**)

wake for Re = 6.5 × 10^4 is shown in Fig. 36.3a. The results of numerical analysis for cylinder under vortex induced vibrations and for B/D = 0.6 are shown in Figs. 36.3b and 36.4b. On Fig. 36.3b the dimensionless amplitudes of motion A/D for different damping parameter η_c for dimensionless velocity parameter $S_t\, v_r$ is presented. The amplitude increases up to maximum value of A/D = 0.56 for $\eta_c = 0.05, 0.22$ for $\eta_c = 0.10$, and 0.13 for $\eta_c = 0.15$. On Fig. 36.4a the phase angle between displacement $y(t)$ and the force $F(t)$ is shown. The phase angle is changed from 0 to π. The dependence of dimensionless air velocity $S_t v_r$ and response frequency $\frac{f_s}{f_s^0}$ is shown in Fig. 36.4b. The frequency of motion is increasing with the increase of air velocity and is equal to frequency given by Strouhal law. The obtained results are consistent with the results presented in [8]. Figure 36.5a shows that dimensionless amplitude increases with increase of Reynolds number. Figure 36.5b presents the numerical simulations performed for fixed aspect ratio for

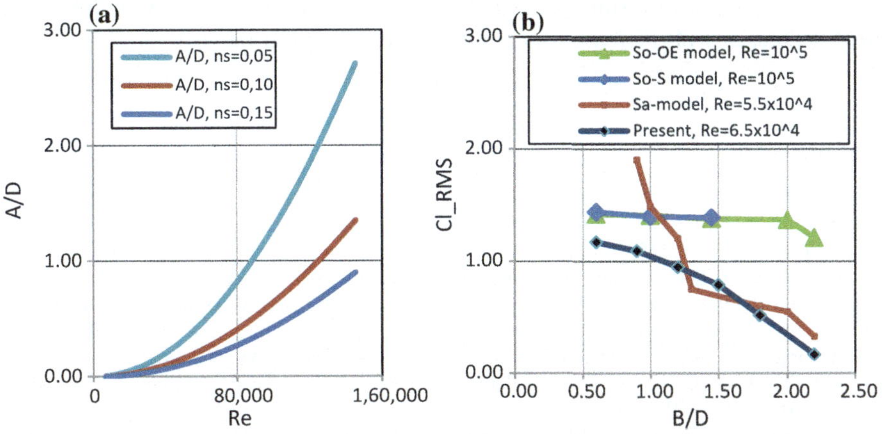

Fig. 36.5 Amplitude of motion for different Re (**a**) variation of r.m.s. lift coefficient with aspect ratio, So-OE and So-S models from [11], Sa-model from [10] (**b**)

the purpose of presenting the general trend of relations between B/D and C_{LRMS} and to compare with other results [10, 11], i.e. for B/D = 0.6, 0.9, 1.2, 1.5, 1.8, 2.2. The r.m.s. lift coefficient decreases with increase aspect ratio (Fig. 36.5b). A similar trend of r.m.s. lift coefficient for rectangular section is found in the study of the aspect ratio dependency presented by Sakamoto [10] and Sohankar for one equation model (So-OE) and Smagorinsky model (So-S) [11].

36.4 Conclusions

In the paper computations were performed for subcritical regime from the range of Re = 6.5×10^4 to Re = 1.4×10^5. In the analyzed case, the RNG $k - \varepsilon$ method, belonging to the Reynolds-Averaged Navier-Stokes models, was used. The numerical simulations were performed for aspect ratio of the cylinder from the range of B/D = 0.60 to B/D = 2.2. The value of lift coefficient was calculated for the purpose of the analysis in the model of vortex induced vibrations. The spectra density analysis for lift force was presented and the velocity contour around ellipse cylinder was drawn. On the basis of the results from the first model, the dimensionless amplitudes of motion A/D for different damping parameter η_c was presented. Also the phase angle between displacement $y(t)$ and the force $F(t)$ was shown. The dependence of dimensionless air velocity $S_t\, v_r$ and response frequency $\frac{f_s}{f_s^0}$ was presented. The analysis of influence of Reynolds number on dimensionless displacement amplitude was investigated. Amplitude increases with increase of Reynolds number. The numerical simulations were performed for fixed aspect ratio for the purpose of presenting the general trend of relations between B/D and C_{LRMS}

and to compare with results presented by other researchers. The r.m.s. lift coefficient decreases with increase aspect ratio in the analyzed range.

References

1. R.A. Skop, O.M. Griffin, A model for the vortex-excited resonant response of bluff cylinders. J. Sound Vib. **27**, 225–233 (1973)
2. M.H. Warner, D.H. Cooper, Some aspects of oscillations of fullscale piles, in *Flow-induced Structural Vibrations*, ed. by E. Naudascher (Springer, Berlin, 1974), pp. 587–601
3. R.E.V. Bishop, A.Y. Hassan, The lift and drag forces on a circular cylinder oscillating in a flowing fluid, in *Proceedings of the Royal Society of London*, Series A, vol. 277 (1964)
4. R.T. Hartlen, I.G. Currie, Lift-oscillator model of vortex-induced vibration. J. Eng. Mech. Div. **96**(5), 577–591 (1970)
5. R.N. Sainsbury, D. King, The flow induced oscillation of marine structures. Proc. Inst. Civ. Eng. **49**, 269–301 (1971)
6. R. King, M.J. Prosser, D.J. Johns, On vortex excitation of model piles in water. J. Sound Vib. **29**, 169–188 (1973)
7. W.W. Martin, I.G. Currie, E. Naudascher, Streamwise oscillations of cylinders. J. Eng. Mech. Div. Am. Soc. Civ. Eng. **107**(589), 607 (1981)
8. C.H.K. Wiliamson, R. Govardhan, A brief review of recent results in vortex induced vibrations. J. Wind Eng. Ind. Aerodyn. **96**(6–7), 713–735 (2008)
9. E.L. Houghton, P.W. Carpenter, *Aerodynamics for Engineering Students* (Butterworth-Heinemann, 2003)
10. H. Sakamoto, H. Haniu, Y. Kobayashi, Fluctuating force acting on rectangular cylinders in uniform flow (on rectangular cylinders with fully separated flow). Trans. Jpn Soc. Mech. Eng. Ser. B **55**(516), 2310–2317 (1989)
11. A. Sohankar, Large eddy simulation of flow past rectangular-section cylinders: Side ratio effects. J. Wind Eng. Ind. Aerodyn. **96**, 640–655 (2008)
12. B. Hunt, *Fluid Mechanics for Civil Engineers* (Christchurch, New Zealand, 1995)

Chapter 37
Crack Localization in L-Shaped Frames

Gilbert-Rainer Gillich, Zeno-Iosif Praisach, Codruta Hamat, Nicoleta Gillich and Jean Louis Ntakpe

Abstract This paper introduces a method to detect damages in L-shaped frames. The problem of identifying cracks in such structures arises from the low frequency changes due to damage. In the early stage, cracks are hard to be assessed. We present two distinct parts: the assessment of the dynamic behavior of L-shaped and description of the damage evaluation method. In the first part, the mathematical relations contrived for the natural frequencies and mode shapes respectively curvatures are introduced. Afterward, the distribution of the modal energy is found and associated with effects produced by a crack with a given depth in different locations. Findings are used to identify the crack location and severity. Accurate frequency evaluation, a crucial aspect since low frequency changes are expected, was made by employing an algorithm developed by the authors. The damage assessment method with the integrated frequency evaluation algorithm is employed to evaluate the health of an L-framed structure.

G.-R. Gillich (✉) · Zeno-Iosif Praisach · Codruta Hamat · Nicoleta Gillich · J. Louis Ntakpe
Universitatea "Eftimie Murgu" Resita, P-ta Traian Vuia 1-4, 320085 Resita, Romania
e-mail: gr.gillich@uem.ro

Zeno-Iosif Praisach
e-mail: zpraisach@yahoo.com

Codruta Hamat
e-mail: c.hamat@uem.ro

Nicoleta Gillich
e-mail: n.gillich@uem.ro

J. Louis Ntakpe
e-mail: Jean-Louis.Ntakpe@alphabet.com

N. Herisanu and V. Marinca (eds.), *Acoustics and Vibration of Mechanical Structures—AVMS-2017*, Springer Proceedings in Physics 198,
https://doi.org/10.1007/978-3-319-69823-6_37

37.1 Introduction

Structures subjected of damage changes their dynamic behavior, since the modal parameters are affected. The changes are deterministic, thus the correlation between these features and damage parameter can be used to assess damage [1–6]. Among the modal parameters, natural frequencies are most promising, while these can be easily measured and imply basic equipment [7–9]. Despite numerous existing damage detection methods, precise assessing damage in the early state remains a challenging issue, especially for complex structures [10, 11].

Researches developed by the authors lead to establishing an advanced detection method, valid for beam-like structures with transverse cracks [12]. It makes use of the modal strain energy distribution and the way how the stored energy loss due to a crack affects the modal frequency components. Based on this property, a predictive model that is suitable for any beam boundary conditions was contrived [13]. This paper proves that the method developed for beams works well even for complex structures. For exemplification, the predictive model was contrived for an L-shaped frame and the detection process was performed on a real structure.

37.2 Assessment of the Dynamic Behavior of L-Shaped Frames

To find the predictive model that describes the dynamic behavior of an L-shaped frame request knowing the mathematical relations for the frequencies, mode shapes and modal curvatures. Note that herein just the weak-axis bending modes are considered. The L-frame, presented in Fig. 37.1a, was decomposed into two elements, one vertical (nominated *column*) and one horizontal (nominated *beam*). The loads induced by the beam on the column are illustrated in Fig. 37.1b, these being the constant axial compression force N due to the beam weight, the variable share force T depending on the joint acceleration and a variable bending moment

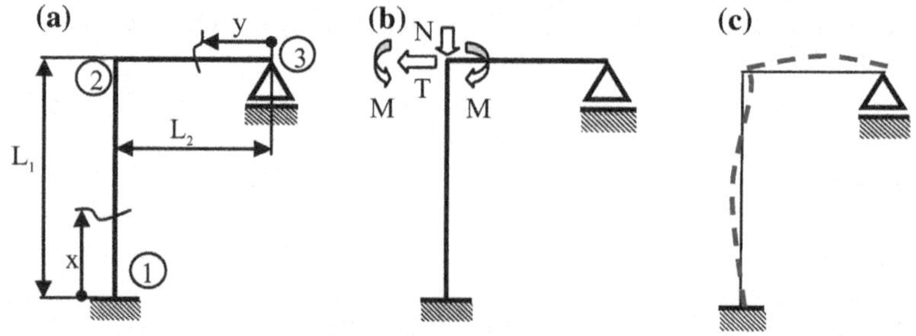

Fig. 37.1 Geometry of the L-shaped frame (**a**), loads on the joint (**b**) and representation of the analytically contrived shape for vibration mode two (**c**)

Table 37.1 Boundary conditions for the two frame elements

Point 1 (fixed end)	Point 2 (joint—column)	Point 2 (joint—beam)	Point 3 (free end)
No displacement	Imposed share force	No displacement	No bending moment
No rotation	Similar rotation	Similar bending moment	No displacement

M. Because N has negligible influence on the frequencies and mode shapes comparing to T and M, see reference [14], it is not considered in this analysis.

From the Euler–Bernoulli beam theory we have the well-known solution for the equation of motion. Written for the two frame elements, it becomes

$$\begin{cases} W(x) = A_1 \sin(\alpha x) + B_1 \cos(\alpha x) + C_1 \sin\mathrm{h}(\alpha x) + D_1 \cos\mathrm{h}(\alpha x) \\ V(y) = A_2 \sin(\alpha y) + B_2 \cos(\alpha y) + C_2 \sin\mathrm{h}(\alpha y) + D_2 \cos\mathrm{h}(\alpha y) \end{cases} \tag{37.1}$$

where $W(x)$ and $V(y)$ are the transverse displacements, x and y are the distances to point 1 respectively point 3. Hence, the two variables take the values: $x = 0\ldots L_1$ and $y = 0\ldots L_2$. The constant α in (37.1) is the dimensionless wave number calculated from the relation $\alpha^4 = \rho A\omega^2 (EI)^{-1}$. Here, ω is the angular frequency, ρ the mass density, A the column respectively beam cross-section, E the Young's modulus. Since the column and the beam are two elements belonging to the same frame, they have the same natural frequencies resulted from the characteristic transcendental equation, which is found by imposing proper boundary conditions.

For this structure, the boundary conditions are presented in Table 37.1. Solving (37.1) in accordance to this conditions, the characteristic equation results in

$$\frac{(\Omega_1(\alpha L_1))^2 + \Pi_1(\alpha L_1)\cdot \Gamma_1(\alpha L_1)}{\Sigma_1(\alpha L_1)\Pi_1(\alpha L_1) - \Omega_1(\alpha L_1)\Gamma_1(\alpha L_1)} \cdot \frac{\sin(\alpha L_2)\sin h(\alpha L_2)}{\cos(\alpha L_2)\sin h(\alpha L_2) - \sin(\alpha L_2)\cos h(\alpha L_2)} - \frac{1}{2} = 0 \tag{37.2}$$

where following functions are defined:

$$\begin{array}{ll} \Gamma() = \sin() + \sin h & \Pi() = \sin() - \sin h() \\ \Sigma() = \cos() + \cos h & \Omega() = \cos() - \cos h() \end{array} \tag{37.3}$$

The coefficients α_i derived from (37.2) are indicated in Table 37.2 for the first $i = 3$ modes. Frequencies are found from the well-known frequency relation, for the mechanical properties of *structural steel* from the ANSYS library. Analytically achieved results and those found from the FEM analysis also are presented in Table 37.2 and compared via relative error; one observes the low error range.

Modal curvatures are derived from (37.1) by replacing the coefficients $A_1 \div D_2$. These are involved in the damage assessment procedure.

Table 37.2 Wave numbers and natural frequencies for the first three bending vibration modes

Mode i	Coefficient α_i	Frequency (Hz)		Error ε_i (%)
		f_i analytic	f_i FEM	
1	6.920	55.532	55.599	0.119647
2	9.053	95.028	95.341	0.328397
3	12.468	180.255	180.705	0.249051

$$\overline{W}_i''(x) = \eta_i \left[\frac{\Omega_1(\alpha_i L_1)}{\Pi_1(\alpha_i L_1)} (-\sin(\alpha_i x) - \sin h(\alpha_i x)) + (\cos(\alpha_i x) + \cos h(\alpha_i x)) \right] \quad (37.4)$$

$$\overline{V}_i''(y) = \kappa_i \left[-\sin(\alpha_i y) - \frac{\sin(\alpha_i L_2)}{\sin h(\alpha_i L_2)} \cdot \sin h(\alpha_i y) \right] \quad (37.5)$$

Normalization for each mode curvature is necessary, so that the coefficients η_i and κ_i are introduced in (37.4) and (37.5). These are determined from the condition to have $max\{W_i(x), V_i(y)_i\} = 1$ for each individual vibration mode.

37.3 Description of the Damage Detection Method

Cracks affect structural element's capacity to store energy, which has a consequence the natural frequency drop. We demonstrated that the frequency decrease due to a crack in a specific slice located at x_j is proportional to the square of the normalized modal curvature for that location [15]. As a consequence, a sequence of normalized mode shape curvature squares for one location form patterns indicating the crack depth and location. Normalizing again these values the damage severity is eliminated and therefore the results indicate the damage location alone. We nominated the individual values

$$\Phi_1(x_j) = \frac{\left[\bar{W}_1''(x_j)\right]^2}{\max\left\{\left[\bar{W}_i''(x_j)\right]^2\right\}}, \ldots, \Phi_n(x_j) = \frac{\left[\bar{W}_n''(x_j)\right]^2}{\max\left\{\left[\bar{W}_i''(x_j)\right]^2\right\}} \quad (37.6)$$

Damage Location Coefficients (DLC), while the Damage Location Indicator (DLI) nominates the entire sequence $\Phi(x_j)$. The left diagram in Fig. 37.2a presents the strain energy distribution in the column for the first three bending vibration modes. By extracting the DLC for locations x_j the damage pattern is attained. Figure 37.2b presents patterns for $x_1 = 0.06$ m and $x_2 = 0.54$ m by considering six bending vibration modes.

The relative frequency shift (RFSh) of vibration mode i, denoted Δf_i, is the ratio of the frequency shift and the frequency of the undamaged beam, that is

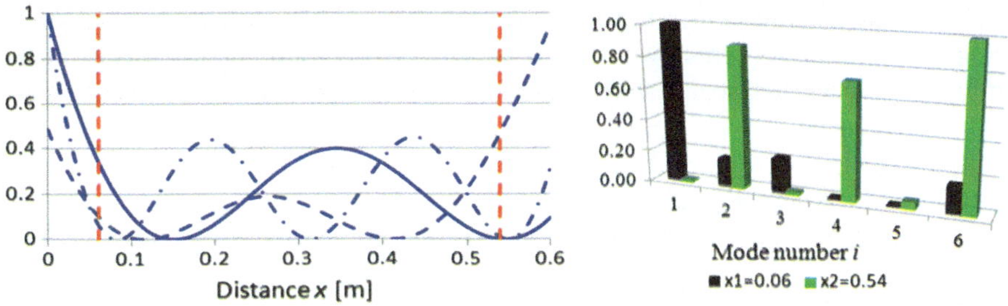

Fig. 37.2 Strain energy distribution for the first three bending vibration modes (left) and patterns indicating the damage location for several values of *x/L* (right diagram)

$$\Delta f_i = \frac{f_{i-U} - f_{i-D}}{f_{i-U}} \tag{37.7}$$

These values can be normalized by dividing each to the highest value of the series, resulting in

$$\Psi_1 = \frac{\Delta f_1^m}{\max\{\Delta f_i^m\}}, \ldots, \Psi_n = \frac{\Delta f_n^m}{\max\{\Delta f_i^m\}} \tag{37.8}$$

which has the condensed form $\Psi = \{\Psi_1, \Psi_2, \ldots, \Psi_n\}$ and represents the damage signature DS, also independent from the damage severity.

The DS has the same connotation as DLI, the difference consisting in the way how the values are obtained. Comparing the two sequences one can identify the damage location x_j. The similarity between DS and created DLIs can be made involving the Minkowski Distance d_{L_2}, see for instance [16], the crack location being found at the distance x_j where the d_{L_2} takes the lowest value.

37.4 The Experimental Research

Experiments were made to proof that the original damage assessment method is available for complex structures. For exemplification, we employed the steel structure presented in Fig. 37.1, having the column length L_1 = 600 mm, beam length L_2 = 400 mm and cross-sectional area $B \times H = 20 \times 5$ mm. The material's relevant physical and mechanical characteristics are: the mass density ρ = 7850 kg/m^3, the Young's modulus $E = 2.0 \times 10^{11}$ N/m^2 and Poisson's ratio ν = 0.3.

The frequencies, measured for the undamaged frame and that acquired if the column has a crack of depth h = 2.5 mm at distance x = 300 mm from the fixed end, are present in Table 37.3. The crack, with the position randomly chosen, was made by a saw cut with width 2 mm.

To get the resolution of 2 mm we have chosen j = 299 evenly distributed points along the column, for which we derived the DLIs. Afterward, the DS was extracted from the relative frequency shifts, obtained from frequency measurements made with an advanced tool [17] developed in LabView, see Fig. 37.3.

By applying the Minkowski Distance we found that the DS fitting the DLI is for $x_j \approx 300$ mm where the global minimum was found, see Fig. 37.4. Since the damage location was correctly found, the method proved its availability.

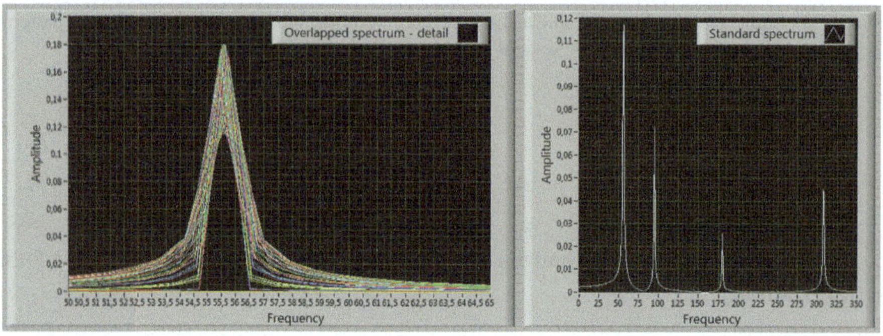

Fig. 37.3 The overlapped-spectrum VI developed for high-precision frequency measurements

Table 37.3 Measured frequencies from the healthy and damaged frame identification process

Mode i	Frequency (Hz)		RFSh (%)	DS Ψ_i	DLC $\Phi_i(0.3)$
	Undamaged	Damaged			
1	55.5986	54.4112	2.135656	0.896586	0.702407
2	95.3414	94.7152	0.656880	0.275770	0.324345
3	180.7055	180.5471	0.087647	0.036796	0.031431
4	307.8279	304.7196	1.009767	0.423918	0.415456
5	381.3240	372.2409	2.381987	1.000000	1.000000
6	592.5542	591.4657	0.183689	0.077116	0.078114

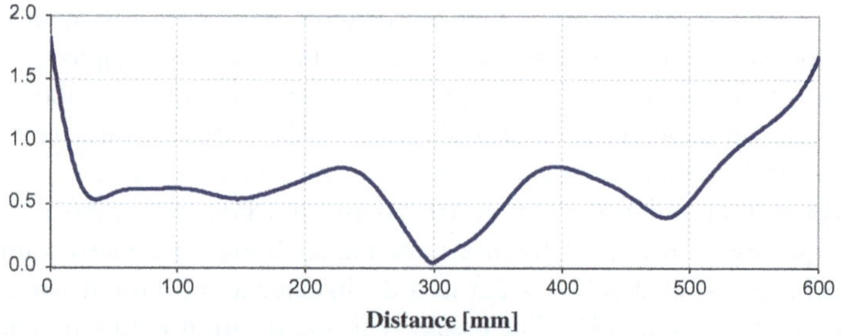

Fig. 37.4 The Minkowski distance represented for the column with damage at x = 300 mm

37.5 Conclusion

This paper presents researches made by the authors to define damage patterns that can be used as a reference in damage detection for complex structures. The developed patterns were successfully employed in damage assessment of complex structures as L-shaped frames are. The method can be applied for structures made of any isotropic material, because it bases on the modal curvatures which do not depend on the material physical and mechanical properties.

References

1. S.W. Doebling, C.R. Farrar, M.B. Primc, A summary review of vibration-based damage identification methods. Shock Vibr. Dig. **30**, 91–105 (1998)
2. E.P. Carden, P. Fanning, Vibration based condition monitoring: a review. Struct. Health Monit. **3**, 355–377 (2004)
3. J.V.A. dos Santos, N.M.M. Maia, C.M.M. Soares, C.A.M. Soares, Structural damage identification: a survey, in trends in computational structures technology. ed. by B.H.V. Topping, M. Papadrakakis (Saxe-Coburg Publications, UK, 2008), pp. 1–24
4. W. Fan, P. Qiao, Vibration-based damage identification methods: a review and comparative study. Struct. Health Monit. **10**, 83–111 (2011)
5. E. Mesquita, P. Antunes, F. Coelho, P. Andre, A. Arede, H. Varum, Global overview on advances in structural health monitoring platforms. J. Civil Struct. Health Monit. **6**(3), 461–475 (2016)
6. Z.B. Yang, M. Radzienski, P. Kudela, W. Ostachowicz, Damage detection in beam-like composite structures via Chebyshev pseudo spectral modal curvature. Compos. Struct. **168**, 1–12 (2017)
7. N.M.M. Maia, J.V.A. dos Santos, R.P.C. Sampaio, C.M.M. Soares, Damage identification using curvatures and sensitivities of frequency-response-functions, in *Proceedings of the Third European Workshop Structural Health Monitoring* (2006), pp. 547–554
8. M. Chandrashekhar, R. Ganguli, Damage assessment of composite plate structures with material and measurement uncertainty. Mech. Syst. Signal Process. **75**, 75–93 (2016)
9. Z.B. Yang, M. Radzienski, P. Kudela, W. Ostachowicz, Fourier spectral-based modal curvature analysis and its application to damage detection in beams. Mech. Syst. Signal Process. **84**(A), 763–781 (2017)
10. G.R. Gillich, I.C. Mituletu, M. Tufoi, V. Iancu, F. Muntean, A method to enhance frequency readability for early damage detection. J. Vibr. Eng. Technol. **3**(5), 637–652 (2015)
11. P. Van Phung, Q.X. Lieu, H. Nguyen-Xuan, M. Abdel Wahab, Size-dependent isogeometric analysis of functionally graded carbon nanotube-reinforced composite nanoplates. Compos. Struct. **166**, 120–135 (2017)
12. G.R. Gillich, Z.I. Praisach, Robust method to identify damages in beams based on frequency shift analysis, in *Proceedings of SPIE 8348*, article number 83481D (2012)
13. G.R. Gillich, N.M.M. Maia, I.C. Mituletu, M. Tufoi, V. Iancu, Z. Korka, A new approach for severity estimation of transversal cracks in multi-layered beams. Lat. Am. J. Solids Struct. **13** (8), 1526–1544 (2016)
14. I. Negru, G.R. Gillich, Z.I. Praisach, M. Tufoi, E.V. Gillich, Nondestructive evaluation of piers, in *Proceedings of SPIE 9438*, article number 943817 (2015)
15. G.R. Gillich, Z.I. Praisach, Modal identification and damage detection in beam-like structures using the power spectrum and time-frequency analysis. Signal Process. **96**(A), 29–44 (2014)

16. V. Iancu, G.R. Gillich, C.P. Chioncel, About the effectiveness of several dissimilarity estimators used in damage assessment. TEM J. Technol. Educ. Manag. Inform. **5**(3), 253–262 (2016)
17. G.R. Gillich, D. Frunzaverde, N. Gillich, D. Amariei, The use of virtual instruments in engineering education. Procedia Soc. Behav. Sci. **2**(2), 3806–3810 (2010)

Chapter 38
Assessments on Operational Modal Identification Refining of a Structural Element

Silviu Nastac and Carmen Debeleac

Abstract This study deals with operational modal identification techniques area, presenting a group of experimental and computational approaches about refining of modal characteristics for structural parts during their dynamic exploitation regime. Hereby, it can be included both into the modal experimental/operational analysis domain—with application in dynamic state evaluation in order to provide essential information to vibration control measures, and into the structural health monitoring area—providing a feasible tool for compiling the referenced state and estimating the failure imminence. The analyses were developed based on the laboratory setup, taking into account a simple structural element with constant mechanical and geometrical characteristics. The theoretical approaches contain both analytical evaluations, and computational simulations with the help of the finite element method. The results comparison between the classical modal identification techniques and the proposed method reveals an improved capability of the last for modal characterization of a singular structural element within its dynamic evolution, also taking into account that the number of experimental measurements was constantly maintained. Future developments will take into consideration the applicability of these assessments for structural ensembles or variable characteristic elements.

S. Nastac (✉) · C. Debeleac
Engineering and Agronomy Faculty, Research Center for Mechanics of Machines and Technological Equipments, "Dunarea de Jos" University of Galati, Calea Calarasilor 29, 810017 Braila, Romania
e-mail: silviu.nastac@ugal.ro

C. Debeleac
e-mail: carmen.debeleac@ugal.ro

N. Herisanu and V. Marinca (eds.), *Acoustics and Vibration of Mechanical Structures—AVMS-2017*, Springer Proceedings in Physics 198,
https://doi.org/10.1007/978-3-319-69823-6_38

38.1 Introduction

Experimental identification of modal parameters has been developed during the last decades. Nowadays, the structural vibration testing and analysis provide a useful tool for many industries, including aerospace, auto-making, manufacturing, materials production, power generation, and many others. It is useful to reduce the effects of unwanted vibration because those can cause fatigue or degrade the performance of the structure. Vibration can be unavoidable or even desirable, hereby the goal may be to understand the effect on the structure, or to control or modify the vibration, or to isolate it from the structure and minimize structural response. Dynamic testing and continuous monitoring based on Operational Modal Analysis (or Output-only Modal Analysis—OMA) supply a very useful experimental tool, which allow periodic/continuous assessment of structures through the analysis of their response to ambient excitation, while they are in normal operation regime. For small structures tested in a controlled environment, inside a laboratory, it is available the Experimental Modal Analysis (EMA), which is a method based on the measurement of the structure response to one/several also measured dynamic forces. Combining EMA and OMA techniques recently has been explored and developed so called Operational Modal Analysis with eXogenous inputs (OMAX), which supposed an artificial force used in operational conditions. Major difference between OMAX and EMA derived from the operational forces that are included in the model of the analyzed system instead of being assumed as residual noise.

A new idea of EMA that only use the structural free vibration response due to initial conditions was presented by Wang and Cheng [1]. This idea can also be extended and applied to the general structure with non-proportional damping case and the proposed methodology can therefore enhance the structural modal analysis technique [1]. In the study [2], six aluminum beams of different configurations (with/without cuts of various lengths) were used for conducting experiments regarding impact test, shaker test, and operational modal analyses. Within the thesis [3] the uncertainty impacting OMA was grouped into four categories, in order to provide an insight into relative impacts of different uncertainty sources in OMA.

A novel procedure to perform OMA on a rotating cantilever beam is described in [4], using Digital Image Correlation to measure the deformation of the beam from images captured with a pair of high-speed digital cameras. Modal parameters are determined from the deformation data using Ibrahim Time Domain method. DaSilva and Pereira [5] discuss the effect of the presence of harmonic components in the process of identification of modal parameters, and use the concept of probability density function as a tool to separate the operational modes (harmonic) from the structural ones. Practical problems to be dealt with when planning an ambient vibration based system identification test on a structure are discussed by Cantieni in [6].

In the paper [7], the comparison between two different vibration testing techniques is presented. The first approach takes advantage of the frequency domain decomposition, of the response cross power spectral densities (CPSD) to estimate both the natural frequencies and the "unsealed" mode shapes, whereas the second one, starting from the Hilbert Transform of auto power spectral densities and taking account of the CPSD, allows one to get the frequency response functions [7]. Ulriksen et al. [8] demonstrates an application of a proposed modal and wavelet analysis-based damage identification method to a wind turbine blade. OMA was conducted to obtain mode shapes for undamaged and damaged states of the blade.

This work is a part of a large study regarding simplified EMA and OMA procedures available for modal identification of various structural elements. The authors present some experimental assessments dealing with the improvement of the modal information database based on acquiring and processing additional dynamics of explored structural element.

38.2 Experimental Basics

A general view of the experimental setup was depicted in Fig. 38.1, where the first picture (a) presents a snapshot during the experimental tests and the last picture (b) shows a detail of the accelerometer montage on the explored structural element. For laboratory tests was assumed a cantilever montage based on a constant section beam element with 15 mm width, 2 mm thickness and 1000 mm length, of aluminum alloy, with 68,900 MPa Young's Modulus, 0.33 Poisson's Ratio and 2710 kgm^3 mass density. Modal identification procedure was performed based on transversal dynamics of the beam, and comparison with the analytic and finite element method results was done. Rolling hammer technique was used, with five inputs sequence.

Diagrams in Fig. 38.2 present magnitudes of frequency response function evaluated for acquired acceleration of transversal beam direction (y axis—see Fig. 38.1a). It have to be mentioned that EMA method was used in order to provide a validation tool for measuring points through the coherences signals between each input and the output. Goodness of modal identification procedure results from the correlation of the peaks marked on diagrams within Fig. 38.2.

(a) (b)

Fig. 38.1 General view of the experimental setup (**a**) and detail of vibration transducer montage (**b**)

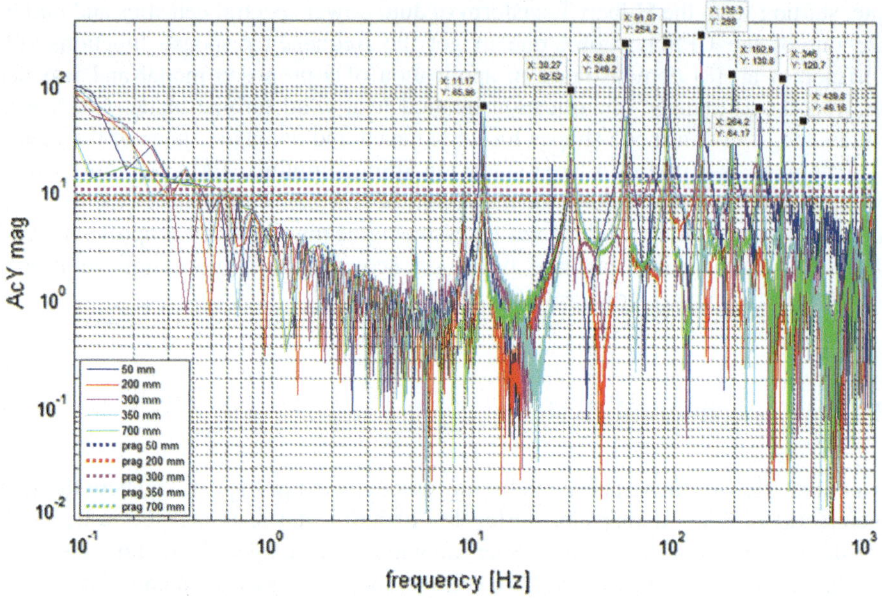

Fig. 38.2 Spectral compositions for transversal vibration of the explored cantilevered beam

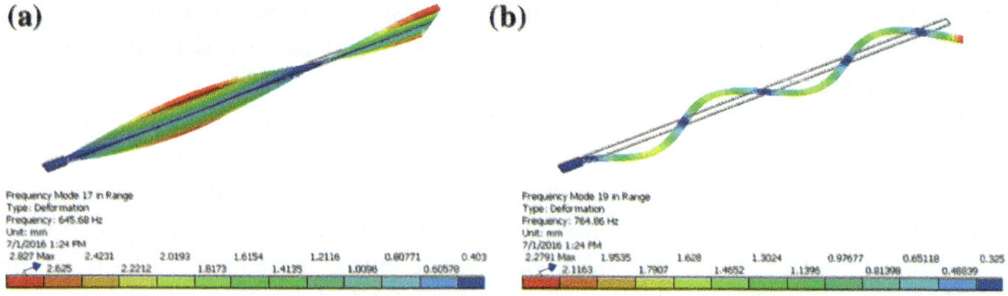

Fig. 38.3 Torsional (**a**) and lateral bending (**b**) mode shapes of the explored cantilevered beam

During the analysis it was observed that the finite element analysis provides additional frequency information and modes besides those specifically for the basic transversal bending motion. Two examples were depicted in Fig. 38.3, where picture (a) denotes the torsional and picture (b) denotes lateral bending shapes.

Taking into account that many of health monitoring methods require evaluations of changes within modal frequencies, and the vibration control techniques based on the resonances shifting, the authors had been proposed to experimentally evaluate the most of the spectral information related to dynamics of the analyzed element, maintaining, if it is possible, the same volume of experimental tests.

38.3 Additional Experimental Research

According with previously mentioned goal, it was used a six degree-of-freedom (6DoF) transducer in order to explore the entire range of analyzed element motions. Picture in Fig. 38.4 depicts a detailed view of glued montage of 6DoF MPU-6050 transducer onto the beam in respect with motion directions—z axis respects the orthogonal rule related on transducer plane. Magnitudes of frequency response functions for the three accelerations and three gyrations, provided by the 6DoF transducer, were depicted in Fig. 38.5. Red dots denote the peaks over the thresholds. The correlation between the frequency values related with these peaks, for all six signals, can be evaluated within the chart in Fig. 38.6, where star symbols was used for main direction of beam vibratory motion (transversal bending).

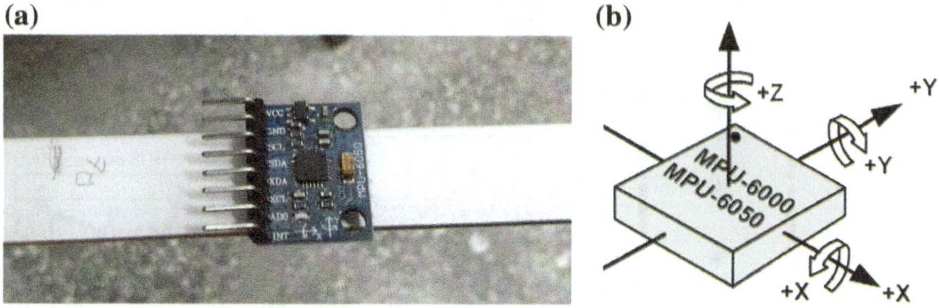

Fig. 38.4 Evaluative experimental setup with 6DoF MPU-6050 transducer (a) and sensor axes (b)

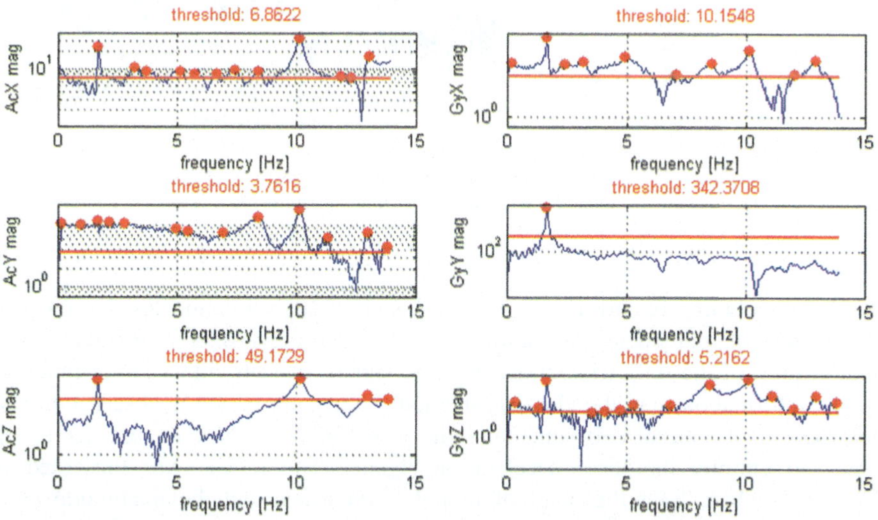

Fig. 38.5 Magnitude spectrums of each signal provided by the 6DoF transducer

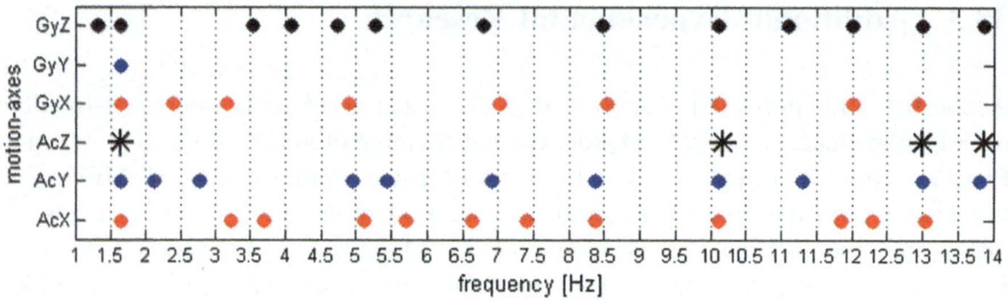

Fig. 38.6 Frequencies chart of 6DoF evaluative experiment. Star symbol denotes main direction

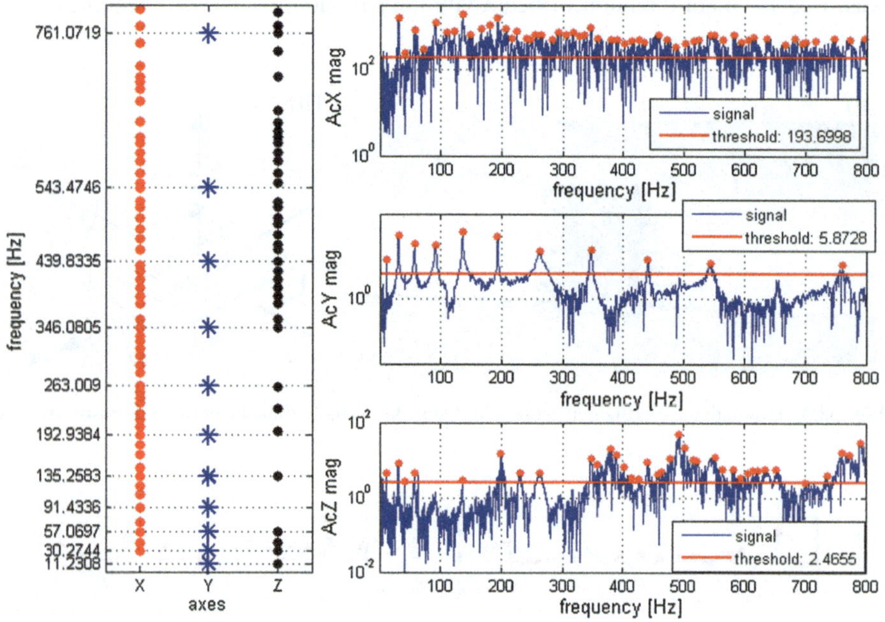

Fig. 38.7 Frequencies chart (left side) and magnitude spectrums (right side) for triaxial acceleration measurements related on basic experimental setup. Star symbol denotes main direction

One concluding remark resulted from this chart analysis indicates that most of the significant frequencies can be discovered within the acceleration signals, thus that acquire of the only triaxial acceleration signals can provide a larger range of modal information than the classical method. Supposing previous observation and returning to the initial experimental setup (see Fig. 38.1), it was considered and analyzed entirely the three acceleration signals. The results was presented in Fig. 38.7, where both the chart of modal frequencies, and the magnitude plots reveal the additional significant information contained within acceleration onto

x and z axes. Similar to diagram in Fig. 38.6, the star symbols denote main vibratory motion in respect with the transversal direction.

38.4 Concluding Remarks

The observations related to the experimental tests performed by the authors and partially presented within this paper lead to the conclusion that using the triaxial accelerometers for exploring the vibratory motion of a structural element provides an improved tool to enlarge the spectral information database and offers a simplest way to refine the modal identification analyses.

References

1. B.-T. Wang, D.-K. Cheng, Modal analysis of mdof system by using free vibration response data only. J. Sound Vib. (2007). doi:10.1016/j.jsv.2007.09.030
2. S. Rudroju, A. Gupta, S. Yandamuri, Operational modal analysis of aluminum beams. J. IEST **50**(1), 74–85 (2007)
3. S.K. Ciloglu, The impact of uncertainty in operational modal analysis for structural identification of constructed systems. A Thesis Submitted to the Drexel University in Partial Fulfillment of the Requirements for the Degree of Ph.D., Aug 2006
4. S.S. Rizo-Patron, J. Sirohi, in *Operational modal analysis of a rotating cantilever beam using high-speed digital image correlation.* 57th AIAA/ASCE/AHS/ASC Structures, Structural Dynamics, and Materials Conference, AIAA SciTech Forum, (AIAA 2016–1957). http://dx.doi.org/10.2514/6.2016-1957
5. B.E.P. da Silva, J.A. Pereira, Effect of harmonic components in an only output based modal analysis, in *Proceedings of COBEM 2009—20th International Congress of Mechanical Engineering*, Gramado, RS, Brazil 15–20 Nov 2009
6. R. Cantieni, Application of ambient vibration testing (operational modal analysis) in practice, http://rcidynamics.ch/site/downloads/
7. A. Agneni, R. Brincker, B. Coppotelli, (2004). On modal parameter estimates from ambient vibration tests, in *Proceedings of the International Conference on Noise and Vibration Engineering: ISMA2004* Leuven, Belgium, 20–22 Sept 2004
8. M.D. Ulriksen, D. Tcherniak, P.H. Kirkegaard, L.Damkilde, Operational modal analysis and wavelet transformation for damage identification in wind turbine blades. Struct. Health Monit. **15**(4), 381–388 (2015). doi:10.1177/1475921715586623

Chapter 39
Gear Drive System Simulation with Different Model of Input Speed

Kamil Řehák, Barbora Kopečková and Aleš Prokop

Abstract Presented paper deals with investigation of the basic gearbox parameter effect on the surface normal velocity of the gearbox top cover. First the single stage gearbox was designed and manufactured for the inputs and results validation. Afterwards a dynamic model was built and each part of it was validated by an appropriate technical experiment. The presented part of investigation focuses on comparison of different kind of input speed—ideal smooth increasing input speed and variable input speed. The evaluation is done by amplitude of gearbox surface normal velocity. The main goal is to describe dynamic behaviour of whole gearbox by numerical simulation and find the numerical model recognisability. The simulation results show that the variable input speed has strong influence on dynamic behaviour.

39.1 Introduction

Combustion engines, electric motors, compressors or transmissions are designed to have maximum possible efficiency, wide variety of running conditions, complexity and low production costs. The increasing number of such devices closely relates to the prevailing popular health issue topics. Since the combustion engines are absent in this regard, other noise components come into focus, mainly the transmissions. Currently, the gearboxes in the locomotive area are becoming a hot topic because trains are usually associated with densely inhabited district areas. In addition, transmissions in wind power plants, which are examined in papers [1, 2], as a

K. Řehák (✉) · B. Kopečková · A. Prokop
Faculty of Mechanical Engineering, Brno University of Technology, Technicka 2896/2, 61669 Brno, Czech Republic
e-mail: rehak@fme.vutbr.cz

B. Kopečková
e-mail: Barbora.Kopeckova@vutbr.cz

A. Prokop
e-mail: prokop.a@fme.vutbr.cz

N. Herisanu and V. Marinca (eds.), *Acoustics and Vibration of Mechanical Structures—AVMS-2017*, Springer Proceedings in Physics 198,
https://doi.org/10.1007/978-3-319-69823-6_39

renewable power sources, are dealt with very often when the noise in surrounding areas is concerned, significantly influencing life quality there. Automotive industry with an extensive number of products is another part of this issue, where the transmission is one of the investigated topic [3].

In cases where the transmission is expected to be the dominant noise source, usually it is the housing, which is responsible for more than 80% of the whole transmission emitted noise. Housing NVH improvement is performed predominantly by a modal analysis, which can provide basic information about frequencies and mode shapes. This approach has some disadvantages, the frequency does not have to be, necessarily, excited by operation conditions or, on the other hand, the higher mode (over the operation mode) can be excited. Moreover, this approach does not provide any information about surface velocity, which is closely connected to the emitted noise. Modal analysis has an essential potential in a design phase when validating the differences between two possible options takes place [4, 5]. The next level of gearbox investigation lies in using harmonic analysis, which can be performed by experimental or numerical approach. Harmonic analysis can provide information about surface normal velocity, but the exciting forces have to be known, which is sometimes difficult to obtain.

For that purpose, the Multibody dynamics system can be used. Inputs causing excitation here can be formulated in an easier form—input torque and rotational speed, which are transferred into related forces acting on each model part. The numerical simulation of transmission is frequently examined and also extensive topic, but all models have to be at the end verified by an experiment. The most advantageous variant of the transmission seems to be the single stage gearbox, because of easy modification to different gearwheels, bearings, shafts and covers. This kind of transmission is widely used for examination of oil behaviour, the influence of contact behaviour, unbalance, backlash, torque value and transmission error [6–8]. This paper deals with dynamic simulations of the single stage transmission, where the surface normal velocity on the top cover is mainly observed.

39.2 Analytical Approach

Unlike vector dynamics that works with vector variables, analytical dynamics works with scalar variables (work, energy). Basics of analytical analysis are formulated using principles of mechanics that work with generalized coordinates, because of it, analytical approach is universal approach for solution of dynamic problems. Thanks to this approach is possible to derive Lagrange's equations of motion.

Using generalized forces and the principle of virtual work, we can write virtual work as

$$\delta A = \sum_{j=1}^{N} \boldsymbol{p}_{\boldsymbol{F}_j} \cdot \delta \boldsymbol{r}_j = \sum_{j=1}^{N} \boldsymbol{p}_{\boldsymbol{F}_j} \cdot \sum_{i=1}^{n} \frac{\partial \boldsymbol{r}_j}{\partial q_i} \cdot \delta q_i = 0, \tag{39.1}$$

where $\boldsymbol{p}_{\boldsymbol{F}}$ are workforces, δq_i, $\delta \boldsymbol{r}_j$, represents virtual displacement, respectively written in vector approach. Using principle of minimum energy, D'Alembert's principle for body system and mathematical adjustments, we obtain from (39.1) basic form of Lagrange's equations of motion of second kind

$$\frac{d}{dt}\frac{\partial L}{\partial \dot{q}_i} - \frac{\partial L}{\partial q_i} = 0, \tag{39.2}$$

where L represents Lagrange's function expressed by kinetic and potential energy of the form

$$L = E_k - E_p. \tag{39.3}$$

Equation (39.2) can be extended by adding others variables like for example Rayleigh's dissipative function R_d and Lagrange's multipliers λ_s, which are used to express dependent variables. Using above mentioned extensions we get extended Lagrange's equations of motion of second kind

$$\frac{d}{dt}\frac{\partial L}{\partial \dot{q}_i} - \frac{\partial L}{\partial q_i} + \frac{\partial R_d}{\partial \dot{q}_i} + \sum_{s=1}^{K} \lambda_s \frac{\partial f_s}{\partial q_i} = \tilde{Q}_i. \tag{39.4}$$

For the applications in specific cases, (39.4) can be rewritten with help of an expression kinetic energy using generalized mass matrix $\boldsymbol{M}(\boldsymbol{q})$, potential energy using generalized stiffness matrix $\boldsymbol{K}$, generalized gravitational force $\boldsymbol{f}_g$ and Rayleigh's dissipative function using damping matrix $\boldsymbol{B}$. Using this extensions we obtain the following equation

$$\boldsymbol{M}\ddot{\boldsymbol{q}} + \dot{\boldsymbol{M}}\dot{\boldsymbol{q}} - \frac{1}{2}\left[\frac{\partial \boldsymbol{M}}{\partial \boldsymbol{q}}\dot{\boldsymbol{q}}\right]^T \cdot \dot{\boldsymbol{q}} + \boldsymbol{K}\boldsymbol{q} + \boldsymbol{f}_g + \boldsymbol{B}\dot{\boldsymbol{q}} + \left[\frac{\partial \boldsymbol{f}}{\partial \boldsymbol{q}}\right]^T \cdot \lambda = \tilde{\boldsymbol{Q}}. \tag{39.5}$$

39.3 Numerical Methods

The numerical simulations are widely used for study to compare dynamic behaviour of two transmission design or sensitive study to see the trend. The main advantage is, no prototype has to be created, which significantly save money. For that reason, it is widely used in research and at developing phase of transmission [9]. The presented approach is based on dynamic model with multiple degree of freedom, which is shown in Fig. 39.1. The whole model consists of gears mounted on the flexible shafts, flexible gearbox housing, shaft mounting is through bearings

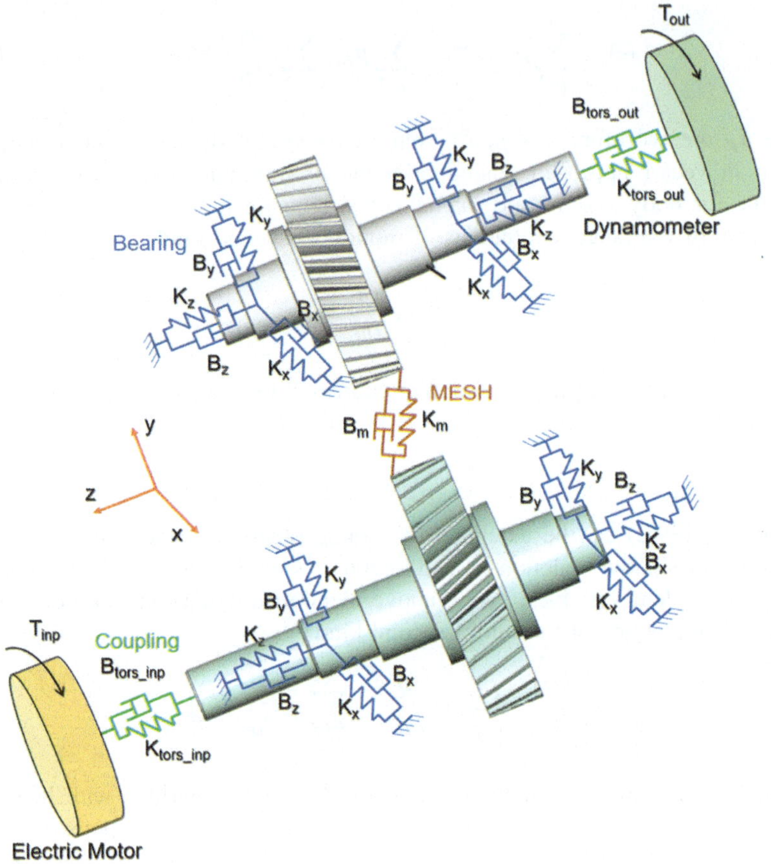

Fig. 39.1 Dynamic model with multiple degree of freedom

specified by radial and axial stiffness. The variable gear mesh stiffness is used for interaction between gears.

The whole approach is based on the basic equation of motion, which corresponds to the analytical (39.5).

The presented method combines the finite element method (FEM) and multibody software ADAMS. The whole methodology is validated with technical experiment on single stage gearbox, the gearbox parameters and process of validation is described in detail in [10]. The FEM is used for getting input parameters for multibody simulation. The whole virtual prototype in ADAMS is performed as open code. That enables to include and simple change all substantial parameters for different transmission as basic dimensions, positions, flexible/rigid bodies, bearings stiffness, gear mesh stiffness, backlash, variable input speed and torque.

The bearing stiffness is calculated on one rolling segment and afterwards converted to the whole bearing [11]. The gear mesh stiffness is very important because it significantly change dynamic behaviour of whole system. The type, real tooth geometry significantly affect the course of variable gear mesh stiffness function. In the presented paper the function is calculated during contact of one tooth pair, that corresponds to three teeth gear pairs, and afterwards repeated periodically. No misalignment and manufacturer error is presented. The stiffness shown in Fig. 39.2 corresponds to the results mentioned in [12].

To include the dynamic behaviour of bodies, the flexible variant is necessary. It can be performed by the Craig-Bampton reduction, which replaces real modal properties with simplified approximation established from the two variants of degree of freedom and multiplied by the special Craig-Bampton transform matrix, see (39.6):

$$\{\boldsymbol{u}_A\} = \left\{ \begin{matrix} \boldsymbol{u}_b \\ \boldsymbol{u}_L \end{matrix} \right\} = \begin{bmatrix} \boldsymbol{I} & 0 \\ \Phi_R & \Phi_L \end{bmatrix} \left\{ \begin{matrix} \boldsymbol{u}_b \\ \boldsymbol{q} \end{matrix} \right\} = \Phi_{CB} \left\{ \begin{matrix} \boldsymbol{u}_b \\ \boldsymbol{q} \end{matrix} \right\}, \tag{39.6}$$

where $\boldsymbol{u}_A$ is the original vector of deformation, $\boldsymbol{u}_b$ are the boundary degrees of freedom (DOFs), $\boldsymbol{u}_L$ stands for interior DOFs, $\boldsymbol{q}$ represents modal DOFs, $\boldsymbol{I}$ is the identity matrix element, Φ_R is rigid body matrix element and in the analogical way Φ_L are the fixed base mode shapes matrix element.

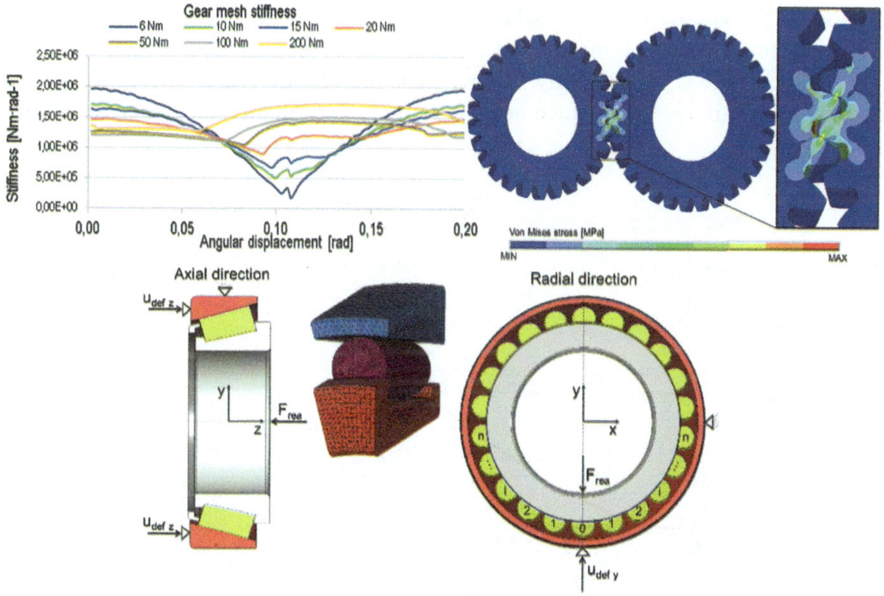

Fig. 39.2 Gear mesh stiffness and definition of bearings

39.4 Results

The main aim of this comparison is to find the effect of different input speed parameters. The input speed is changing in range 0–3000 RPM. First the ideal smooth profile is used. The second kind of input speed profile represents four-cylinder stroke engine, where the second, fourth and eight harmonic functions are presented. The smooth profile is in virtual prototype represented by special STEP function, which enables smooth change of designed RPM parameter, see Fig. 39.3. Whole RPM ramp respects following equation:

$$\omega_1 = STEP(time,\ 0,\ 0,\ 9,\ 3000) + \sum_{k=2;\ 4;\ 8} A_k \sin(\varphi_1 k) \qquad (39.7)$$

where A_k is the amplitude of k-th harmonic function and φ_1 represents the angular displacement of the input gear.

The backlash is related to the original one, which is measured on the physical single stage gearbox prototype. That is marked as 100%. The backlash range is from 0%, which represents ideal (unreal), to 200%.

The input speed has influence on the angular deformation, where there is no local amplitude increase in case of non-variable input speed, Fig. 39.4. The effect of variable input speed is getting stronger with increasing backlash. On the other side the torque is extreme without backlash, Fig. 39.5.

The effect of different input speed is evaluated on the surface normal velocity of top cover, which is significantly thinner than the rest of gearbox housing for the reason of sensitivity. The surface normal velocity is evaluated for top cover middle

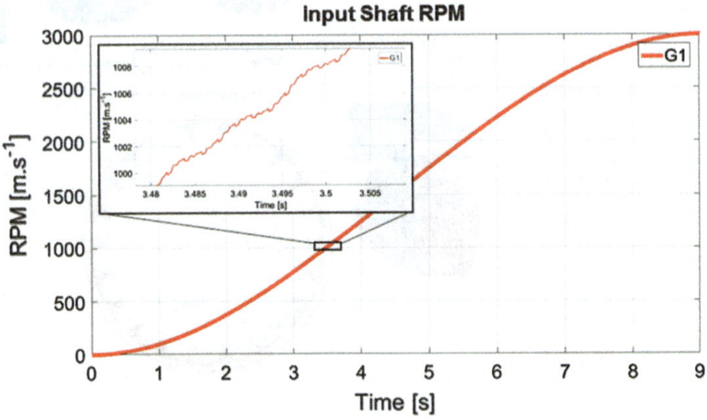

Fig. 39.3 Input shaft speed profile

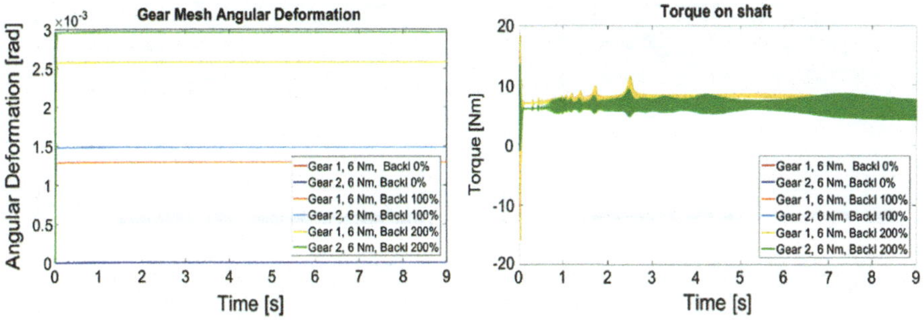

Fig. 39.4 The course of angular deformation and torque for non-variable input speed

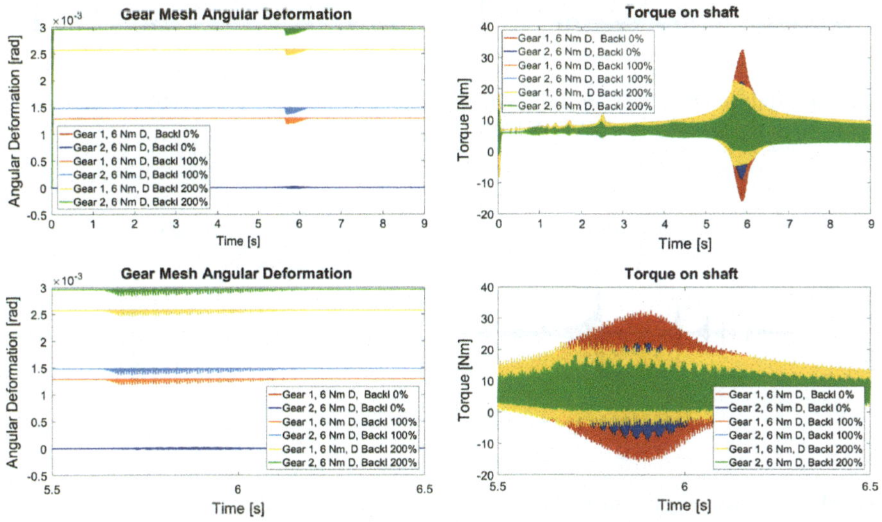

Fig. 39.5 The course of angular deformation and torque for variable input speed

point, Fig. 39.6. The effect of the variable input speed is noticeable in Fig. 39.6a, c. The results of variable input speed shows that higher harmonic orders frequencies are able to significantly affect surface normal velocity. When the backlash is presented, the rattle can occur, which is visible in Fig. 39.6c, d. In these figures the critical area occurs out of tooth frequency and the gearbox housing eigenfrequency intersection. This is noticeable also in Fig. 39.7, where the local increase of the amplitude occurs close the time 6 s.

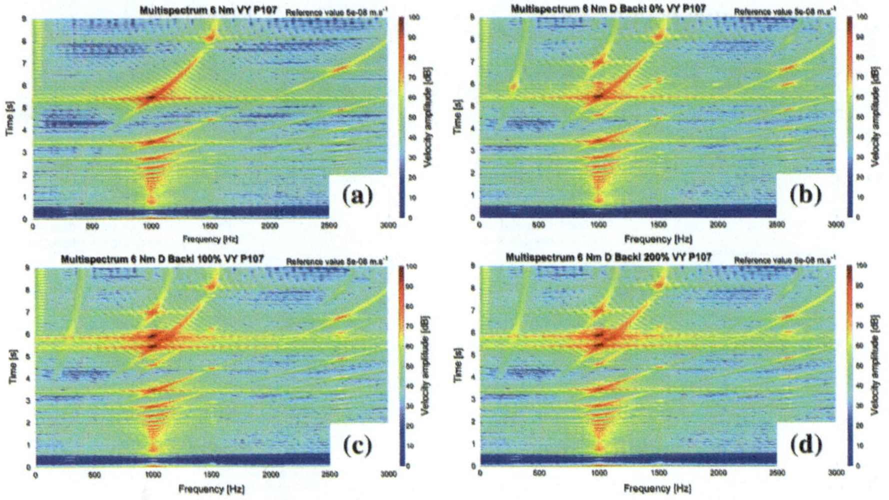

Fig. 39.6 Multispectrum diagram for **a** non-variable input speed backlash 100%; variable input speed with backlash **b** 0%; **c** 100%; **d** 200%

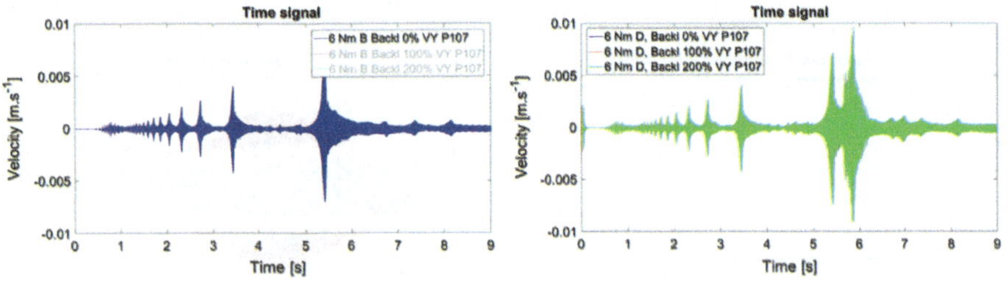

Fig. 39.7 Surface normal velocity in time domain, top cover middle point

39.5 Conclusions

For the study the model, which consists of connection FEM and MBS, is used. The virtual prototype was developed and validated on the single stage gearbox with the weakest part on the top, for the reason of research. Presented study is focused on the effect of different input speed profile. The surface normal velocity is used for comparison. The results show, that for qualitative description of gearbox dynamic behaviour the virtual prototype should contain possibility to include variable input speed. The variability, which is due to functional principle of combustion engine, is the main excitation for the rattle effect initiation. The type of variability should consider real application.

Acknowledgements The research leading to these results has received funding from the Ministry of Education, Youth and Sports under the National Sustainability Programme I. (Project LO1202) and with help of the project FSI-S-17-4104 granted by specific university research of Brno University of Technology. The authors gratefully acknowledge this support.

References

1. L. Feng, Q. Yumo, G. Linshan, P. Zhao, L. Shaokang, L. Donglong, Influences of planetary gear parameters on the dynamic characteristics—a review. J. Vibroeng. **17**(2) (2015) ISSN 1392-8716
2. X. Jin, L. Li, W. Ju, Z. Zhang, X. Yang, Multibody modeling of varying complexity for dynamic analysis of large-scale wind turbines. Renew. Energy **90**, 336–351 (2016)
3. Y. Kadmiri, E. Rigaud, J. Perret-Liaudet, L. Vary, Experimental and numerical analysis of automotive gearbox rattle noise. J Sound Vib **331**(13), 3144–3157 (2012). ISSN 0022-460X
4. L. Yongxiang, J. Lihong, S. Wenquan, N. Liwen, Z. Youjia, An efficient optimal design methodology for abnormal noise control of automobile transmission in the neutral idle condition. J. Vibroeng. **16**(1) (2014). ISSN 1392-8716
5. P. Novotny, V. Pistek, New Efficient Methods for Powertrain Vibration Analysis. Proc. Inst. Mech. Eng. Part D J. Automob. Eng. **224**(5), 611–629 (2010)
6. M. Razpotnik, T. Bischof, M. Boltežar, The influence of bearing stiffness on the vibration properties of statically overdetermined gearboxes. J. Sound Vib. **351**, 221–235 (2015). ISSN 0022-460X
7. T. Loutas, G. Sotiriades, I. Kalaitzoglou, V. Kostopoulos, Condition monitoring of a single-stage gearbox with artificially induced gear cracks utilizing on-line vibration and acoustic emission measurements. Appl. Acoust. **70**(9), 1148–1159 (2009). ISSN 0003-682X
8. A. Andersson, L. Vedmar, A dynamic model to determine vibrations in involute helical gears. J. Sound Vib. **260**, 195–212 (2003). ISSN 0022-460X
9. A. Kumar, H. Jaiswal, R. Jain, P.P. Patil, Free vibration and material mechanical properties influence based frequency and mode shape analysis of transmission gearbox. Procedia Eng. **97**, 1097–1106 (2014)
10. A. Prokop, K. Rehak, M. Zubik, P. Novotny, Experimental validation of the gearbox NVH parameters. J. Middle Eur. Constr. Des. Cars **13**(2), 16–21 (2015)
11. I. Atanasovska, The mathematical phenomenological mapping in non-linear dynamics of spur gear pair and radial ball bearing due to the variable stiffness. Int. J. Non-Linear Mech. [online] **73**, 114–120 (2015). ISSN 00207462
12. T. Kiekbush, D. Sppok, S. Bernd, H. Ian, Calculation of the combined torsional stiffness of spur gears with two- and three-dimensional parametrical FE models. J. Mech. Eng. **57**, 810–818 (2011)

Chapter 40
Dynamic Analysis of the Inverted Tooth Chain Plates Moving Towards Sprocket

Karim Shalaby, Simona Lache and Radu Plamadeala

Abstract The investigation of the dynamic behaviour of the plates of an inverted tooth chain (used as driving system in a wide range of automotive applications) is still a research issue. The contact between the chain plates and the sprocket generates vibrations, with negative effect on the overall driving system performance. This paper aims to perform an in depth analysis of the behaviour of the chain plates contacting a sprocket at different moments, considering the Analogous Huygens Pendulum model and the plates movement towards the sprocket. A frequency domain analysis is performed. The study leads to a better understanding of various scenarios that a plate contacting the sprocket can produce and will allow future studies comparing similarities between a chain plate in a pendulum motion system and a chain drive motion system.

40.1 Introduction

The study of contacts between chain plates and sprocket or chain plates and joints has been always a challenging subject for research [1, 2]. Contacts always put the problem of how one could protect a system from reaching failure and maintain a long life with great performances [3, 4]. Their analysis will always raise the question of "how can one diminish stress increase over certain limits in contacts?" Attempts have been done for a better understanding of the contact phenomena but there are still unsolved issues related to the behaviour of chain plates towards sprockets.

K. Shalaby (✉) · S. Lache
Transilvania University of Brasov, Brasov, Romania
e-mail: karim.shalaby@unitbv.ro

S. Lache
e-mail: slache@unitbv.ro

R. Plamadeala
Schaeffler Romania, Cristian, Romania
e-mail: raduplam@yahoo.com

N. Herisanu and V. Marinca (eds.), *Acoustics and Vibration of Mechanical Structures—AVMS-2017*, Springer Proceedings in Physics 198,
https://doi.org/10.1007/978-3-319-69823-6_40

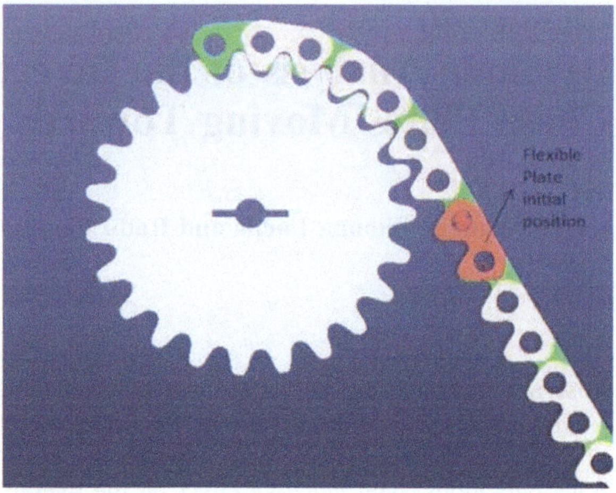

Fig. 40.1 Sprocket and the ITC chain with the flexible body highlighted

Considering an Analogous Huygens Pendulum is an approach for such a research which can lead to a better understanding and finding common behaviours of an Inverted Tooth Chain (I.T.C.) in a chain drive system. The main objective of the paper is to perform a frequency analysis of an ITC plate element in order to identify the correct way for avoiding critical modes. The chain under study has a 2-2-1 lacing, meaning that there are two outer plates, two inner plates and one middle plate. The movement of the elements is reversed, that means the sprocket is fixed and the chain is movable [5]. The model setup in MSC ADAMS and the initial position of the flexible body is shown in Fig. 40.1.

40.2 The Numerical Analysis

In order to perform the dynamic analysis, the multibody approach was used for modelling the flexible plate. The model has been developed with MSC ADAMS, the flexible bodies being considered with a small linear deformation at the local reference frame [6]. The natural frequencies and their corresponding eigenmodes are determined at the location of the flexible plate in the overall system (Fig. 40.1). Figures 40.2 and 40.3 present some results of mode shapes obtained in this way.

The model was set to run for 10 s. This gave the possibility of calculating the maximum and minimum principle stresses of the flexible body and indicating the "hotspot", the highest node element affected at the plate. The analysis gave the possibility to show the most affected zones caused by the contact, as illustrated in Fig. 40.4.

According to the calculations performed, the maximum principal stress was determined: 628.1 MPa at t = 0.674 s, in node 4089. This gave the possibility to measure the contact forces between the plates of interest and the sprocket and the

Fig. 40.2 Mode 7: 69,607.92170 Hz

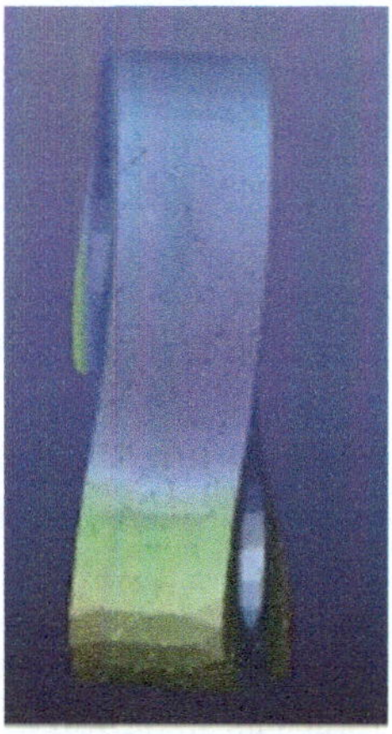

Fig. 40.3 Mode 8: 71,939.727064 Hz

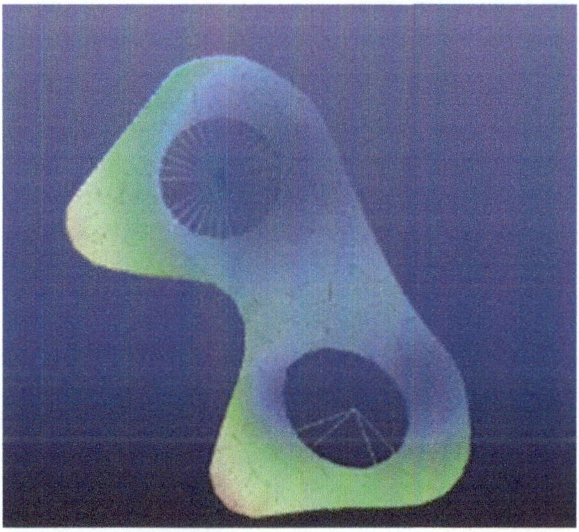

Fig. 40.4 The most affected stress zones due to contacts

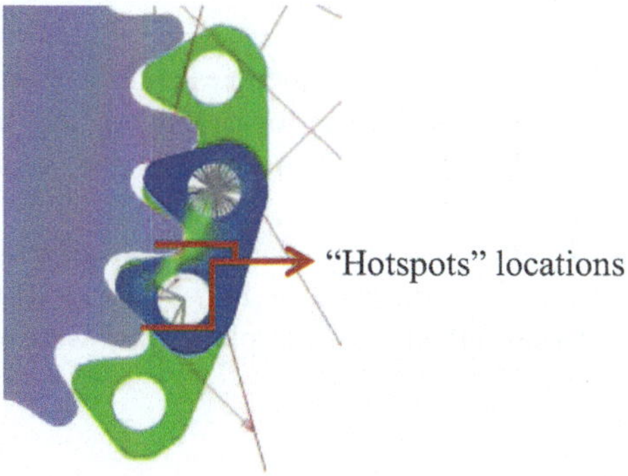

Table 40.1 Forces in joints

Joint	Magnitude of force at the joint [N]
Plate 12–Plate 13	192.79259
Plate 13–Plate 15	241.66594

forces at the joints. The outer plates are not contacting the sprocket in question in contacts, the forces in the joints being divided by three, as three bodies hold that joint. The magnitude of the forces at the joints, considered at the same moment in time, are listed in Table 40.1.

The exact position of each plate was calculated at that exact position in the frame. These results were further used for the finite element analysis, performed with ABAQUS software. For developing the finite element model, the precise position of the model at the specific frame time was set in order to consider the forces as static forces [7]. The FEM model developed for plate 13 was investigated under the boundary conditions, frame by frame. In the next step, the joints between the plate holes were added, by the means of introducing kinematic couplers, the couplers being totally constrained. The couplers, representing the pins joining the chain, are considered not deformable.

For simplifying the calculations, assumptions were made that pins were rigid and had low value of surface roughness; consequently, it was assumed that the friction forces between the joints were low enough. The upper hole at plate 12 and the lower hole at plate 15 were fully constrained, yet leaving them to rotate freely around the Z-axis, representing the reactions of the plates due to the imposed forces.

The outside left and outside right surfaces of plates 13 and 14, illustrated in Fig. 40.5, are translational constrained on the Y-axis plane to replicate the teeth of the sprocket reactions on the chain plates. Preloaded forces are added on the holes of plates 13 and 14, as they had been previously calculated in the MSC ADAMS analysis, at the specific moment. The plate was mashed with hexa-form elements.

The aim of the dynamic analysis was to calculate the eigenfrequencies for which a steady state dynamics of the system was obtained.

40.3 Research Results

The first six eigenfrequencies and the corresponding eigenmodes resulted from the dynamic analysis performed using the finite element method are presented in Table 40.2. The frames from the transient response were used for their identification.

Random nodes of plate 13 were considered for further interpretation of results. The nodes presented in Fig. 40.6 are distributed between the two teeth and the external left plan of the plate; they were selected in order to see the evolution of

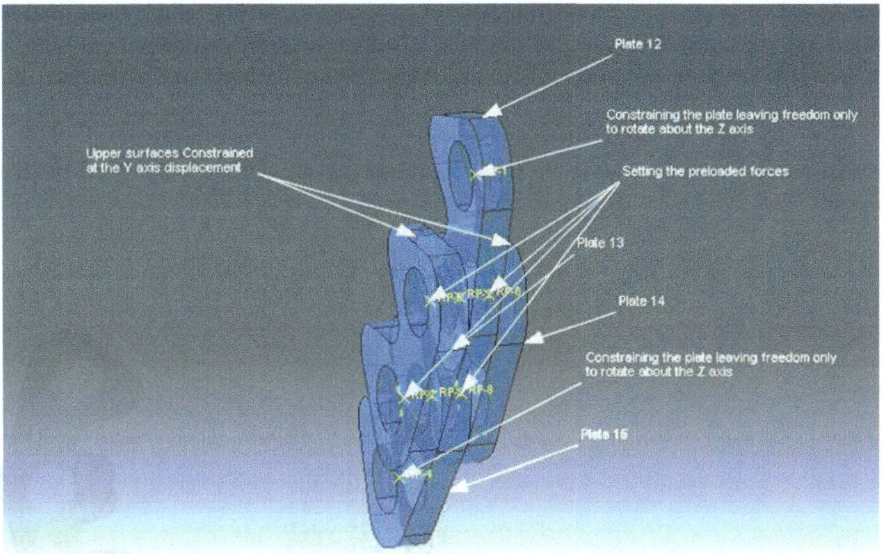

Fig. 40.5 The setup of the model in ABAQUS

Table 40.2 Value of the first six eigenfrequencies

Mode number	Frequency [Hz]
1	26,022
2	29,755
3	30,430
4	55,985
5	59,635
6	65,143

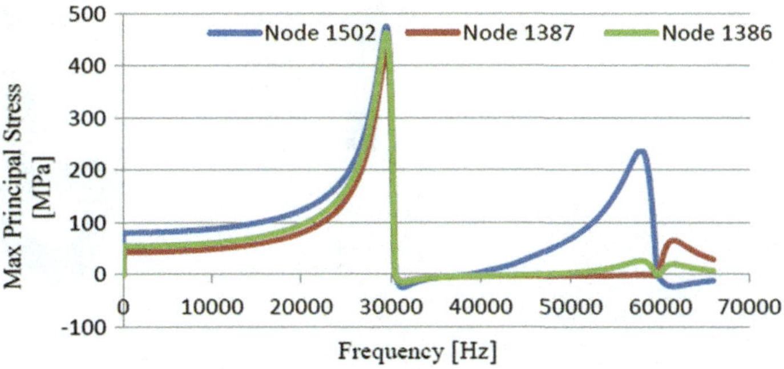

Fig. 40.6 The magnitude of maximum principal stress and the corresponding frequencies

stresses in terms of frequencies applied to the plate. From the graph presented in Fig. 40.6, it can be observed that the most effective frequencies range could be for modes 2, 3 and 5. It is important to notice the distribution of the maximum principal stresses and the displacement of the nodes on the plates at the specified modes.

At the same time, as the frequencies increase, the maximum principal stresses and displacement of nodes decrease, as seen in Figs. 40.7, 40.8, 40.9, 40.10, 40.11 and 40.12.

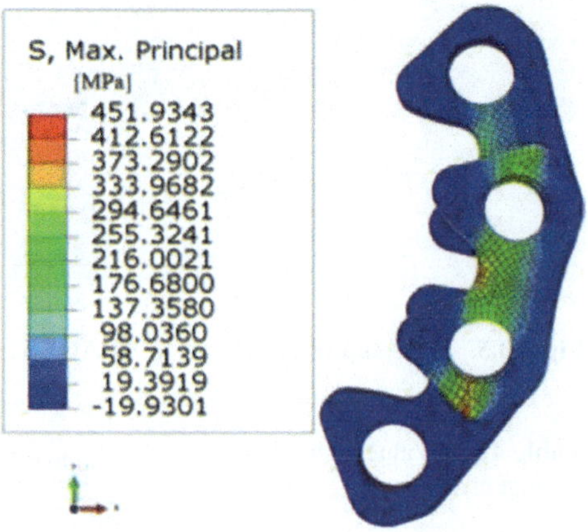

Fig. 40.7 Distribution of the maximum principal stress for mode 2 (f = 27,955 Hz)

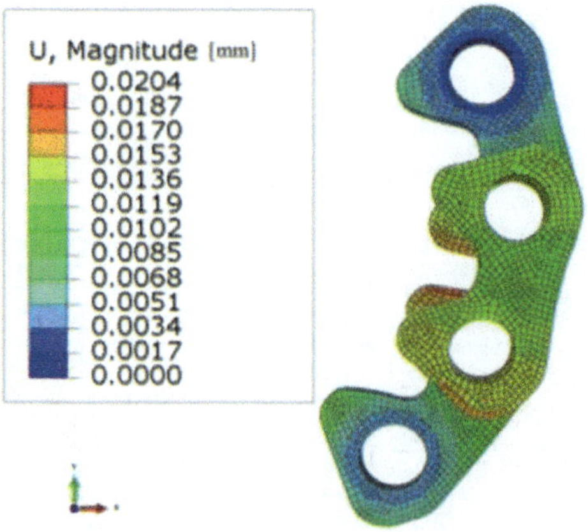

Fig. 40.8 Distribution of displacement field for mode 2 (f = 27,955 Hz)

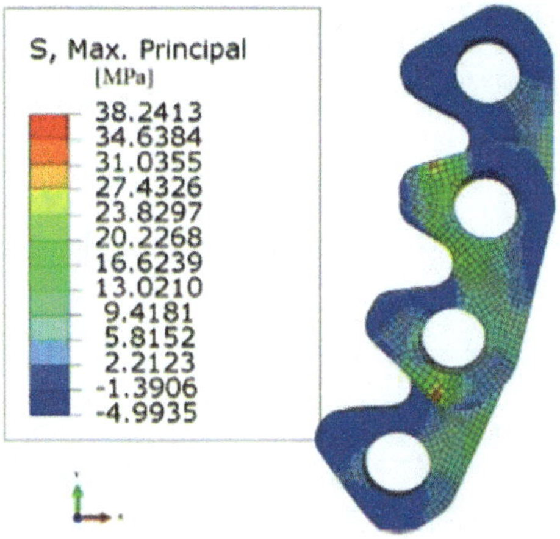

Fig. 40.9 Distribution of the maximum principal stress for mode 3 (f = 30,430 Hz)

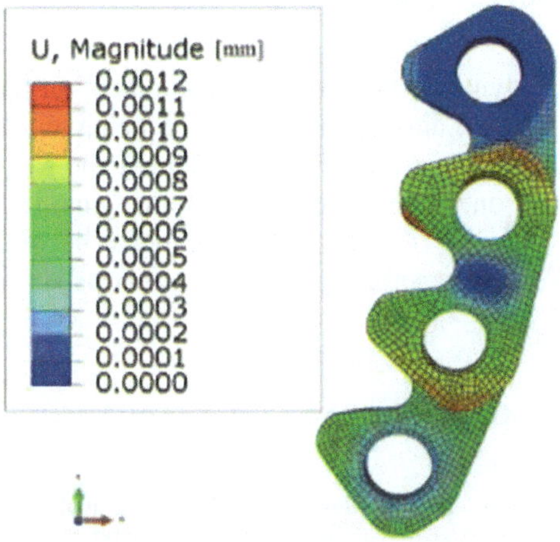

Fig. 40.10 Distribution of displacement field for mode 3 (f = 30,430 Hz)

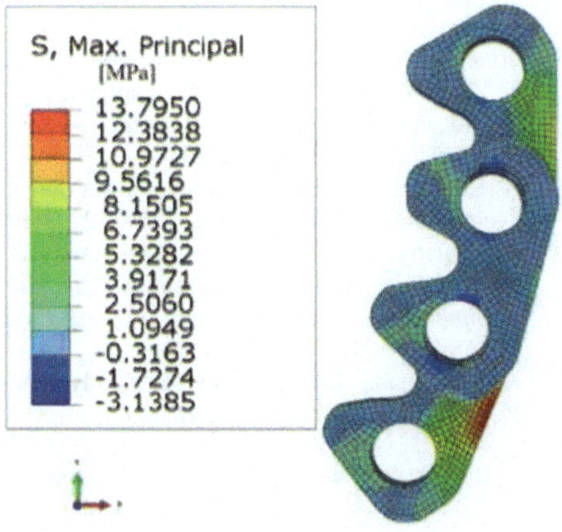

Fig. 40.11 Distribution of the maximum principal stress for mode 5 (f = 59,635 Hz)

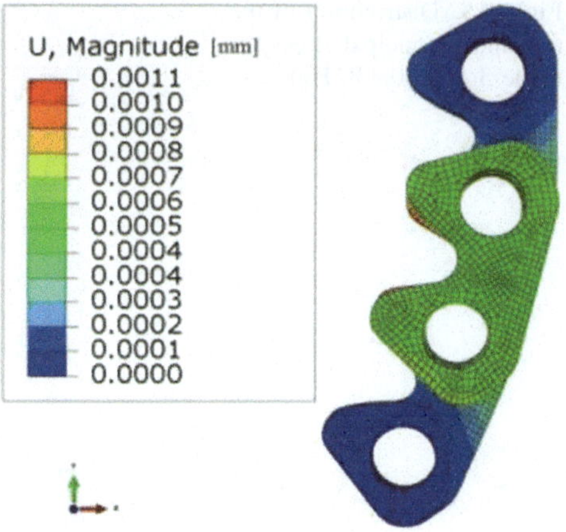

Fig. 40.12 Distribution of displacement field for mode 5 (f = 59,635 Hz)

40.4 Conclusions

A dynamic analysis of the contacts on an ITC plate towards a sprocket was performed, aiming to identify the behaviour of an ITC plate in the dynamic transient state. The results allow to identify the frequency ranges that are harmful, as well as the ones useful for the system dynamic behaviour. In order to have a complete overview of contact phenomena between chains elements and sprockets, the research method presented in this paper can be extended to more chain elements.

References

1. R.V. Mulik, M.M. Joshi, Dynamic analysis of timing chain system of a high speed three cylinder diesel engine. Res. Invent Int. J. Eng. Sci. **4**(5), 21–25 (2014)
2. M. Ishihama, H. Watanabe, *Analysis and control of inverted tooth chain vibration*, World Automotive Congress FISITA 2010, Budapest, 30 May–4 June 2010, pp. 852–861
3. I. Troedsson, L. Vedmar, A method to determine the dynamic load distribution in a chain drives. J. Mech. Eng. Sci. **215**(5), 569–579 (2001)
4. F. Zeng Ming, L. Junlong, L. Gangwu, Dynamic analysis of silent chain drive system for hybrid car, in *Advanced Materials Research*, vol. 694–697 (Trans Tech Publications, Switzerland, 2013), pp. 84–89
5. A. Emmerson, Things are seldom what they seem Christiaan Huygens, the pendulum and the cycloid. NAWCC Bull (362), 295–312 (2006)
6. K. Shalaby, S. Lache, F. Radu, The analysis of an analogous huygens pendulum connected with I.T.C. using flexible multibody dynamic simulations, in *6th International Conference on Computational Mechanics and Virtual Engineering*, 15–16 Oct 2015, Braşov, Romania, pp. 65–71
7. A.A. Shabana, Dynamics of Multibody Systems, 4th edn. (University of Illinois, Chicago, Cambridge University Press, 2013)

Chapter 41
Theoretical and Experimental Analysis for a Pendulum-Based Absorber System Used in Vibration Isolation

Aurora Potîrniche and Gigel Căpățână

Abstract This paper proposed to realize a theoretical and experimental analysis of a vibration isolation system using a dynamic pendulum absorber. The analysis was made from simple to complex. At first, it was approached the case of the pendulum absorber system, considered as an individual system, then the case of a structural system in free vibration, without pendulum absorber, and finally, the case of the insulator-structure assembly in dynamic varied working regime. For this system was developed a set of tests and, based on this tests, was established a methodology of approach for testing in dynamic regime. This methodology allows the determining of essential dynamic parameters of the proposed pendulum system and the evaluating of their behavior in dynamic working regime.

41.1 Introduction

The antiseismic designing and of protection to vibrations of the buildings aims the achieving of mechanical devices that are designed as energy absorbers or insulators and that can favorably influence the dynamic response of the equipments subjected to random actions [1, 2]. The applicability domain of antiseismic protection systems is very wide, so that the range of specific necessary parameters is characterized by a great flexibility, so as to ensure the functionality and capability of these systems in terms of isolation against harmful effects of disturbances. In the context described above, the present work aims to analyze the behavior in dynamic regime of a dynamic pendulum absorber system [3–5].

A. Potîrniche (✉) · G. Căpățână
Engineering and Agronomy Faculty, Research Center for Mechanics of Machines and Technological Equipments, "Dunărea de Jos" University of Galați, Galați, Romania
e-mail: aurora.potirniche@ugal.ro

G. Căpățână
e-mail: gcapatana@ugal.ro

N. Herisanu and V. Marinca (eds.), *Acoustics and Vibration of Mechanical Structures—AVMS-2017*, Springer Proceedings in Physics 198,
https://doi.org/10.1007/978-3-319-69823-6_41

41.2 Theoretical Approaches

In the first part of the study was carried out the mathematical modeling of a system consisting of a rod hinged at the top and a lumped mass at the end of the bar attached to its underside (Fig. 41.1). The mathematical calculation was based on several simplifying assumptions: the hypothesis of the small oscillations (the amplitude of the pendulum system does not exceed 4–5°); the movement of the considered system occurs in the *XOY* plan; the rod component of the dynamic pendulum absorber system was considered a rigid rod (without deformations); the mass placed on the end of the bar was considered rigid too, without relative movements to the rod. It has pursued the calculation of the eigenvalues of the frequencies and of the amplitudes corresponding to degrees of freedom of the system, two parameters of interest in obtaining results that highlight the opportunity of the structures isolation using additional systems of dynamic pendulum absorber type. In order to determine the above-mentioned interest parameters it was considered a system with a single degree of freedom of pendulum dynamic absorber type. The degree of freedom is a rotation and is represented by the angle located between the bar with number 1 and the vertical direction, denoted by θ [6, 7].

At the writing of the differential equation that defines the motion of the system have been taken into account the mass of the rod, denoted by m_1 and articulate on the top side and the concentrated mass, denoted by m_2 and attached at the bottom of the bar. The pendulum joint includes a visco-elastic element with a linear characteristic. The moment of inertia for the bar in rotation motion was denoted by J_1 and the length of the bar was denoted by l. It was denoted the stiffness coefficient of elastic element of the hinge by k and the damping coefficient by b.

After applying the Lagrange's equation II for the present case it results the eigenvalue of pulsation, denoted by p. For the proposed system presents also interest the obtaining of the amplitude of the motion corresponding to the considered degree of freedom. In this respect it was considered that on the system acts an harmonic excitation whose amplitude is F_0. The amplitude of the motion is A.

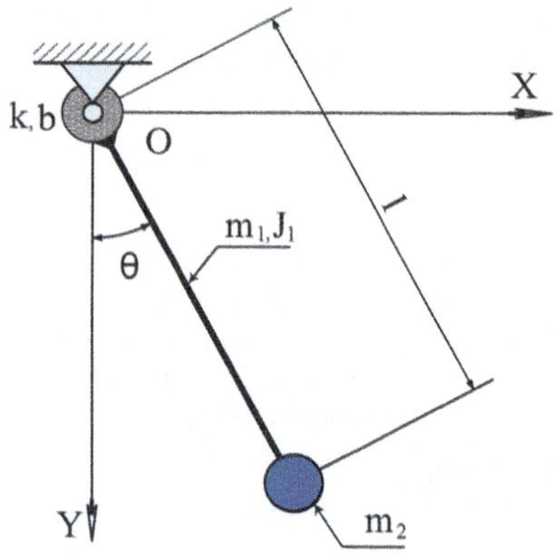

Fig. 41.1 The pendulum dynamic absorber, with visco-elastic element in the hinge

$$p = \frac{1}{l}\sqrt{\frac{0.5m_1gl + m_2gl + k}{m_1/3 + m_2}};$$
$$A = F_0\left(-p^2l^2\left(\frac{m_1}{3} + m_2\right) + \frac{1}{2}m_1gl + m_2gl + k\right)^{-1} \quad (41.1)$$

Further in the present study is treated the classic case of the movement of the base mass of the complex assembly in which the dynamic absorber is only an additional system (Fig. 41.2). The reason for the treating of this simple case is the following: it is wanted in this paper a complete study, progressively, from simple to complex, of the behavior in dynamic regime of the pendulum system, harmonic excited in the horizontal direction. The above assumptions remain valid, and there is an additional hypothesis: the base mass of the complex system which must to be isolated has only one degree of freedom (a movement in the horizontal direction, denoted by x). The mass m_1 simulates the behavior of a structure type system and it is considered that the system is double restricted. In this respect, the rigidities of the two elastic elements were denoted by k_1, respectively by k_2 [8, 9].

The eigenvalue of pulsation and the amplitude (considering that on the system is acting a harmonic excitation whose amplitude is F_0) in this case are:

$$p = \sqrt{(k_1 + k_2)/m_1}; \quad A = F_0(k_1 + k_2)^{-1} \quad (41.2)$$

Based on the two cases above, further is approached the case of the system with dynamic absorber pendulum restricted with elastic link in the hinge (Fig. 41.3).

The model developed is that of a complex system made of a structure and an absorber pendulum, the connection of elastic elements with the system is on top of this one. In the hinge point denoted by k_3 was introduced an elastic element of torsion which produced the restriction of the pendulum absorber movement [10]. Given the specific assumptions for the present case, the eigenvalues of pulsations are:

$$p_1 = \sqrt{\frac{(C_1C_5 + l^2C_3C_4) + \sqrt{(C_1C_5 + l^2C_3C_4)^2 - 4C_3C_5\left(C_1C_4l^2 - l^2C_2^2\right)}}{2\left(C_1C_4l^2 - l^2C_2^2\right)}} \quad (41.3)$$

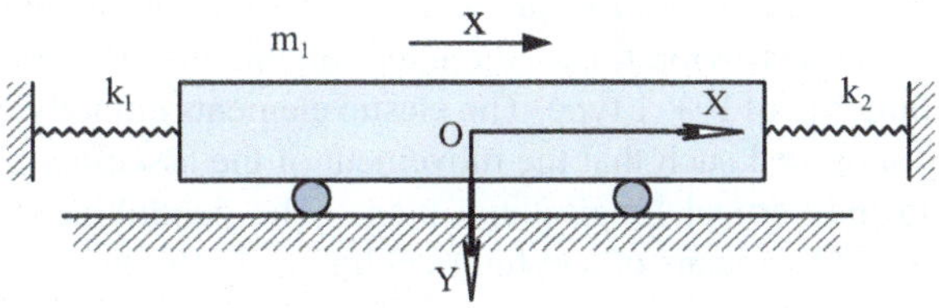

Fig. 41.2 The base mass of the complex assembly, considered as a system with SDoF

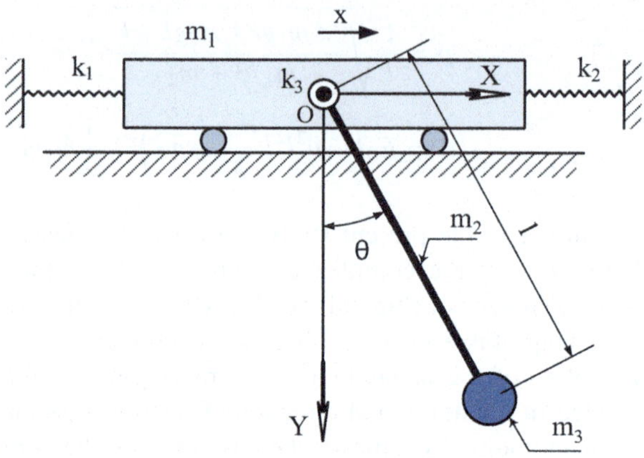

Fig. 41.3 The model with dynamic absorber pendulum, restricted within the joint

$$p_2 = \sqrt{\frac{(C_1C_5 + l^2C_3C_4) - \sqrt{(C_1C_5 + l^2C_3C_4)^2 - 4C_3C_5(C_1C_4l^2 - l^2C_2^2)}}{2(C_1C_4l^2 - l^2C_2^2)}} \tag{41.4}$$

In the above pulsations were made the following notations:

$$\begin{cases} C_1 = m_1 + m_2 + m_3;\ C_2 = 0.5m_2 + m_3;\ C_3 = k_1 + k_2 \\ C_4 = m_2/3 + m_3;\ C_5 = 0.5m_2gl + m_3gl + k_3 \end{cases} \tag{41.5}$$

Considering that on the system is acting an harmonic excitation whose amplitude is F_0, it is obtained the amplitudes:

$$A = \frac{-p^2l^2C_4 + C_5}{p^4l^2(C_1C_4 - C_2^2) - p^2(l^2C_3C_4 + C_1C_5) + C_3C_5}F_0 \tag{41.6}$$

$$B = \frac{p^2lC_2}{p^4l^2(C_1C_4 - C_2^2) - p^2(l^2C_3C_4 + C_1C_5) + C_3C_5}F_0 \tag{41.7}$$

41.3 Case Study

To identify and analyze the working regimes of the dynamic absorber pendulum, was available an experimentally set which is a laboratory stand, in a compact construction, on a small scale, containing the equipment which generate the unidirectional vibrations, mounted on an elastic structure materialized through a building of P + 1 type. The elastic elements embodying the vertical studs have been configured such that the movement of the assembly to be restricted to develop only in the vertical longitudinal plane. The pendulum absorber system is fixed to the upper by means of a cylindrical hinge. The joint is made by a releasable connection so that it can be changed if are required higher or lower values of length than the capable domain of values of the configuration initially used. The pendulum mass is mounted on the lower end of the rigid rod by means of a detachable joint so as to be able to adjust anytime the pendulum length. In Fig. 41.4 are presented constructive details of the structure and of the pendulum absorber system. The dynamic exciter applied to the top level is an unidirectional vibration generator, mounted such that the movement of the system to be in horizontal direction, in the longitudinal plane of the assembly. The experimental plan involved three sets of tests, namely: (a) the structural system in free vibration, without the pendulum absorber in operation; (b) the structural system in forced vibration, without the pendulum absorber in operation and with the dynamic exciter in varied regime of operation; (c) the assembly formed from structure—pendulum absorber—dynamic exciter in varied regime of operation. For all three cases listed above, the parameters of the structure and of the pendulum absorber were kept constant, with the following values: (i) the pendulum mass $m_2 = 0.8$ kg; (ii) the mass of the pendulum rod $m_1 = 0.4$ kg; (iii) the length of the pendulum rod $l = 0.252$ m.

The first stage corresponds to the analysis of the structural system in free vibration, without pendulum absorber in operation and without the dynamic exciter in operation. In order not to change the value and the distribution of masses in the considered assembly, the dynamic unidirectional exciter was kept mounted in the working position, blocking the movement of its mobile equipment. The pendulum absorber was also blocked so as not to affect the movement of the structure. The removing of the steady state of the whole assembly was carried out by the moving of the structure in the horizontal direction and then the release of this one (a value different of zero for the initial displacement). This case is particularly useful in

Fig. 41.4 Constructive details of the structure and of the pendulum absorber system

assessing the specific parameters of the structure, namely: the natural frequencies, with the estimation of the significant spectral component and the global dissipative feature of the structure.

In Fig. 41.5 is presented the set of acquired signals. The three diagrams mean: (1) excitation signal—the acceleration recorded on the mobile equipment of the exciter; (2) response signal—the acceleration recorded on the higher level of the structure; (3) excitation signal unanswered—the acceleration recorded on the mobile equipment of the exciter of which has been removed the component due to the movement of the structure. This last component should be theoretically zero, because the exciter does not work in this situation. In fact, it is observed on the third diagram of Fig. 41.5 that there are differences between the structure movement and the exciter movement; this is justified by the existence of some elements of the

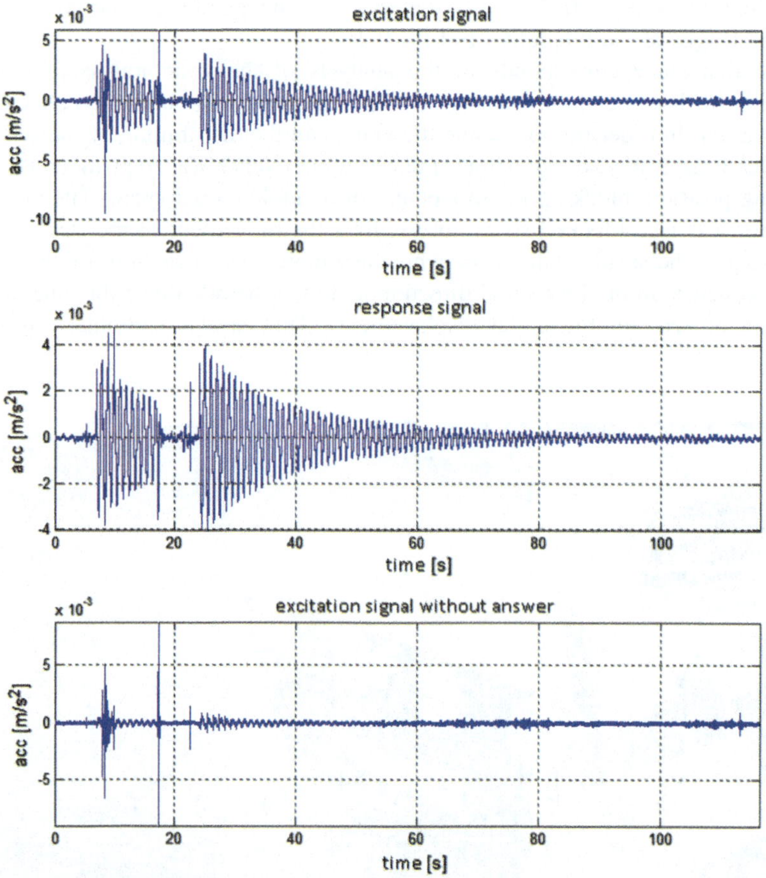

Fig. 41.5 The acceleration signals recorded for the first situation

exciter which are highly mobile and require additional components in the global dynamics of the assembly. The free evolution of the system for the analyzed period, from the point of view of the significant spectral components, can be observed in the spectrograms of the input or output signals. These diagrams are shown in Fig. 41.6.

In the second case analyzed, the structural system is in forced vibration, without pendulum absorber in working and with the dynamic exciter in varied regime of operation. In reality, in this situation it does not exist an isolation system to vibrations or seismic waves. The structure moves under the effect of external disturbing dynamic actions. In Fig. 41.7 is presented the set of acquired signals for the second stage. The significance of the three diagrams in the figure is identical to that set out in the previous case. The two spectrograms corresponding to the excitation signals and, respectively, to the structural response are shown in Fig. 41.8. There is a constant of the fundamental frequency of the structure throughout the analyzed period, and, at the same time, a deviation in frequency of the other components presented in the evaluated spectrum. In fact, are temporarily excited, depending on parameters of the dynamic instantaneous regime, one or the other components of the structure or of the vibrations generator.

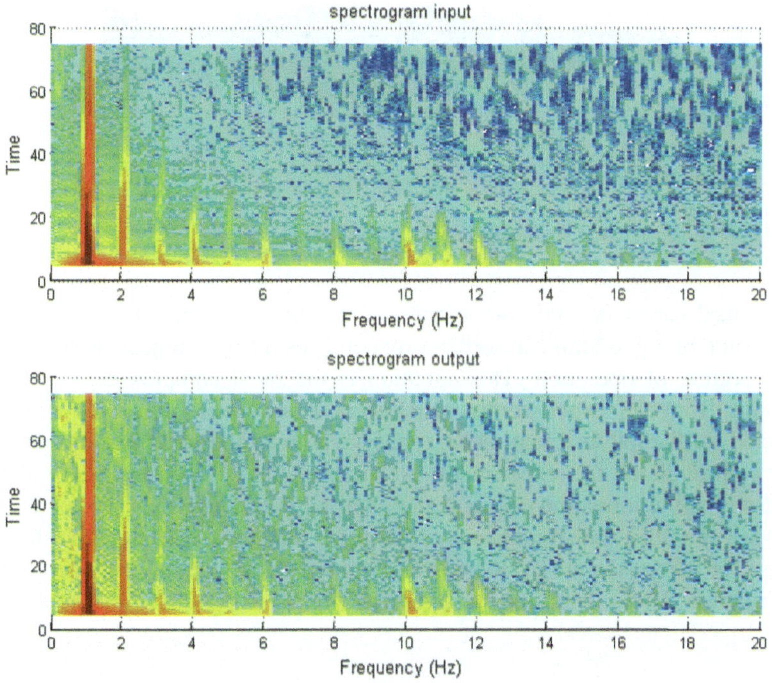

Fig. 41.6 The spectrograms of the recorded accelerations for the first situation

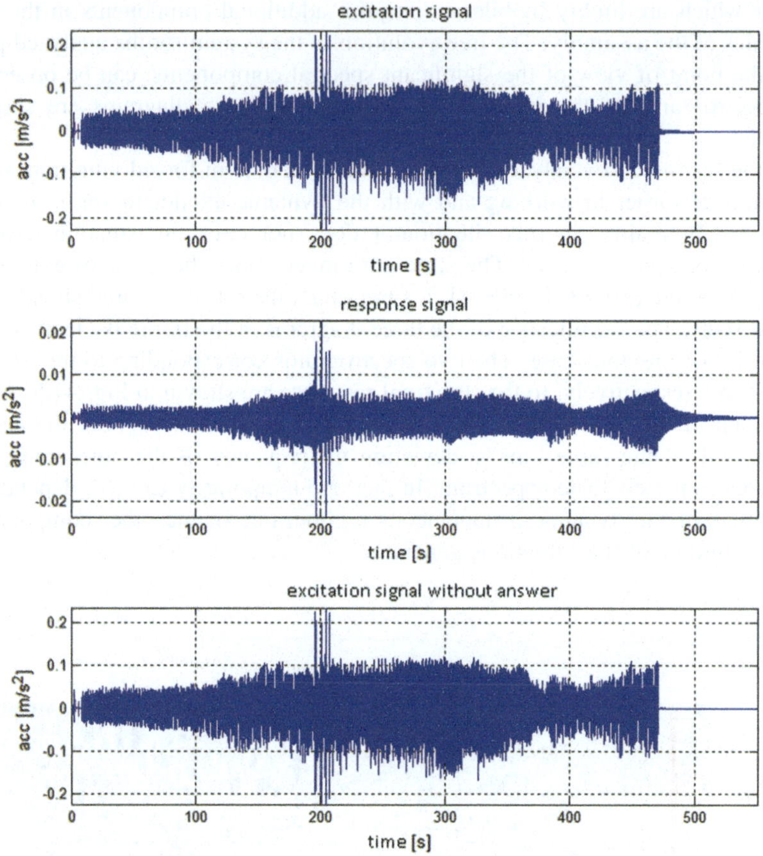

Fig. 41.7 The acceleration signals recorded for the second situation

The third case analyzed corresponds to the situation where the system formed from structure—pendulum absorber—dynamic exciter is subjected to a dynamic varied regime of operation. This situation is the most complex case of all three situations from the experimental plan, because supposes both structural system subjected to disruptive action and the intervention of the pendulum absorber type system, throughout the experiment. As in previous cases, the temporal global evolutions of the input and output signals for the entire period of test are shown in Fig. 41.9, in which the significance of the three diagrams is identical to the above.

Also for this case were evaluated the spectrograms of the two signals: excitation and response. These diagrams are shown in Fig. 41.10.

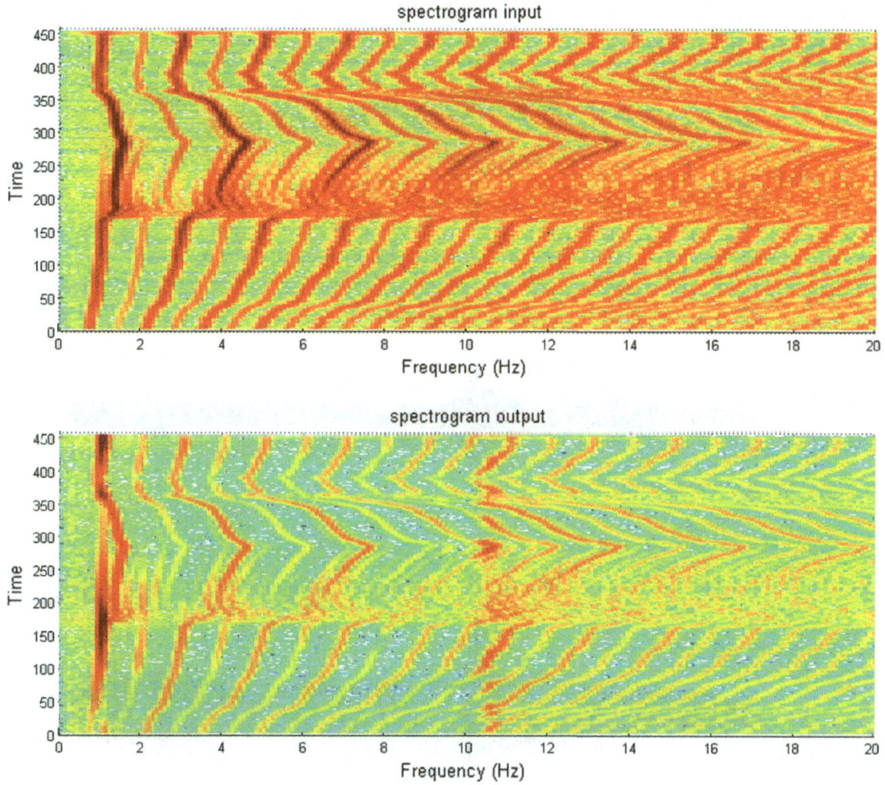

Fig. 41.8 The spectrograms of the recorded accelerations for the second situation

41.4 Conclusions

The comparative analysis of these results is based on the assumption that the first set of tests was conducted solely in order to fully characterize the structure in terms of essential dynamic parameters (stiffness, damping, natural frequencies). The other two sets of tests were proposed and realized in order to ensure the necessary support for evaluate the efficiency of the dynamic pendulum absorber for reducing the dynamics of the base structure, even from qualitative point of view. In this meaning, the second situation was necessary to ensure the point of reference in terms of the structure evolution subjected to dynamic disturbing actions intense and varied. The third situation contributed to obtain information relating to the behavior in dynamic regime of the system formed from structure—pendulum absorber.

In these conditions, from the comparative analysis of the cases II and III were observed the following relevant aspects: the dynamic pendulum absorber has the ability to provide the reduction of the dynamic effects induced in the basic structure by the action of an external disturbance; the acceptable efficiency domain as effect

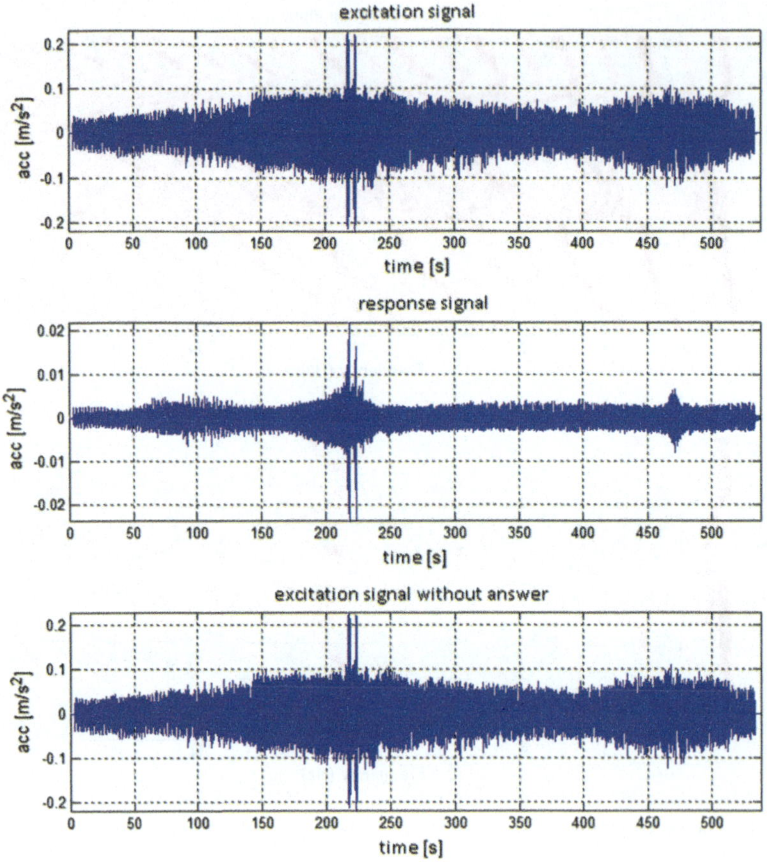

Fig. 41.9 The acceleration signals recorded for the third situation

is determined by both structural and functional configuration of the base structure and of the pendulum system; there is a danger that this pendulum system, initially implemented to reduce the structural dynamics, to increase this in certain conditions depending on the nature of disturbing excitation and on the specific values of these parameters; the behavior of such a system which reduces dynamic effects induced by an external disturbance in a structure must be characterized on the basis of some information related on the behavior in time and on the spectral evolution of involved components, so as to identify as possible the danger of some structural resonances with negative results in the evolution of the assembly dynamics.

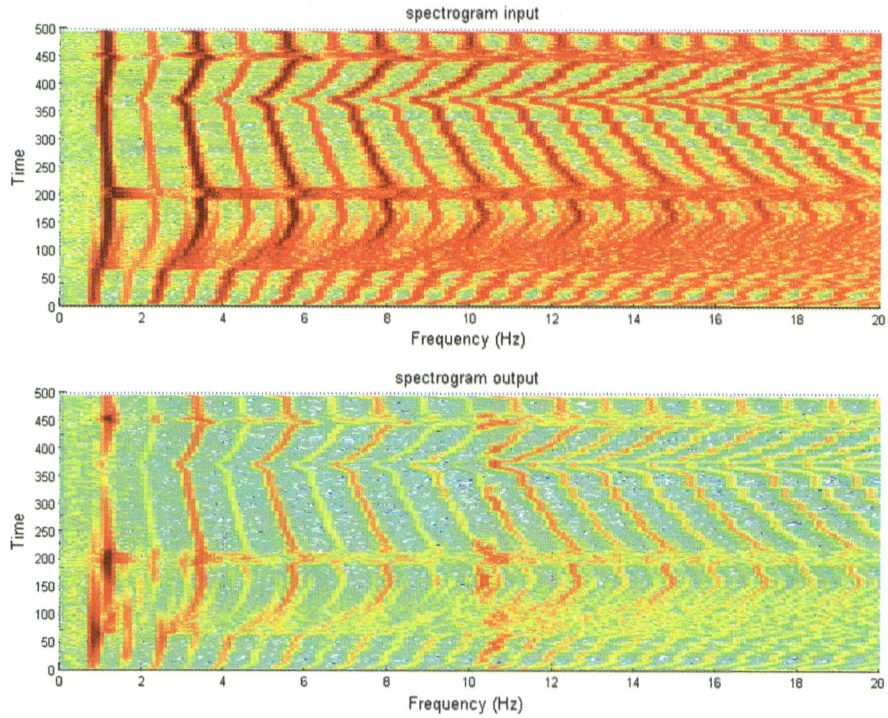

Fig. 41.10 The spectrograms of the recorded accelerations for the third situation

References

1. P. Bratu, *The Analysis of Elastic Structures. The Behaviour to Static and Dynamic Actions* (Impuls Publ, Bucureşti, 2011)
2. P. Bratu, Antivibrating systems having amplified deflection under loading, vibration control and transmissibility, in *The Internaţional Conference of RSA on Sound and Vibration*, Bucharest, România, 14–17 Oct 2004, p. 271
3. C.M. Harris, A.G. Piersol, *Harris' Shock and Vibration Handbook*, 5th edn (The McGraw-Hill Book Co, USA, 2002). ISBN 0-07-137081-1
4. M.J. Kelly, A.D. Konstantinidis, *Mechanics of Rubber Bearings for Seismic and Vibration Isolation* (Wiley, 2011)
5. Y.J. Jiang et al., Experimental study on seismic isolation bearing of large aqueduct. Appl. Mech. Mater. **226–228**, 1693–1696 (2012)
6. S. Năstac, Advances in Computational Dynamics of Passive Vibration Isolation Devices, in *The CD-Proceedings of the 1st EAA-EuroRegio 2010 Congress on Sound and Vibration*. ISBN 978-961-269-283-4, Ljubljana, Slovenia, paper 230, with abstract in Acta Acustica united with Acustica, vol. 96, Supplement 1—2010, E21 466, ISSN 1610-1928, S5-14, pp. S48
7. F.S. Oliveira, A.L.Z. Gomez, S.M. Avila, J.L.V. Brito, Design criteria for a pendulum absorber to control high building vibrations. Int. J. Innov. Mater. Sci. Eng. (IMSE) **1**(2)
8. A. Potîrniche, Aspects on the dynamic isolation for buildings using pendulum systems, in *4th International Conference on Advanced Engineering in Mechanical Systems*, 24–25 October,

2013, ACTA TECHNICA NAPOCENSIS—Series: Applied Mathematics and Mechanics, vol IV, no. 56, pp. 829–834, Technical University of Cluj-Napoca, ISSN 1221-5872
9. A. Potîrniche, G. Căpăţână, C. Simionescu, Assessments in experimental dynamics of a pendulum system with application in vibration isolation, Applied Mechanics and Materials, vol. 2015, *Proceedings of XIII-th International Symposium "Acoustics and Vibration of Mechanical Structures*" AVMS 2015, 28–29 May 2015, Timişoara, România
10. A. Potîrniche, G. Căpăţână, Experimental analysis of a vibration isolation system based on constrained pendulum device, in *The 40th International Conference on Mechanics of Solids, Acoustics and Vibrations & The 6th International Conference on "Advanced Composite Materials Engineering*" ICMSAV2016&COMAT2016 Brasov, Romania, 24–25 Nov 2016

Chapter 42
Universal Absorber Applied to NVH in EV's Powertrain

Calin Itu, Sorin Vlase, Maria Luminita Scutaru, Consuella Sofia Pena, Paul Nicolae Borza and Mircea Mihalcica

Abstract Inertial dampers ensure, through the structural design, mechanical energy dissipation due to vibration or noise generating source. Also these dampers allow, as result of vibration energy dissipation in volume of plastics on the one hand and as result of dissipative higher power efficiency and stability on the other hand (due to the high energy absorption distributed system) the use of corresponding non-linear zone flexibility based plastics. Moreover, the spatial distribution of inertial masses ensure a more uniform thermal loading plastic base and can be achieved without breaking conditions thereby avoiding pregnancy and the development of concentrators zones in basic plastic. The same spatial distribution in conjunction with inertial masses form, their surface roughness and especially the flexibility of plastics allow obtaining anisotropic features extremely useful. In this paper a study of a such type of damper is made.

C. Itu (✉) · S. Vlase · M. L. Scutaru · P. N. Borza · M. Mihalcica
Transilvania University of Braşov, Braşov, Romania
e-mail: calinitu@unitbv.ro

S. Vlase
e-mail: svlase@unitbv.ro

M. L. Scutaru
e-mail: lscutaru@unitbv.ro

P. N. Borza
e-mail: borzapn@gmail.com

M. Mihalcica
e-mail: mihmi_1@yahoo.com

C. S. Pena
Extremadura University, Badahoz, Spain
e-mail: cgragera@unex.es

N. Herisanu and V. Marinca (eds.), *Acoustics and Vibration of Mechanical Structures—AVMS-2017*, Springer Proceedings in Physics 198,
https://doi.org/10.1007/978-3-319-69823-6_42

42.1 Introduction

In [1] are shown the theoretical equations of motion corresponding inertial mass immersed in basic plastic. In the paper [2] such dampers are included in the class of 'meta-materials', mechanical hybrid structures respectively (composite) that allow "predicting the mechanical properties of the design phase" and damping properties in relation to specific needs in system which the damper is included. Such systems are found in the living world that can fulfill both dissipation function but also amplification and therefore accelerations of vibration transmitted through the damper. First, lower, corresponding to the fundamental frequency applied to the electric motor that is dependent on variable angular speed. The second frequency range of vibration is due to electrical switchgear and circuit characteristics and correspond to higher frequencies such as those produced by electrical switching circuit. This second field will vary very little in relation to the angular speed of the engine or the velocity of the vehicle. This second area is intended to be covered by the damping characteristics (band pass filter) that can perform inertial damper. Alternative solutions, involving the development of an absorber of oscillations in electric commutation type Wiener filter [3], which both in terms of reliability, energy consumption, especially high computing power necessary to implement satisfy only partially the needs of applications involving a rotating as uniform for electric vehicle drive system. Furthermore, if for some reason practically disappears the power control system of the torsional vibration, disappears too the possibility to decrease the vibrations.

42.2 Mechanical Model and Analysis Scenario

In Fig. 42.1 a standard model of an inertial damper is presented. The damper is made by rubber reinforced with balls. In the paper two cases are considered: the balls are made by steel and, alternately, by aluminum (Al). The results are similar, which is why we only present the case where steel balls are. The properties for rubber and steel are those currently used in engineering. A finite element model for the damper is presented in Fig. 42.2. The main goal of the paper is to determine the eigenvalues for a functional structure and for the damaged structure. Making an analysis in the eigenfrequencies changes is possible to identify the moment when the damage is produced. The method was presented in previous paper as [3, 4, 7]. Analytical methods in the field of the eigenvalues can be found in many paper (see [5, 6, 8–14]) (Figs. 42.3 and42.4).

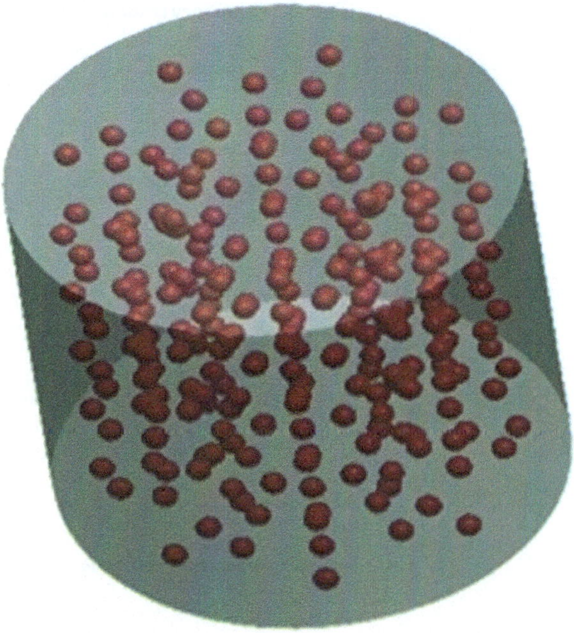

Fig. 42.1 Physical model of the inertial damper

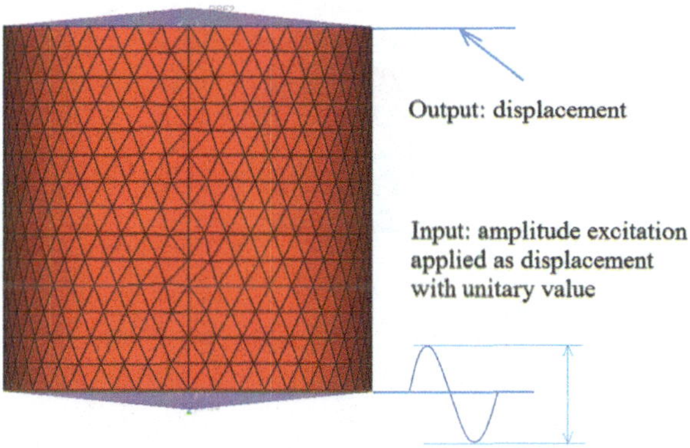

Fig. 42.2 Finite element model of the damper

42.3 Eigenvalues for the Damaged Damper

The eigenvalues and eigenmodes were computed using a finite element soft. The result for the eigenvalues computed for the scenarious are presented in Fig. 42.5. Symmetric modes gives equal eigenvalues [15–18] but a crack in the material makes one of the eigenmodes to keep his value while the other will change. That is why it is necessary to consider various eigenmodes each time to make comparison. Some eigenvalues, despite the emergence of crack and damage of the damper, do not change. If we analyze one of the cases, for example, the last case (version 10—

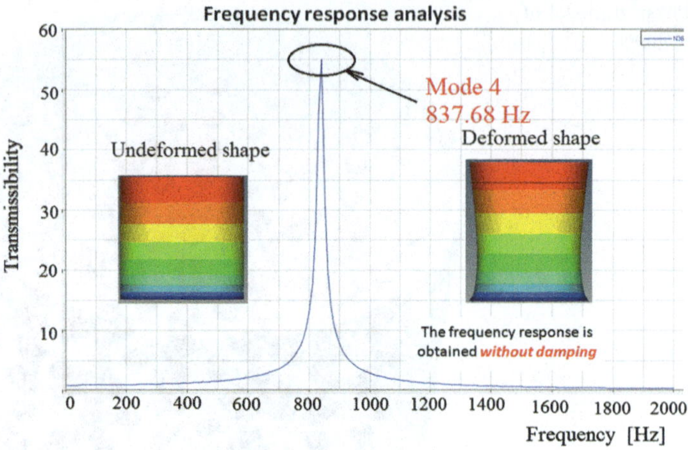

Fig. 42.3 Frequency response for a rubber model

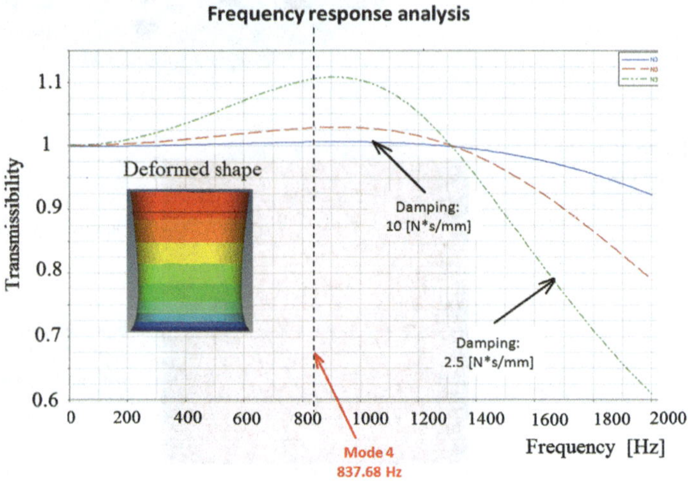

Fig. 42.4 Frequency reponse for an inertial damper with steel balls

Fig. 42.5) you can see that the eigenmode 7, for example, is not changed although the case study shows a consistent fissure. In contrast to most other eigenvalues are severely affected. It will result so that some eigenvalues are not altered by a change in material continuity while others give some consistent differences. It follows therefore that for detecting a beginning of damage is necessary to consider several vibration modes to be comparable to the initial situation.

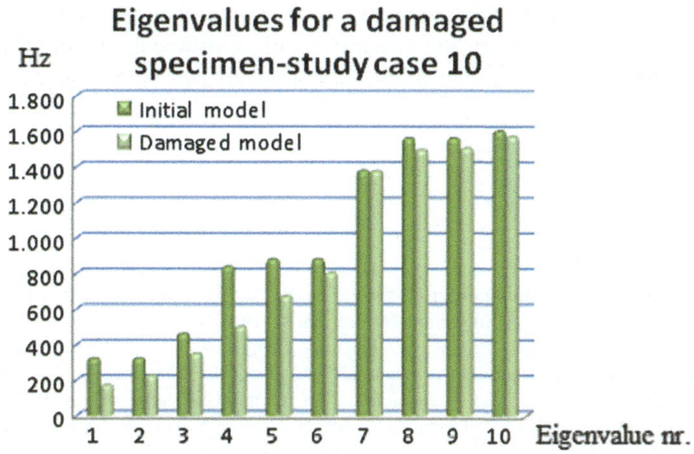

Fig. 42.5 The first 10 eigenvalues for the scenario nr. 10 (initial and damaged model)

When we use balls made by aluminum the results offer similar conclusions as for the values presented for the steel balls, from a qualitative point of view. The density of aluminum being smaller as the density for steel, the quantitative results differ.

42.4 Conclusions

In the operation of a vibration damper, insert mode of the balls inertial in the mass damper rubber makes it vulnerable to damage. It is, of course, be very useful to detect early failures of this damper before damage. It can thus avoid significant material expenses. A periodic visual inspection requires expenses and waste of time. For this reason it can be very useful to find a method of early detection of cracks in the rubber body. In this paper we propose the use of their eigenfrequencies damper measurement and comparison to baseline. If there are significant differences the damper may be replaced. As seen in the presented figures, some eigenfrequencies can differ more or less, depending on the place and dimension of the fissure. For this reason it should be analyzed and compared multiple frequencies simultaneously. We must not confine ourselves to the study of a single frequency of failure because there may be cases where the eigenfrequency corresponding to a particular mode of vibration, not be influenced by the mechanism of destruction, but other frequencies are affected. The analysis was carried out for a single shock absorber, attached at one end. In practice it may be more useful to consider the damper mounted in the mechanical system. This arrangement makes its eigenfrequencies of the damper to change. They can be determined from a more complex calculus but the findings of this study do not change. From the results presented it can be concluded that a change of eigenfrequency more than 5% can pull the alarm on the possibility of cracks in the structure while a difference of more than 10% should lead to a revision of the whole assembly to determine the cause of this change.

References

1. A. Carotti, E. Turci, A tuning criterion for the inertial tuned damper. Design using phasors in the Argand-Gauss plane. Appl. Math. Model. **23**(3), 199–217 (1999)
2. C.C. Claeys, B. Pluymers, P. Sas, W. Desmet, Design of a resonant metamaterial based acoustic enclosure, in P*roceedings of ISMA2014*, pp. 3351–3358 (2014)
3. N. Shimkin, Estimation and Identification in Dynamical Systems. Israel Institute of Technology, Department of Electrical Engineering, Lecture Notes, 2009, Chapter 3, Fall 2009
4. G.-R. Gillich, Z.-I. Praisach, C.M. Iavornic, Reliable method to detect and assess damages in beams based on frequency changes, in *ASME 2012 International Design Engineering Technical Conferences and Computers and Information in Engineering Conference*. pp. 129–137 (2012)
5. G.-R. Gillich, Z.-I. Praisach, Damage-patterns-based method to locate discontinuities in beams, in *SPIE Smart Structures and Materials*, pp. 869532–869538 (2013)
6. Z.-I. Praisach, G.-R. Gillich, D.E. Birdeanu, Considerations on natural frequency changes in damaged cantilever beams using FEM, in *International Conference on Engineering Mechanics, Structures, Engineering Geology, International Conference on Geography and Geology-Proceedings*. pp. 214–219, 2010
7. V.H. Neubert, Design of damping and control matrices by modification of eigenvalues and eigenvectors. Shock Vibr. **1**(4), 317–329 (1994)
8. A.S. Nobari, D.J. Ewins, On the effectiveness of using only eigenvalues in structural model updating problems. Shock Vibr. **1**(4), 339–348 (1994)
9. S. Vlase, A method of eliminating lagrangian-multipliers from the equations of motion of interconnected mechanical systems. J. Appl. Mech. Trans. ASME **54**(1), 235–237 (1987)
10. S. Vlase, Dynamical response of a multibody system with flexible elements with a general three-dimensional motion. Rom. J. Phys. **57**(3), 676–693 (2012)
11. S. Vlase, P.P. Teodorescu, Elasto-dynamics of a solid with a general "rigid" motion using FEM model. Part I., Theoretical approach. Rom. J. Phys. **58**(7–10), 872–881 (2013)
12. M.L. Scutaru, S. Vlase, Some properties of motion equations describing the nonlinear dynamical response of a multibody system with flexible elements. J. Appl. Math. (Article Number: 628503) (2012)
13. S. Vlase, Eigenvalues and eigenmodes of an inclined homogeneous truss in a rotational field. Rom. J. Phys. **59**(7–8), 699–714 (2014)
14. A. Modrea, S. Vlase, H. Teodorescu-Draghicescu et al., Properties of advanced new materials used in automotive engineering. Optoelectron. Adv. Mater. Rapid Commun. **7**(5–6), 452–455 (2013)
15. S. Vlase, C. Danasel, M.L. Scutaru, M. Mihalcica, Finite element analysis of a two-dimensional linear elastic systems with a plane "rigid motion". Rom. J. Phys. **59**(5–6) (2014)
16. S. Vlase, P.P. Teodorescu, C. Itu, M.L. Scutaru, Elasto-dynamics of a solid with a general "rigid" motion using FEM model. Part II. Analysis of a double cardan joint. Rom. J. Phys. **58**, 7–8 (2013)
17. S. Vlase, A method of eliminating lagrangian-multipliers from the equation of motion of interconnected mechanical systems. J. Appl. Mech. Trans. ASME **54**(1), 235–237 (1987)
18. S. Vlase, M.V. Munteanu, M.L. Scutaru, On the Topological Description of the Multibody Systems. Annals of DAAAM, Book Series, pp. 1493–1494, 2008

Chapter 43
The Analysis of a Distribution Mechanism for the Miller-Atkinson Cycle

Ionuţ Dragomir, Bogdan Mănescu and Nicolae-Doru Stănescu

Abstract In this paper the authors consider a classic distribution mechanism designed to implement the Miller-Atkinson cycle for an internal combustion engine. The mechanism was presented in a previous paper and it is characterized by the spherical head of valve and the asymmetrical motion of the roller of the lever on the spherical head. Only the contact between the roller of the lever and the valve is considered here, the synthesis of the cam being the object of another study. In our paper we extend the discussion for the vibration of the lever considering its angular velocity and acceleration with respect to the rotational angle of the shaft. The analysis of the non-linear vibration of the lever is performed with respect to different geometric parameters: radius of the spherical head of the valve, radius of the roller of the lever, dimensions and positions of different elements. The maximum values of the angular velocity and acceleration of the lever are highlighted in each case by numerical simulation. A great attention is paid to the relative motion between the roller and the spherical head of the valve, which is a measure of the wear of these two elements. The same problem of wear is discussed for the vibrations of the lever in its joint with the aid of the angular velocity and acceleration. The authors also discuss the particular case of the flat head of the valve. In this paper we consider different laws of displacement of the valve in function of the angle of rotation of the shaft; for each such a law we perform a complete discussion using the results obtained by numerical methods. The most important characteristic of these laws of motion of the valve is that they simulate the Miller-Atkinson cycle of the internal combustion engine. The analysis is performed from the kinematic point of view, the authors highlighting the advantages and disadvantages of the laws of motion considered in their study.

I. Dragomir · B. Mănescu
AKKA ROMSERV, Bucharest, Romania
e-mail: ionut.dragomir1988@gmail.com

B. Mănescu
e-mail: bogdan.manescu04@gmail.com

N.-D. Stănescu (✉)
University of Piteşti, Piteşti, Romania
e-mail: s_doru@yahoo.com

N. Herisanu and V. Marinca (eds.), *Acoustics and Vibration of Mechanical Structures—AVMS-2017*, Springer Proceedings in Physics 198,
https://doi.org/10.1007/978-3-319-69823-6_43

43.1 Introduction

In [1] it was proved that the internal combustion engines which work after the Miller cycle lead to the diminishing of the NO_x reduction, but the diminishing depends on the angle of the late intake valve closing. The reference [2] proved that the Miller cycle increases the efficiency based on longer expansion ratio. In [3] the performances of a standard Atkinson engine were analyzed with the aid of the finite-time thermodynamics. Early Inlet Valve Closure offers benefits at certain range of loads and speeds for a Ford engine [4]. Both Late Inlet Valve Closure and Early Inlet Valve Closure lead to the reduction of soot [5] and NO_x [6]. Other authors [7] realize a complex theoretical and experimental study using a 3 cylinder engine in which 2 cylinders realize 4 strokes cycles, while the third one performs a 5 stroke cycle. Some authors [8] consider a turbocharged Dual cycle model studying, with the aid of a Matlab simulation, the effects of different parameters. The study of the Miller internal combustion engine is still at the beginning, many aspects remaining unsolved. Nowadays, there are some companies (Mazda, Toyota, Ford etc.) which use this type of engines for their hybrid models, especially for the USA market. The general synthesis of a distribution mechanism with general contact curve is described in [9]. The vibrations of the engine considering different types of non-linear suspensions are discussed in [10, 11].

43.2 Mechanical Model

The mechanical model was considered in a previous paper of us. It is drawn in Fig. 43.1. It consists in the bar (which symbolizes the lever) OC having the length equal to l_1, and having at its end a roll of radius R_2. In the initial position the angle between the bar and the horizontal direction is equal to β_0, which is considered to be known. The roll of radius R_2, supports at any moment of the motion, on an arc of circle of radius R_1; the center of this circular arc is situated at the distance d from the vertical axis Oy. The displacement of the valve in vertical direction with the distance s leads to the displacement of the circular arc of radius R_1, so that the center C of this arc moves from the position C_0 to the new position C, but remaining situated at the distance d from the axis Oy. The bar of length l_1 rotates such that the roll of radius R_2 remains tangent to the circular arc of radius R_1, while the angle between the bar OC_2 and the axis Ox takes the value β. The systems must assure a required maximum displacement of the valve, s_{max}. The implicit equation that gives the angle β depending on the position y_C (and, consequently, the displacement s) was found in [12] as

Fig. 43.1 Mechanical model

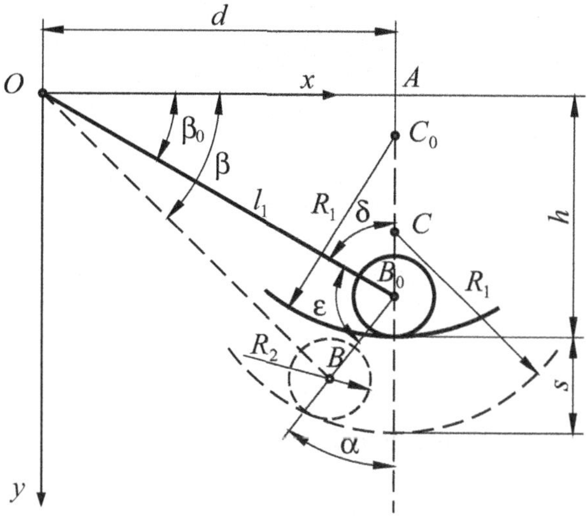

$$\left\{2x_C(l_1\sin\beta - y_C)^2 + (l_1\cos\beta - x_C)[C_1 - 2y_C(l_1\sin\beta - y_C)]\right\}^2$$
$$- \left(l_1^2 + x_C^2 + y_C^2 - 2l_1x_C\cos\beta - 2l_1y_C\sin\beta\right)\left\{4x_C^2(l_1\sin\beta - y_C)^2\right. \tag{43.1}$$
$$\left. + [C_1 - 2y_C(l_1\sin\beta - y_C)]^2 - 4R_1^2(l_1\sin\beta - y_C)\right\} = 0,$$

where

$$C_1 = R_1^2 - R_2^2 + l_1^2 - \left(x_C^2 + y_C^2\right), \quad x_C = d, \quad y_C = h + s - R_1 \tag{43.2}$$

Writing (43.1) as $f(y_C, \beta) = 0$ and applying the implicit functions theorem, we get

$$\frac{\mathrm{d}\beta}{\mathrm{d}y_C} = -\frac{\partial f/\partial y_C}{\partial f/\partial \beta}, \quad \frac{\mathrm{d}\beta}{\mathrm{d}s} = \frac{\mathrm{d}\beta}{\mathrm{d}y_C}, \quad \frac{\mathrm{d}\beta}{\mathrm{d}\varphi} = \frac{\mathrm{d}\beta}{\mathrm{d}y_C}\frac{\mathrm{d}s}{\mathrm{d}\varphi} \tag{43.3}$$

in which φ is the rotational angle of the crankshaft. Similarly, we may write

$$\frac{\mathrm{d}^2\beta}{\mathrm{d}y_C^2} = -\frac{\left(\frac{\partial^2 f}{\partial y_C^2} + \frac{\partial^2 f}{\partial y_C \partial \beta}\frac{\mathrm{d}\beta}{\mathrm{d}y_c}\right)\frac{\partial f}{\partial \beta} - \frac{\partial f}{\partial y_C}\left(\frac{\partial^2 f}{\partial y_C \partial \beta} + \frac{\partial^2 f}{\partial \beta^2}\frac{\mathrm{d}\beta}{\mathrm{d}y_C}\right)}{\left(\frac{\partial f}{\partial \beta}\right)^2} \tag{43.4}$$

$$\frac{\mathrm{d}^2\beta}{\mathrm{d}s^2} = \frac{\mathrm{d}^2\beta}{\mathrm{d}y_C^2}, \quad \frac{\mathrm{d}^2\beta}{\mathrm{d}\varphi^2} = \frac{\mathrm{d}^2\beta}{\mathrm{d}s^2}\frac{\mathrm{d}s}{\mathrm{d}\varphi} + \frac{\mathrm{d}\beta}{\mathrm{d}s}\frac{\mathrm{d}^2 s}{\mathrm{d}\varphi^2} \tag{43.5}$$

An important case is that of the flat head of the valve, that is $R_1 = \infty$. In this situation, the angle β is given by

$$\beta = \arcsin\left(\frac{s + l_1 \sin\beta_0}{l_1}\right) \tag{43.6}$$

The angle α (necessary to calculate the relative velocity) is obtained as

$$\delta = \arctan\left(\frac{h + s - R_1}{d}\right), \quad \varepsilon = \arctan\left(\frac{\sqrt{l_1^2 - R_1^2/2}}{R_1/2}\right), \quad \alpha = \frac{\pi}{2} - \delta - \varepsilon \tag{43.7}$$

43.3 Numerical Study

Two laws for the motion of the valve are chosen:

$$s_1(\varphi) = \begin{cases} 0 \text{ for } \varphi \in [0, \varphi_1), \\ s_{\max} \sin^2\left(\frac{90(\varphi - \varphi_1)}{\varphi_2 - \varphi_1}\right) \text{ for } \varphi \in [\varphi_1, \varphi_2], \\ h \text{ for } \varphi \in (\varphi_2, \varphi_3), \\ s_{\max} \sin^2\left(\frac{90(\varphi + \varphi_4 - 2\varphi_3)}{\varphi_4 - \varphi_3}\right) \text{ for } \varphi \in [\varphi_3, \varphi_4], \\ 0 \text{ for } \varphi \in (\varphi_4, 360], \end{cases} \tag{43.8}$$

$$s_2(\varphi) = \begin{cases} 0 \text{ for } \varphi \in [0, \varphi_1), \\ s_{\max} \frac{(\varphi - \varphi_1)^2 (\varphi - 2\varphi_2 + \varphi_1)^2}{(\varphi_2 - \varphi_1)^4} \text{ for } \varphi \in [\varphi_1, \varphi_2], \\ h \text{ for } \varphi \in (\varphi_2, \varphi_3), \\ s_{\max} \frac{(\varphi - \varphi_4)^2 (\varphi - 2\varphi_3 + \varphi_4)^2}{(\varphi_3 - \varphi_4)^2} \text{ for } \varphi \in [\varphi_3, \varphi_4], \\ 0 \text{ for } \varphi \in (\varphi_4, 360]. \end{cases} \tag{43.9}$$

The following values were selected for the numerical study: $l_1 = 0.05$ m, $R_1 = 0.025$ m, $R_2 = 0.005$ m, $\beta_0 = 30°$, $s_{\max} = 0.012$ m; the angles of the crank at which the valve starts to open, is open, starts to close, and is closed are $\varphi_1 = 80°$, $\varphi_2 = 100°$, $\varphi_3 = 200°$, $\varphi_4 = 220°$; the angular step of iterations $\Delta\varphi = 0.1°$; the angular velocity of the crankshaft $\omega = 40\pi$ rad/s. Different values of these parameters selected in order to highlight their influences are $\varphi_1 = 85°$, $\varphi_2 = 95°$, $\varphi_3 = 205°$, $\varphi_4 = 215°$, $l_1 = 0.06$ m, $s_{\max} = 0.018$ m, $\beta_0 = 60°$, $R_1 = 0.028$ m, $R_2 = 0.006$ m. The results of the simulation are given in Tables 43.1 and 43.2 (Fig. 43.2).

Table 43.1 Variation of different parameters for the first law of motion of the valve

	Standard	β_0	$s_{1_{max}}$	l_1	φ_i	R_1	R_2
β_{max} (°)	44.99832	78.05304	51.25186	42.41806	44.99832	45.27013	44.89266
$(d\beta/d\varphi)_{min}$	−0.00265	−1.50549	−0.00389	−0.00131	−0.00529	−0.00397	−0.00300
$(d\beta/d\varphi)_{max}$	0.00265	1.50549	0.00389	0.00131	0.00529	0.00397	0.00300
$(d^2\beta/d\varphi^2)_{min}$	−0.03793	−59.15634	−0.06385	−0.01878	−0.15172	−0.05982	−0.03717
$(d^2\beta/d\varphi^2)_{max}$	0.02169	19.39334	0.02429	0.01063	0.08678	0.03425	0.02405
$v_{r_{max}}$ (m/s)	4.42571	3.52616	6.42890	3.58590	8.85143	5.0673	4.52262

Table 43.2 Variation of different parameters for the second law of motion of the valve

	Standard	β_0	$s_{2_{max}}$	l_1	φ_i	R_1	R_2
β_{max} (°)	44.99832	78.05304	51.25186	42.41806	44.99832	45.27013	45.27013
$(d\beta/d\varphi)_{min}$	−0.00260	−1.54665	−0.00385	−0.00128	−0.00519	−0.00389	−0.00389
$(d\beta/d\varphi)_{max}$	0.00260	1.54665	0.00385	0.00128	0.00519	0.00389	0.00389
$(d^2\beta/d\varphi^2)_{min}$	−0.03666	−65.31096	−0.05500	−0.01839	−0.14455	−0.05432	−0.05432
$(d^2\beta/d\varphi^2)_{max}$	0.01758	30.92266	0.01986	0.00861	0.07034	0.02776	0.02776
$v_{r_{max}}$ (m/s)	4.36819	3.54977	6.38030	3.53940	8.73638	4.93252	4.93252

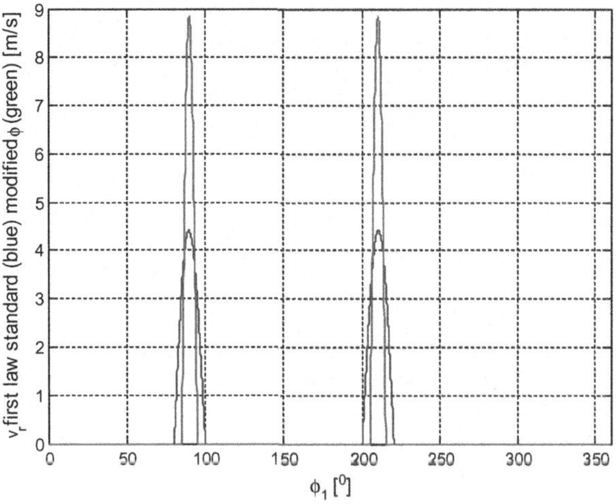

Fig. 43.2 Variation $v_r = v_r(\varphi_1)$ using the standard parameters (blue) and modifying the angles at which the inlet valve opens and closes (green). The moving law of the valve is the first one

43.4 Conclusions

The wear at the contact between the lever and the head of the valve is determined by the relative velocity v_r and the preload. One may see that the parameter s_{max} (the maximum displacement of the valve) is the most important factor. The reduced acceleration is important in order to determine the real acceleration and, consequently, it may characterized the inertial forces and moments and determine the necessary preload. In addition, the reduced velocity is a measure of the wear in the revolute joint. The reader may see that the factor with the greatest influence on the reduced velocity and acceleration is the initial angle β_0.

References

1. G. Gonca, B. Sahin, A. Parlak, Y. Ust, V. Ayhan, I. Cesur, B. Boru, Theoretical and experimental investigation of the Miller cycle diesel engine in terms of performance and emission parameters. Appl. Energy **138**, 11–20 (2015)
2. M.E.S. Martins, T.D.M. Lanzanova, Full-load Miller cycle with ethanol and EGR: potential benefits and challenges. Appl. Therm. Eng. **90**, 274–285 (2015)
3. R. Ebrahimi, Effects of mean piston speed, equivalence ratio and cylinder wall temperature on performance of an Atkinson engine. Math. Comput. Model. **53**, 1289–1297 (2011)
4. J.P. Zammit, M.J. McGhee, P.J. Shayler, T. Lawa, I. Pegg, The effects of early inlet valve closing and cylinder disablement on fuel economy and emissions of a direct injection diesel engine. Energy **79**, 100–110 (2015)
5. Y. Murata, J. Kusaka, M. Odaka, Y. Daisho, D. Kawano, H. Suzuki et al., Achievement of medium engine speed and load premixed diesel combustion with variable valve timing. SAE paper 2006-01-0203 (2006)
6. W. De Ojeda, Effect of variable valve timing on diesel combustion Characteristics. SAE paper 2010-01-1124 (2010)

7. A. Kéromnčs, B. Delaporte, G. Schmitz, L. Le Moyne, Development and validation of a 5 stroke engine for range extenders application. Energy Convers. Manag. **82**, 259–267 (2014)
8. S. Zhu, K. Deng, S. Liu, S. Qu, Comparative analysis and evaluation of turbocharged Dual and Miller cycles under different operating conditions. Energy **93**, 75–87 (2015)
9. S. Mihalcea, N.-D. Stănescu, D. Popa, Synthesis and kinematic and dynamic analysis of a variable valve lift mechanism with general contact curve. Proc Int. Inst. Mech. Eng. Part K J Multi-body Sys. **229**(1), 65–83 (2015)
10. N.-D. Stănescu, D. Popa, The vibrations of the engine with Neo-Hookean suspension. Acoust. Vib. Mech. Struct. Appl. Mech. Mater. **430**, 53–59 (2013)
11. N.-D. Stănescu, D. Popa, Stability of the equilibrium positions of an engine with nonlinear quadratic springs. CEJE **4**(2), 170–177 (2014). doi:10.2478/s13531-013-0138-1
12. N.-D. Stănescu, I. Dragomir, N. Pandrea, A. Clenci, D. Popa, Geometric constraints at the distribution mechanism with spherical contact between the lever and the head of the valve. in *Proceedings of International Automotive Congress, CONAT 2016* (2016), pp. 43–51. ISBN 2069-0401

Chapter 44
The Transitory Vibrations for a Variable Compression Ratio Mechanism

Bogdan Mănescu, Ionuț Dragomir and Nicolae-Doru Stănescu

Abstract This paper considers the case of a Variable Compression Ratio (VCR) mechanism for an internal combustion engine. The mechanism consists in the crank, the shaft, a triangular plate which links the previous two parts one to another and connected to a lever. One end of the lever has a vertical translation motion, the cause of its displacement being the practical conditions of driving. For constant conditions the lever acts as a constraint which states that an apex of the triangular plate must be situated on a circle. In this case one deals with a mechanism with one degree of freedom, this degree being given by the angle of rotation of the shaft. If the conditions change, then the end of the lever moves on the vertical direction, the mechanism having now two degrees of freedom: the angle of rotation of the shaft and the vertical displacement of the lever. When an optimal condition of functioning for the engine is obtained, then the vertical motion of the end of the lever stops and the mechanism becomes again a one degree of freedom one. In the cases of one degree of freedom mechanism the vibrations of the mechanism may be studied similar to the vibrations of a simple one degree of freedom system, the parameters of vibrations being different and depending on the position of the end of the lever. The main goal of this paper is to present the transitory vibrations for the two degrees of freedom system which occur during the vertical motion of the one end of the lever. The study is realized using a multibody approach and the numerical results are presented and discussed. The transitory vibrations start when the end of the lever starts its motion and they end after the lever stops its motion. The dependence of the length of the interval of time of transitory vibrations as function of the geometric dimensions of the elements and the law of motion of the end of the lever is also highlighted by the authors.

B. Mănescu · I. Dragomir
AKKA ROMSERV, Bucharest, Romania
e-mail: bogdan.manescu04@gmail.com

I. Dragomir
e-mail: ionut.dragomir1988@gmail.com

N.-D. Stănescu (✉)
University of Pitești, Pitești, Romania
e-mail: s_doru@yahoo.com

N. Herisanu and V. Marinca (eds.), *Acoustics and Vibration of Mechanical Structures—AVMS-2017*, Springer Proceedings in Physics 198,
https://doi.org/10.1007/978-3-319-69823-6_44

44.1 Introduction

Different methods to obtain VCR engines are presented in [1]. A type of an articulated engine block engine is the Hara engine developed by professor Hara from University of Pitesti. Professor Hara's research leads to registering several patents numbers 111863B, 109770C, 104027, 96876, in Romania. The maximum compression ratio 12.5:1 obtained at the idle speed; when the engine load increases, the upper part of the engine is moving and the volume of combustion chamber increases, while the compression ratio decreases down to 8.5:1 at full load. Regulation of compression ratio is self-made by the level of cylinder pressure. This type of engine keeps an unmodified combustion chamber and a several series parts [2]. The SVC engine's cylinder block is divided into two parts. The lower part is the crankcase, where crankshaft, con-rods, pistons etc. are mounted. The upper part is what SAAB calls the "Monohead", is where the combustion chamber, valves, cylinder liners etc. are situated. The Monohead can be tilted relative to the crankcase through a pivot shaft. The pivot shaft location is a compromise between the previous requirements avoiding too much cylinder axial variation and keeping the engine width down. Tilting of the Monohead (enabling the variation of the CR) is controlled by a device which is called the eccentric shaft mechanism by SAAB. The eccentric shaft is turned to the chosen position by a hydraulic actuator situated at the front of the engine. A sensor indicates the position of the eccentric shaft and thereby the actual CR. A separate hydraulic system supplies hydraulic pressure between 60 and 100 bars to the hydraulic actuator [3]. Ford patented a method to modify combustion chamber volume by using a secondary piston. The motion of the piston creates an optimum compression ratio according to engine load. The volume of combustion chamber is increased to reduce the compression ratio by moving secondary piston witch communicates with combustion chamber. The cylinder head cooling needs to be improved and the auxiliary piston needs proper lubrication [4]. The piston crow deflects according cylinder pressure with modifications of compression ratio. This mechanism limits peak pressure at high loads without an additional control device, allowing engine to operate with a high compression ratio at low load conditions. It can be adapted to convectional engines with only changes of piston and connecting road design. The main reason to apply the VR/LE concept to SI car engines is to gain the engine efficiency at partial loads by increasing the compression ratio up to that limited by the occurrence of the knock [5]. Using the method of separation of kinematic structure and function, Freudenstein and Maki [6] developed a search procedure in order to determine the possible structures of the VCR mechanisms.

44.2 Mechanical Model

The mechanical model is presented in Fig. 44.1. It consists in the shaft *OC*, intermediate triangular element *ABC*, crank *BD*, piston situated at point *D*, and control element *CE*. The crankshaft has an angular velocity ω. The control element has a determined motion considered to be known $Y_E = Y_E(t)$. The system has five elements and, that is, it has maximum 15 possible degrees of freedom. The position of the mechanism is described by the positions of the centers of weight of the five elements, X_{C_i}, Y_{C_i}, and their rotational angles φ_i, $i = \overline{1,5}$. The following constraint functions may be written: point *A* belongs to the elements 1 and 2, point *B* to the elements 2 and 3, point *C* to the elements 2 and 4, point *D* to the elements 3 and 5,

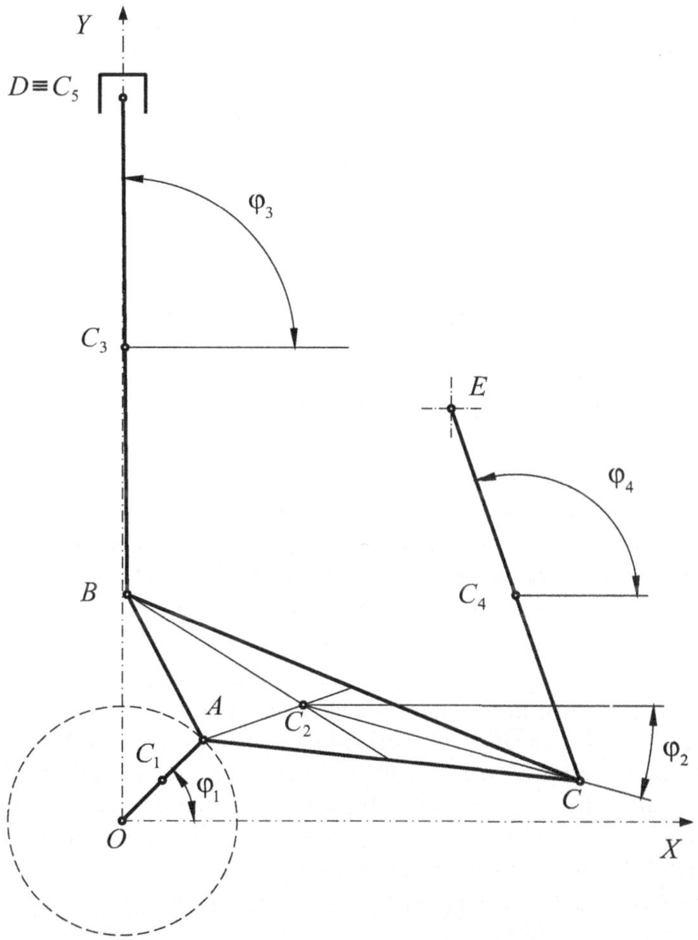

Fig. 44.1 Mechanical model

the coordinates X_E and X_D are known, coordinates X_{C_1} and Y_{C_1} are harmonic functions in φ_1 (12 functions), the rotation of the element 5 is equal to zero (1 function). It results maximum 2 possible degrees of freedom. In addition, if the point E is fixed (no control), then a new constraint function appears (coordinate Y_E is known) and the mechanism has only one degree of freedom. Another variant of the mechanism is characterized by a displacement e between the Y-axis and the motion direction of the piston. In addition, the coordinate X_E is always fixed and denoted by d.

44.3 Numerical Simulation

The following values were selected for the numerical simulation: $\omega = 40\pi\,\text{rad/s}$, $AB = 0.043$ m, $BC = 0.128$ m, $CA = 0.099$ m, $CE = 0.103$ m, $OA = 0.030$ m, $BD = 0.130$ m, $d = 0.086$ m, $e = 0$ m, $Y_E = 0.108$ m, maximum displacement of the end of control lever $\Delta Y_{E_{\max}} = 0.05$ m, maximum control time $t_c = 2$ s, time when control starts $t_{sc} = 2$ s,. Alternative values are as follows $t_c = 4$ s, $CE = 0.120$ m, $BD = 0.150$ m, $AB = 0.060$ m, $Y_E = 0.125$ m.

Two laws of motion are used for the end of the control lever:

$$\Delta Y_{E_1}(t) = \begin{cases} 0 \text{ for } 0 < t_{sc}, \\ \Delta Y_{E_{\max}} \frac{t-t_{sc}}{t_c} \text{ for } t_{sc} \le t \le t_{sc} + t_c, \\ \Delta Y_{E_{\max}} \text{ for } t_{sc} + t_c < t, \end{cases} \tag{1}$$

$$\Delta Y_{E_2}(t) = \begin{cases} 0 \text{ for } 0 < t_{sc}, \\ \Delta Y_{E_{\max}} \left(\frac{t-t_{sc}}{t_c}\right)^2 \text{ for } t_{sc} \le t \le t_{sc} + t_c, \\ \Delta Y_{E_{\max}} \text{ for } t_{sc} + t_c < t. \end{cases} \tag{2}$$

Results of the simulation may be summarized in the following tables. The head of each table marks the parameter which was modified (Tables 44.1 and 44.2).

The initial values are obtained by using a Newton-Raphson procedure and consider as known the initial value for the angle φ_1, that is, $\varphi_1^0 = 45°$. The step for the time is taken $\Delta t = \frac{\pi}{180\omega}$, which correspond to $\Delta\varphi_1 = 1^0$. The derivatives are obtained by numerical calculation using the well known formulae.

The values presented in the tables are the maximum values for the all period of simulation (10 s). The authors also verified these data for the time between t_{sc} and $t_{sc} + t_c$ obtaining only very small differences for some parameters. It results that these maximum values are generally obtained during the transitory regime.

The reader may easily observe that the parameter with the greatest influence is the length AB, following by the length of the control lever CE.

Figure 44.2 presents two diagrams of the transitory regime.

Table 44.1 Variation of parameters for the first law of motion of the end of the control lever

	Standard	*AB*	*BD*	*CE*	t_c	Y_E
a_2 [m/s^2]	78.267	127.181	78.267	79.274	78.271	78.307
ε_2 [rad/s^2]	1049.082	1748.190	1049.082	1062.338	1049.032	1049.579
a_3 [m/s^2]	235.016	381.758	234.981	238.032	234.984	234.921
ε_3 [rad/s^2]	28.706	32.192	24.879	32.165	30.218	26.185
a_4 [m/s^2]	0.028	0.028	0.028	0.030	0.028	0.025
ε_4 [rad/s^2]	0.412	0.412	0.412	0.396	0.408	0.388
a_5 [m/s^2]	235.243	382.044	235.174	238.258	235.265	235.189
ω_{12} [rad/s]	10.482	16.580	10.482	10.437	10.481	10.482
ω_{23} [rad/s]	9.568	15.699	9.544	9.575	9.586	9.687
ω_{24} [rad/s]	9.520	15.618	9.520	9.486	9.520	9.586

Table 44.2 Variation of parameters for the second law of motion of the end of the control lever

	Standard	*AB*	*BD*	*CE*	t_c	Y_E
a_2 [m/s^2]	92.549	127.129	92.549	127.692	78.244	78.282
ε_2 [rad/s^2]	1236.370	1747.465	1236.370	1688.029	1048.685	1049.285
a_3 [m/s^2]	277.941	381.644	277.896	383.090	234.930	235.061
ε_3 [rad/s^2]	28.205	28.926	24.444	28.967	27.789	26.709
a_4 [m/s^2]	0.028	0.028	0.028	0.030	0.028	0.025
ε_4 [rad/s^2]	0.408	0.408	0.408	0.367	0.408	0.387
a_5 [m/s^2]	278.228	381.975	278.139	383.084	235.207	235.289
ω_{12} [rad/s]	10.481	16.579	10.481	13.077	10.481	10.482
ω_{23} [rad/s]	10.126	15.720	10.149	11.941	9.586	9.685
ω_{24} [rad/s]	10.340	15.618	10.340	12.252	9.520	9.586

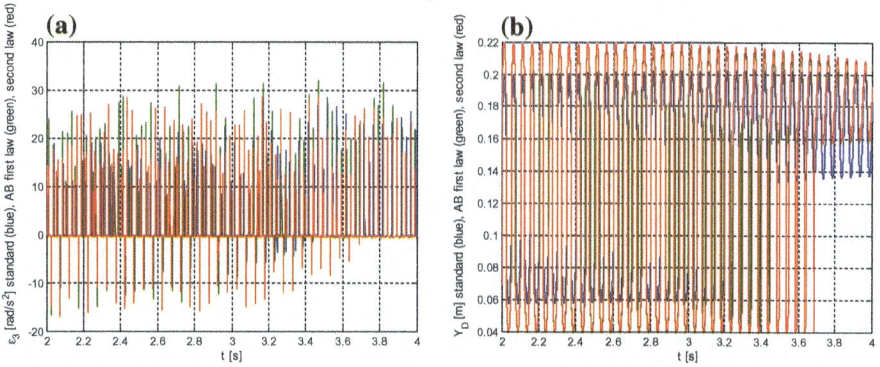

Fig. 44.2 Time history for the **a** angular velocity of the third element; **b** displacement of the piston, during the transitory regime: standard mechanism using first law of motion (blue), mechanism with the modified length of the segment *AB* using the first law of motion for the control element (green) and the second law of motion for the control element (red)

44.4 Conclusions

This paper presents the influence of different parameters on the positions, velocities and accelerations of the elements. Two laws for the displacement of the end of the control lever are used. The influences of different parameters on the transitory vibrations are similar, for the two laws. The most important parameters are the lengths of the elements *AB* and *CE* (the length of the control lever). The time in which the control acts is important for the second law. Increasing this time, the mechanical parameters (velocities and accelerations) decrease. This increasing of the control time can be used at regular cars (family cars, for instance), but it has to be avoided at the sport cars, where the knock phenomenon is expected. Modifying the position of the control lever one may change the working point of the engine. In Fig. 44.2, the reader may observe the modifications of the extreme points of the piston, which influence the compression ratio and the volume of the compression chamber of the engine. These modifications are required in order to avoid the knock phenomena. For each case, one other parameter is modified, the main results being given in tables.

References

1. A. Gabor, Contributions in determining the dynamic response of the variable compression ratio engines, PhD thesis, 2012
2. V. Hara, N. Pandrea, D. Popa, M. Stan, S. Boncea, Motoare termice adaptative, Editura Universitătii din Pitesti (1995)
3. I. Miller, SAAB variable compression motor, available online at the address. www.me.udel.edu/meeg425/SaabVarComp.doc (2001)
4. P.J. Mahlesh, V.K. Aparna, Variable compression ratio engine—a review of future power plant for automobile. Int. J. Mech. Eng. Res. Dev. (IJMERD), **2**(1) (2012)
5. A. Teodorczyk, Variable compression ratio engine VR/LE concept, published on Institute of heat engineering Warsaw in 1995
6. F. Freudenstein, E.R. Maki, Development of an optimum variable-stroke internal-combustion engine mechanism from the viewpoint of kinematic structure, J. Mech. Transm. Autom. Des. **105** (1983)

Chapter 45
Gear Drive System Simulation of Input Parameters Effect on Rattle

Aleš Prokop, Barbora Kopečková and Kamil Řehák

Abstract The understanding of gear dynamics is essential to design the power transmission with low vibration and noise radiation level. The virtual prototype is proposed for transmission dynamic investigation. This article deals with investigation of input parameters effect on gear rattle. The rattle noise is caused by hitting the tooth by each other, which can be noticed by vibrations evaluation on the gearbox housing. The evaluation is done by surface normal velocity comparison, which is connected to the emitted noise. The simulation results show that the torque and backlash has influence on the tooth contact loosening at the starting of rotation.

45.1 Introduction

Gear drive systems are one of the most widespread machinery mechanisms. Vibration and emitted noise of gearboxes significantly affects people at usage of final product. For each manufacturer, it is crucial to design and then manufacture the transmission, which should meet current limits of final product and also for the purpose to be competitive on the market. The understanding of the gear mesh dynamics is key to design the components with the required noise and vibration limits. Manufacturing of whole transmission and its experimental testing is costly in

A. Prokop (✉) · B. Kopečková · K. Řehák
Faculty of Mechanical Engineering, Brno University of Technology, Technicka 2896/2, 61669 Brno, Czech Republic
e-mail: prokop.a@fme.vutbr.cz

B. Kopečková
e-mail: Barbora.Kopeckova@vutbr.cz

K. Řehák
e-mail: rehak@fme.vutbr.cz

N. Herisanu and V. Marinca (eds.), *Acoustics and Vibration of Mechanical Structures—AVMS-2017*, Springer Proceedings in Physics 198,
https://doi.org/10.1007/978-3-319-69823-6_45

terms of both time and material costs, which also significantly affects the final product price. The computational approach is another way how to perform optimization of transmission without manufacturing physical prototypes.

The research of the gear systems is very spread with focused on the description of the dynamic behaviour. The investigation is mainly performed on the single stage gearbox with simplified housing [1]. The main excitation is caused by time-varying gear mesh stiffness [2]. The goal of the gear manufacturer is teeth optimization to reach smooth gear mesh stiffness.

45.2 Methods

With modern computer technologies the numerical simulation enables to perform sensitive study to deep understand dynamic behaviour of gear, for that reason it is widely used in research and at developing phase of transmission [3]. The presented virtual prototype is based on dynamic model with multiple degree of freedom, which is shown in Fig. 45.1. This model incorporates influence of the gear mesh stiffness, which is expressed as time-varying function. Further the elasticity of shafts and the bearings cannot be normally neglected. The equation of motion of a dynamic system is based on Newton–Lagrange method and can be symbolically written by (45.1).

$$\boldsymbol{M}\ddot{\boldsymbol{q}}+\dot{\boldsymbol{M}}\dot{\boldsymbol{q}}-\frac{1}{2}\left[\frac{\partial \boldsymbol{M}}{\partial \boldsymbol{q}}\dot{\boldsymbol{q}}\right]^T\cdot\dot{\boldsymbol{q}}+\boldsymbol{C}\boldsymbol{q}+\boldsymbol{f}_g+\boldsymbol{B}\dot{\boldsymbol{q}}+\left[\frac{\partial \boldsymbol{f}}{\partial \boldsymbol{q}}\right]^T\cdot[\boldsymbol{\lambda}]=\boldsymbol{Q}, \qquad (45.1)$$

where $\boldsymbol{M}$ is the generalized mass matrix, which is in (45.1) related to kinetic energy. $\boldsymbol{C}$ is generalized stiffness matrix, related to potential energy and damping matrix $\boldsymbol{B}$ is related to Rayleigh's dissipative function. Generalized gravitational force is presented by $\boldsymbol{f}_g$, vector $\boldsymbol{q}$ stands for generalised coordinates, $\boldsymbol{Q}$ is vector of forces and $\boldsymbol{\lambda}$ are Lagrange's multipliers.

In the presented paper the combination of finite element method (FEM), which is used for getting input parameters, and multibody software ADAMS is utilized for simulation. Whole methodology is validated with technical experiment, the gearbox parameters and process of validation is described in detail in [4]. The whole virtual prototype in ADAMS is performed as open code to include and also simple change all substantial parameters for different transmission. It enables to include modal properties of the key components, bearings stiffness, gear mesh stiffness, backlash, variable input speed and torque.

The FEM is used to calculate the bearing stiffness on one rolling segment and afterwards converted to the whole bearing, which is used in ADAMS. This approach is used in another research studies [5]. The gear mesh stiffness is very important to get real dynamic behaviour of transmission. To get gear mesh stiffness, the commercial software can be used, but for the aim of open code usage, the

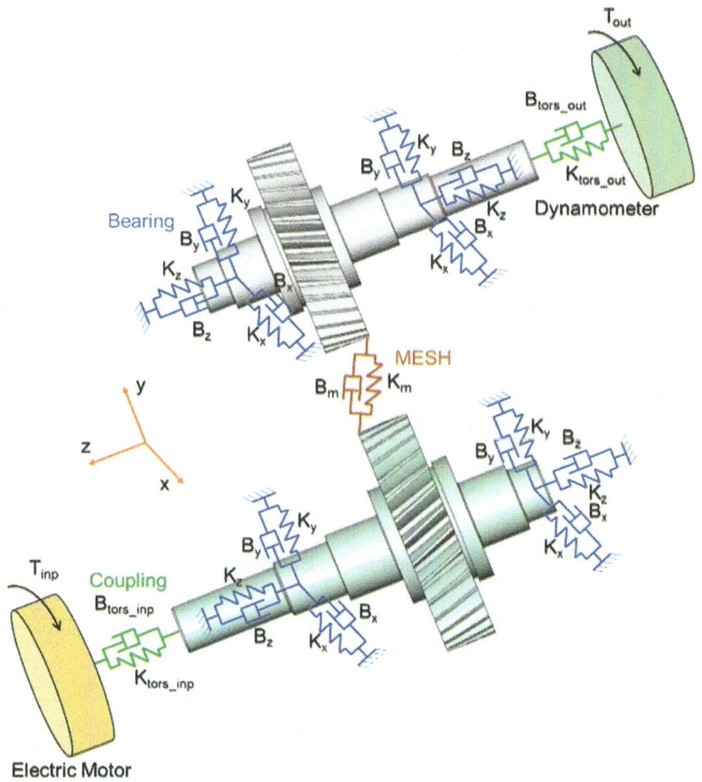

Fig. 45.1 Functional scheme of moving parts of virtual prototype

parametrical FE model is used. The gear tooth real geometry is discretized with focus on fine mesh on four teeth gear pairs. The stiffness is calculated during contact of one tooth angle period and afterwards used periodically. No misalignment and manufacture error are presented. The stiffness shown in Fig. 45.2 corresponds to the results mentioned in [6].

Backlash influence is incorporated by stiffness-based function with dependency on gear mesh deformation in torsional direction, which is stated by angular displacement of both gears. The function for specific load is shown in Fig. 45.3. Backlash has a significant influence on the dynamic of transmissions due to nonlinear behaviour. Because of this phenomenon the gear mesh contact can be in some certain states released periodically, which is accompanied by impact forces. These critical states can occur for example while relatively low level of carried torques.

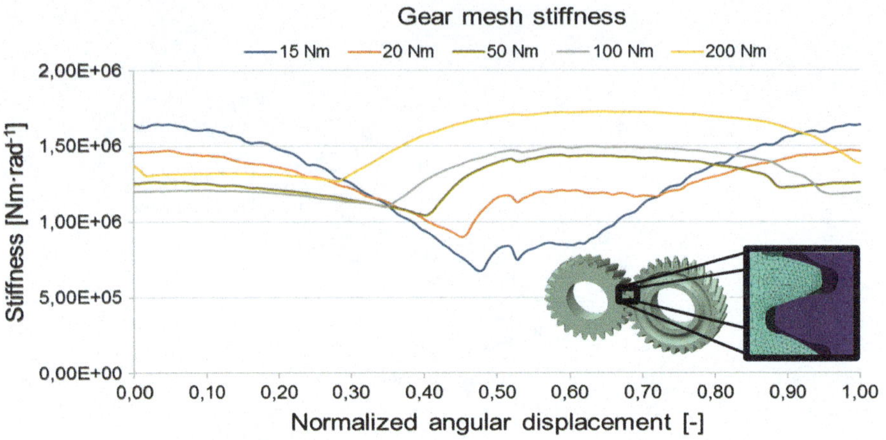

Fig. 45.2 Gear mesh stiffness

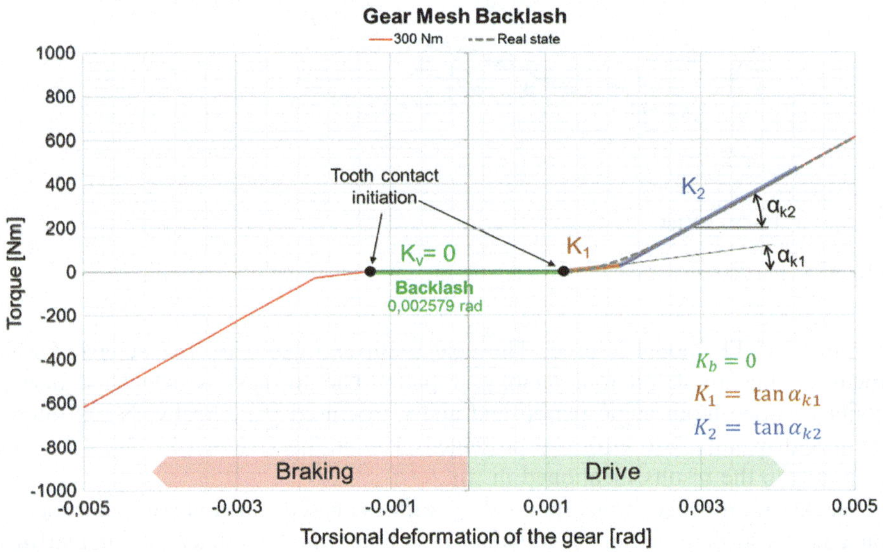

Fig. 45.3 Gear mesh backlash

For the reason to be able to evaluate surface normal velocity, it is necessary to include modal properties of gearbox housing. For that purpose, the Craig-Bampton modal reduction method is used, which replaces real modal properties with simplified approximation established from the two variants of degree of freedom and multiplied by the special Craig-Bampton transform matrix, see (45.2):

$$\{u_A\} = \begin{Bmatrix} u_b \\ u_L \end{Bmatrix} = \begin{bmatrix} I & 0 \\ \Phi_R & \Phi_L \end{bmatrix} \begin{Bmatrix} u_b \\ q \end{Bmatrix} = \Phi_{CB} \begin{Bmatrix} u_b \\ q \end{Bmatrix}, \tag{45.2}$$

where $\boldsymbol{u_A}$ is the original vector of deformation, $\boldsymbol{u_b}$ are the boundary degrees of freedom (DOFs), $\boldsymbol{u_L}$ stands for interior DOFs, $\boldsymbol{q}$ represents modal DOFs, $\boldsymbol{I}$ is the identity matrix element, $\boldsymbol{\Phi_R}$ is rigid body matrix element and in the analogical way $\boldsymbol{\Phi_L}$ are the fixed base mode shapes matrix element.

45.3 Results

The dynamic simulations are performed for different values of torque and backlash to cover whole operating range. The input speed is increasing in range 0 ÷ 3000 RPM without any variation. The backlash is related to the original one, which is measured on the physical single stage gearbox prototype. That is marked as 100%. The backlash range is from 0%, which represents ideal (unreal), to 200%.

The differences between the backlash ratio variants 0%; 100% and 200% at constant loading moment 6 Nm; 30 Nm and 100 Nm are shown in Figs. 45.4 and

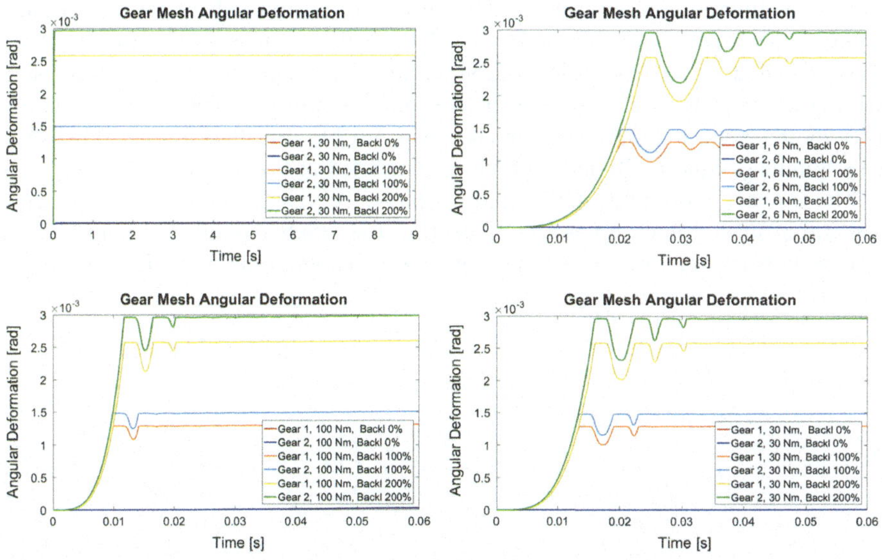

Fig. 45.4 Angular deformation for loading moment 6 Nm; 30 Nm and 100 Nm

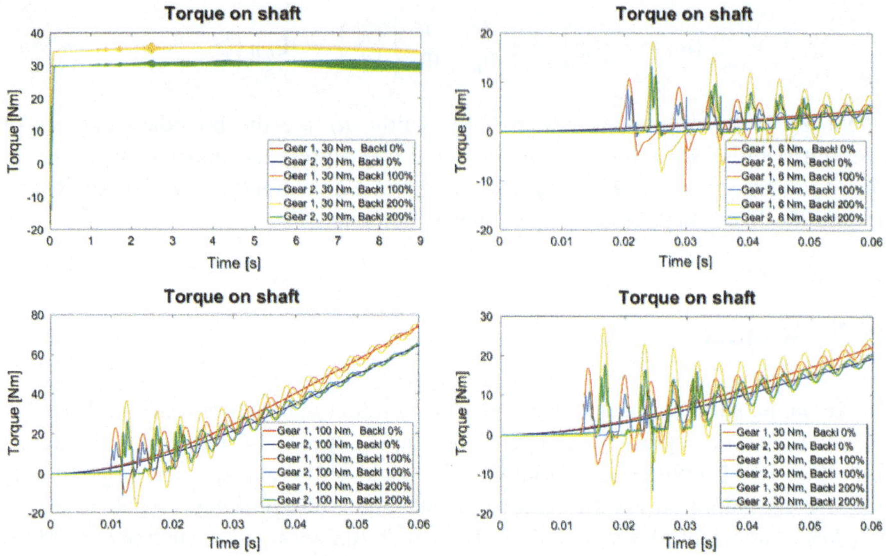

Fig. 45.5 Torque on shaft for loading moment 6 Nm; 30 Nm and 100 Nm

45.5. The difference is only at the beginning of rotation, where with increasing backlash the oscillation value of angular deformation increases and also the peaks number increases. That represents tooth contact loosening. In case of decreasing loading moment the trend of contact loosening is similar to increasing backlash. The tooth contact loosening can be also visible on the torque, Fig. 45.5. The nominal torque 6, 30 and 100 Nm is presented at 0.1 s.

The contact loosening at the starting phase is visible on the multispectrum diagram of surface normal velocity on the top cover middle point. The amplitude of surface normal velocity at the starting phase is increasing with the increasing backlash and torque, see Fig. 45.6.

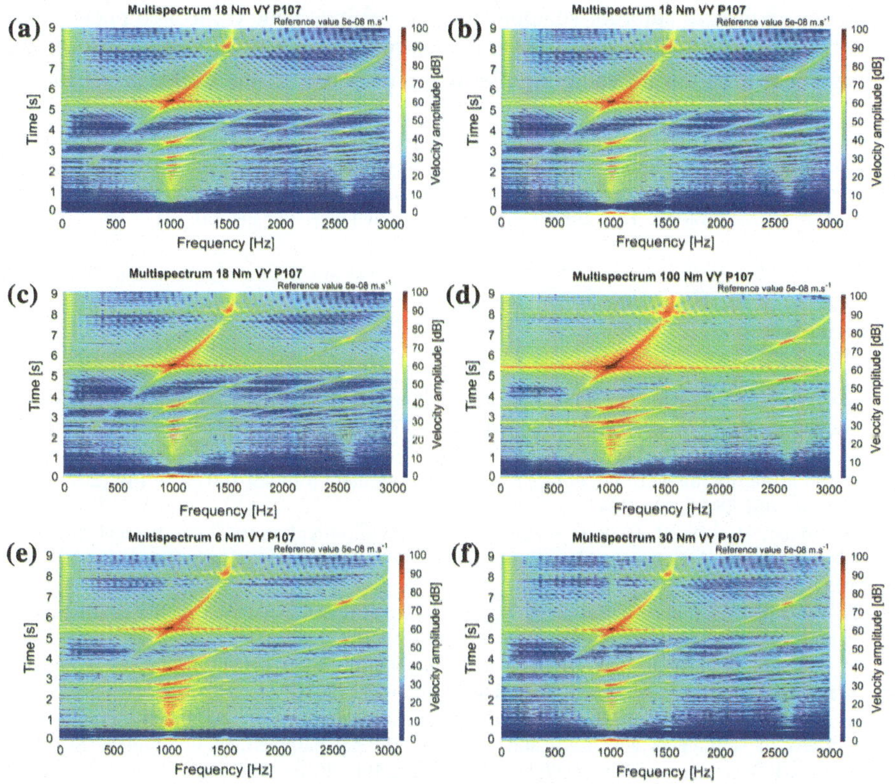

Fig. 45.6 Multispectrum diagram of surface normal velocity of top cover midle point for backlash **a** 0%; **b** 100%; **c** 200%; **d** 200%; **e** 200%; **f** 200%

45.4 Conclusions

Presented methodology, which consists of connection FEM and MBS, is developed on the single stage gearbox. The simple modifiability for use on a wide range of gears is due to open code. That enables to perform sensitive study of wide range of parameters. Presented study is focused on the effect of backlash and torque on the rattle. The course of angular deformation, torque and surface normal velocity is used for evaluation.

The teeth contact loosening, which is connected to the rattle, is presented at the rotation starting and do not have significant effect on dynamic behavior at high speed. The main reason of this statement is because of no variable input speed is applied, which is not typical in real conditions. The backlash has significant effect on the transmission dynamic behavior at the starting phase while the torque is applied. This simulation is useful for cases when the initial starting position of the gears include a backlash. While closing the backlash and initiation of gear mesh contact is mainly in automotive application investigated and controlled phenomena. It is because of possibility of impact excitation of structure in early stage of gear engagement. The torque value highlights the response on the backlash value.

Acknowledgements The research leading to these results has received funding from the Ministry of Education, Youth and Sports under the National Sustainability Programme I. (Project LO1202) and with help of the project FSI-S-17-4104 granted by specific university research of Brno University of Technology. The authors gratefully acknowledge this support.

References

1. M. Byrtus, V. Zeman. On modeling and vibration of gear drives influenced by nonlinear couplings. Mech. Mach. Theory [online] **46**(3), 375–397 (2011). ISSN 0094114X
2. J. Zhan, M. Fard, R. Jazar, A CAD-FEM-QSA integration technique for determining the time-varying meshing stiffness of gear pairs. Measurement [online] **100**, 139–149 (2017) [vid. 2017-03-06]. ISSN 02632241
3. A. Kumar, H. Jaiswal, R. Jain, P.P. Patil, Free vibration and material mechanical properties influence based frequency and mode shape analysis of transmission gearbox. Proc. Eng. **97**, 1097–1106 (2014)
4. A. Prokop, K. Rehak, M. Zubik, P. Novotny, Experimental validation of the gearbox NVH parameters. J. Middle Eur. Constr. Des. Cars **13**(2), 16–21 (2015)
5. P. Novotny, V. Pistek, New efficient methods for powertrain vibration analysis. Proc. Inst. Mech. Eng. Part D J. Automob. Eng. **224**(5), 611–629 (2010)
6. T. Kiekbusch, D. Sappok, B. Sauer, I. Howard, Calculation of the combined torsional mesh stiffness of spur gears with two- and three-dimensional parametrical FE models. Strojniski Vestnik/J. Mech. Eng. [online] **57**(11), 810–818 (2011). ISSN 00392480

Chapter 46
Experimental Study of Road Traffic Vibrations Impact on Heritage Buildings in Braila, Romania

Mihaela Picu and Laurentiu Picu

Abstract In this paper we analyse only the road traffic influence because—on the long-term—these vibrations cause irretrievable damages on heritage buildings. Braila is a true open-air museum; here there are lots of old buildings of undeniable beauty placed near roads with heavy traffic. In traffic, most vibrations occur because of trucks and buses. Cars and minibuses vibrations do not have significant impact, but if intense traffic, they should be taken into account. We have analysed the factors influencing the level and frequency of vibrations: the type and condition of roads, type and condition of the car, etc. We analysed the vibrations transmitted to the sidewalks in front of the houses, in the middle and on the interior walls of the ground floor, 1st floor and 2nd floor, for different speeds of the cars. It was found that accelerations of the buildings resistance structures are higher in case of lighter cars running with the same speed. It was found that accelerations of the buildings resistance structures are higher in case of lighter cars running with the same speed. The frequencies for which maximum accelerations were found (around 2 m/s^2) are below 30 Hz. A few solutions have been proposed for reducing and preventing the destruction of buildings and protecting people: road repairs, reducing traffic, barriers built into the ground.

46.1 Introduction

Ever since traffic increased until today menacing vibrations transmitted to the building became acutely research topic: effects on buildings of vibrations caused by traffic have been studied by a number of researchers, for many years [1–4].

M. Picu (✉) · L. Picu
"Dunarea de Jos" Galați University, Galați, Romania
e-mail: mihaelapicu@yahoo.com

L. Picu
e-mail: Constantin.Picu@ugal.ro

N. Herisanu and V. Marinca (eds.), *Acoustics and Vibration of Mechanical Structures—AVMS-2017*, Springer Proceedings in Physics 198,
https://doi.org/10.1007/978-3-319-69823-6_46

Whiffin et al. [5] found that “vibrations are generated by fluctuations of wheel contact loads as vehicles travel over road surface irregularities. Irregularities of the order of 20 mm in amplitude can cause peak particle velocities in the ground of up to 5 mm/s at which level “architectural damage” may occur in buildings. However, before this level is reached, vibrations become intrusive and even annoying to occupants of buildings (at about 2.5 mm/s) and complaints may result”. Hunt [6, 7] studied the transmission of vibrations generated by passing trains: “The vehicle-track model is used in conjunction with a building model of infinite length. The tracking and building are infinite and parallel to each other and forces applied are statistically stationary in space so that vibration levels at any two points along the building are the same”. After analyzing the traffic in Montreal, Al-Hunaidi and Rainer [8] found that “it appears feasible to significantly reduce vibration induced by transit buses by modifying the characteristics of their suspension systems. Modifications would be made to achieve either a small axle hop amplitude or an axle hop frequency that is below the lowest site cutoff frequency in the city.” “Ground-borne vibrations are produced when vehicles pass over these profiles and in some cases they can reach perceptible levels in adjacent buildings. Measurements of peak particle velocity have been taken alongside a selection of hump and cushion designs using a range of vehicles under controlled driving conditions [9].”

Any external mechanical factor: strong wind, earthquake, auto, rail or air traffic, etc. represents a major danger to buildings and particularly to historical buildings. Also we should consider the vibrations effect on people who are working in these buildings: on their health and on their work (hospitals, research laboratories, libraries, schools, theatres, etc.).

In this paper we analyse only the influence of road traffic because—on the long-term—these vibrations caused irretrievable damages on heritage buildings. Braila is a city in Romania located on the Danube River, before forming the Danube Delta; it is a true open-air museum; here are lots of heritage buildings, of undeniable beauty.

The vibrations produced by cars in traffic travels through the ground towards the buildings. Road condition is extremely important: holes, cracks, bumps, sewers which are not at road level. All this lead to additional load on the road when cars are passing over. These loads propagate through the ground and reach the foundations of buildings, causing them to vibrate. The highest vibrations are produced by heavy cars (buses, trucks, etc.). Cars and minibuses vibrations do not generate significant vibrations, but when intense traffic, they should be taken into account. Also, the smaller the distance between the buildings and the road surface is, the higher the risk is for the buildings. Vehicle traffic around the building produces random, unsteady and intermittent vibrations. The transmission of vibrations from vehicles to buildings is through the ground. The purpose of this paper is to experimentally determine the vibrations transmitted to a heritage building, situated in the proximity of one of the main arteries of the city, with heavy traffic. Also we will suggest solutions for reducing these vibrations.

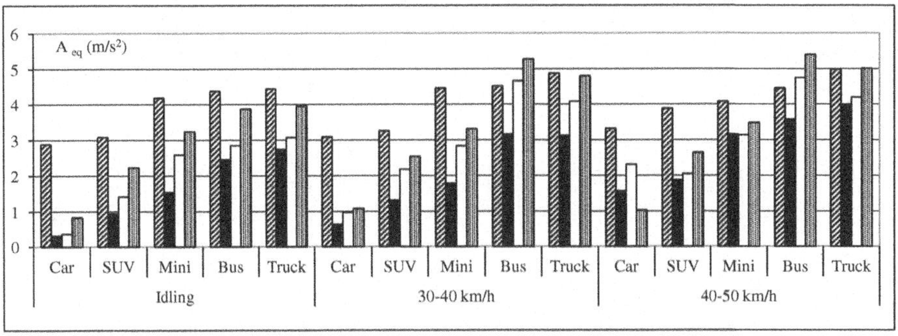

Fig. 46.1 Equivalent acceleration A_{eq} for different vehicles at 3 different situations: idling, speed range 30–40 km/h and speed range 40–50 km/h, in 4 positions: (//) sidewalk; (■) basement; (□) 1st floor; (▩) 2nd floor

46.2 Materials and Methods

The considered building was built in the late 19th century, and was originally the City Court. In this building activities are engaged in basement, ground floor and 1st floor; there are also private offices on the 2nd floor. The total area of all the rooms is 5400 m^2. We analysed the level and frequency of vibrations due to road traffic considering the type and condition of the roads, type and condition of the cars etc. We analysed the vibrations transmitted to the sidewalk in front of the building, in the middle of the ground floor, of the basement and of the 1st floor, for different speeds and type of the cars.

To measure the vibrations we used triaxial accelerometer Type 356A16 from PCB Piezotronics, Net dB 12 acquisition unit from 01 dB and dBFA software suite 4.810 (Fig. 46.1). A handheld Pocket Radar was used to measure the speed of vehicles. The number of vehicles on different time intervals was determined: 6–10AM, 10AM–1PM, 1–6PM from March to July and from September to November 2016. The experimental measurements were made in 5 places: on the sidewalk, in front of the building, in the middle of the hallways in the basement, 1st floor and 2nd floor. The distances from the road axis to each point of determination was measured.

The vibrations generated by different types of vehicles (cars, SUV, minibuses, buses, trucks) were determined for the following situations: Idling and driving on asphalt with two speed ranges: 30–40 km/h and 40–50 km/h, Idling (this was necessary because around the building vehicles are making parking maneuvers, stay at crosswalks, enter or exit the municipal parking garage, etc.). For each of these situations accelerations were measured on all the 3 axes and the weighted r.m.s. acceleration A_w and equivalent acceleration A_{eq} were calculated.

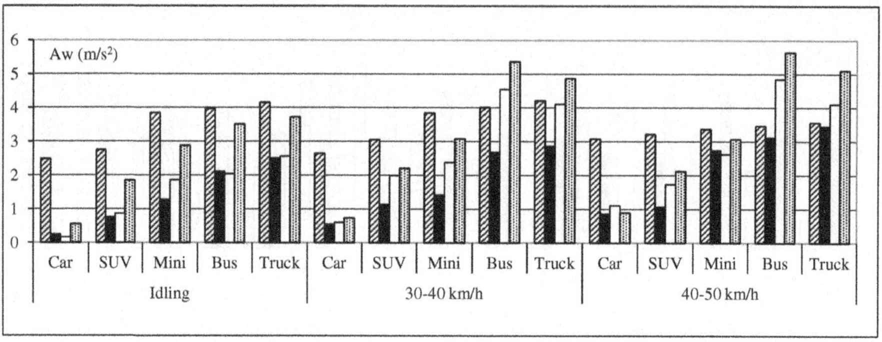

Fig. 46.2 Weighted r.m.s. acceleration Aw for different vehicles at 3 different situations: idling, speed range 30–40 km/h and speed range 40–50 km/h, in 4 positions: (//) sidewalk; (■) basement; (□) 1st floor; (▩) 2nd floor

46.3 Results and Discussions

From the measurements it was found that vibrations transmitted by road traffic have frequencies between 5–20 Hz. The measured accelerations depend on a series of factors: the type of the road (with all kinds of imperfections), the type of the car (big and heavy vehicles or smaller cars), the type of suspensions (pneumatic or steel springs), the load, the weather conditions (dry asphalt or snowy or icy asphalt), the distance from the road axis to the building, etc. It is extremely difficult to find a relation of dependency between all these factors. A special case is when the frequency of the vibrations transmitted by the cars is equal to the frequency of different parts of the building. Then phenomenon of resonance appears and the vibrations are amplified. Figures 46.1, 46.2 and 46.3 shows that the lowest values of accelerations were obtained for the light vehicles: cars and SUVs. Yet their influence is considerable because they outnumber the buses and the trucks. The minibuses generate vibrations with higher accelerations but their number is small enough not to pose problems to the building structure. A worrying situation is the presence of a large number of trucks and buses that generate harmful vibrations to the building structure.

Also, it is seen that the maximum values of the building vibrations occurred, so resonance phenomena occur in all cases where the buses and the trucks have passed in front of the building, as follows (Table 46.1).

Although these values of road traffic accelerations fall within the limits set by the professional standards, the condition of the road must be monitored constantly, because the presence of holes or improper repairs (humps) can lead to an important increase of the acceleration that leads in its turn to the destruction of heritage buildings. Analysing the accelerations depending on their frequency (Fig. 46.4) the peak is clearly seen at the resonance frequency (6.5 Hz) and at the harmonica (12 Hz).

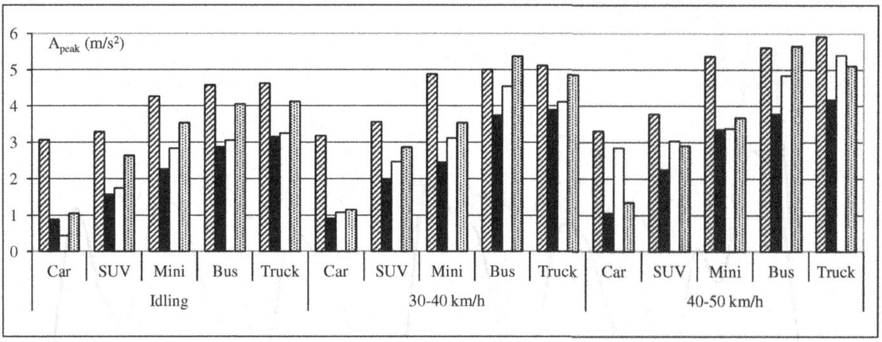

Fig. 46.3 Apeak for different vehicles at 3 different situations: idling, speed range 30–40 km/h and speed range 40–50 km/h, in 4 positions: (//) sidewalk; (■) basement; (□) 1st floor; (▩) 2nd floor

Table 46.1 A_{peak} values measured on the ground floor and on the 1st floor (in m/s^2)

A_{peak}	30–40 km/h		40–50 km/h	
	Buses	Trucks	Buses	Trucks
GF	4.55	4.14	4.85	4.11
1st floor	5.37	4.87	5.65	5.12

For the present case it was also noted that the vibrations level decreases with distance from the road to the point of measurement. Knowing the mechanisms of attenuation of vibrations in the ground:

- Geometric propagation in the case of surface waves: $A_2 = A_1 \cdot (r_1/2r_2)$;
- Geometric propagation in the case of volume waves: $A_2 = A_1 \cdot (r_1/r_2)$;
- Damping by the medium (friction in the ground): $A_2 = A_1 \cdot \exp[\alpha(r_2 - r_1)]$.

From this dependence a model of vibrations propagation by distance could be made, if the soil would be homogeneous and the building regular.

Unfortunately, these conditions cannot be met in reality therefore generally applicable models cannot be made, but only for each particular case.

As opposed to accidental imperfections (holes, sewer caps), road waves or macadam roads generate dynamic loads which act continuously. In most cases, the road has imperfections (potholes, sewer caps, crimps); when a car passes over these imperfections appear an impact load and a load due to oscillation of the vehicle. For example, the vibrations amplitude from busses is higher because of their suspensions; therefore the accelerations will be higher.

Figure 46.4 shows that shocks due to buses are almost 2 times higher than those generated by trucks, but their mitigation time is lower. This is because the trucks suspensions that is more rigid than those of the buses. Another factor which influences the accelerations transmitted to buildings by cars is the weather: it was observed that in winter, when the ground is frozen, the accelerations are lower than

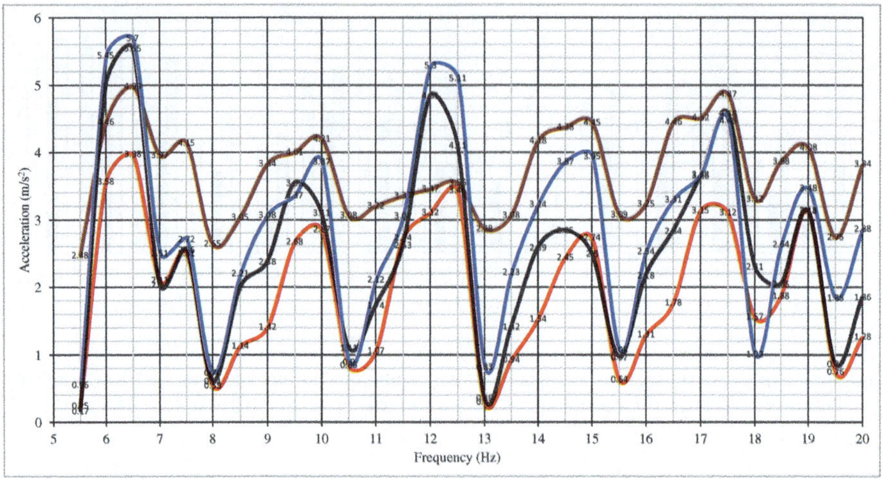

Fig. 46.4 The accelerations spectrum for buses (brown line): Sidewalk; (red line): Basement; (black line): Ground Floor; (blue line): 1st floor

in the summer or fall. During the spring, when the road is covered with water from the melting snow, the accelerations have slightly higher values.

46.4 Conclusions

As we have seen, heavy traffic is one of the causes that lead to irrecoverable damages of heritage buildings: cracks in walls or ceilings, walls peeling, etc. However vibrations are not the only reasons for these deteriorations because they are not high enough to cause them directly. However, a series of measures are necessary to be taken in order to reduce car transmitted vibrations: road maintenance, sewer covers repairs, reducing the stream of cars and their speed, hardening the roads structure, creating barriers which are built into the ground to decrease the amplitude of the Rayleigh waves that propagate close to the ground surface, etc. For these barriers to be effective (i.e. to reduce the vibrations by approximately 25%), they must have a depth equal to a Rayleigh wavelength, at least. Because the vehicle transmitted vibrations have small frequency, the wavelengths are very high, so these barriers are difficult to build. For the heritage buildings where the distance to the road cannot be increased, the most effective measures remain—besides proper maintenance of the roads—heavy traffic restriction, improving soil structure using the deep mixing method and construction of cement barriers columns.

References

1. M. Bata, Effects on buildings of vibrations caused by traffic. Build. Sci. **6**(4), 221–246 (1971)
2. R. Burdzik, Identification of structure and directional distribution of vibration transferred to car-body from road roughness. J. Vibroengineering **16**(1), 324–333 (2014)
3. A.J. Galczynska, R. Jankowski, Traffic-induced vibrations. The impact on buildings and people. The 9th International Conference "Environmental Engineering", Vilnius, Lithuania, 22–23 May 2014
4. A. Camara, A.M. Ruiz-Teran, Multi-mode traffic-induced vibrations in composite ladder-deck bridges under heavy moving vehicles. J. Sound Vib. **355**(27), 264–283 (2015)
5. A.C. Whiffin, D.R. Leonard, *A survey of traffic induced vibrations, Transport and Road Research Laboratory (TRRL)* (Wokingham, Berkshire United Kingdom, 1971)
6. H.E.M. Hunt, Stochastic modelling of traffic-induced ground vibration. J. Sound Vib. **144**(1), 53–70 (1991)
7. H.E.M. Hunt, Modelling rail vehicles and track for calculation of ground-vibration transmission into buildings. J. Sound Vib. **193**(1), 185–194 (1996)
8. M.O. Al-Hunaidi, J.H. Rainer, Control of traffic-induced vibration in buildings using vehicle suspension systems. Soil Dyn. Earthq. Eng. **15**(4), 245–254 (1996)
9. G.R. Watts, V.V. Krylov, Ground-borne vibration generated by vehicles crossing road humps and speed control cushions. Appl. Acoust. **59**(3), 221–236 (2000)

Chapter 47
On Dynamics of a Dual Vibration Insulation Device Based on Elastomeric Kernel and Rolling Friction Dissipation

Fanel Scheaua and Silviu Nastac

Abstract Nowadays, there are several insulation solutions for bridges or viaducts structures that work based on various insulation principles. The basic principles used in the bridge isolation field are represented by vibration insulation and energy dissipation arising from various dynamic loads acting upon the structure, thus that the devices in use are mainly categorized as isolators or energy dissipation devices. An innovative solution for bridge isolation is described in this paper, which uses a composed system able to use both isolation and dissipation characteristics of these two devices types. It has been designed a particular device model, meaning dual vibration insulation device, which is composed of an elastomeric isolation type system, working in tandem with a dissipative device based on rolling friction. This hybrid device has been built and mounted on a bridge structure achieved on a reduced scale. Thus, the isolated structure has been tested at dynamic actions by generating vibrations within the bridge structure using a special exciting device. According with the obtained results higher acceleration values are recorded at the bridge pier support level, while they are more attenuated at the superstructure level, due to interposing of the dual vibration insulation device between the bridge pier and the superstructure. This type of composed isolation device provides additional degrees-of-freedom within the entire ensemble, thus that the bridge support pier can move along with the foundation soil, which induces the dynamic loads, while the superstructure is able to maintain a steady-state dynamic regime without being affected, in a considerable way, by ground movements. In this paper, are presented the results recorded on the experimental model at the bridge support pier and superstructure level regarding the time and the frequency evolutions of the acceleration signals.

F. Scheaua (✉) · S. Nastac
Engineering and Agronomy Faculty of Braila, MECMET Research Center, "Dunarea de Jos" University of Galati, Galați, Romania
e-mail: fanel.scheaua@ugal.ro

S. Nastac
e-mail: silviu.nastac@ugal.ro

N. Herisanu and V. Marinca (eds.), *Acoustics and Vibration of Mechanical Structures—AVMS-2017*, Springer Proceedings in Physics 198,
https://doi.org/10.1007/978-3-319-69823-6_47

47.1 Introduction

The seismic isolation methods used for building structures consist in solutions for controlling and avoiding the maximum transmission of seismic energy from the ground, foundation to the superstructure level in case of considerable magnitude seismic actions.

In order to ensure an adequate stability degree and safety in operation, special isolation systems against the destructive seismic actions, manifested in dynamic regime on the bridge or viaduct infrastructure constructions have been conceived and developed.

Insulation and dissipation systems used are mechanical systems with controlled elastic and damping characteristics, in accordance with building inertial and rigidity characteristics, which by their action can increase the safety level and the lifetime of a bridge.

By attaching an insulation system to the structure base, a higher flexibility degree is ensured for the upper structure, while the vibration period is greatly increased, so the seismic motions effects at the superstructure level are more attenuated.

47.2 Theoretical Aspects for Dual Vibration Insulation System

A dual insulating system is proposed as an innovative solution consisting of a friction pendulum device that works in tandem with an elastomeric support (Fig. 47.1).

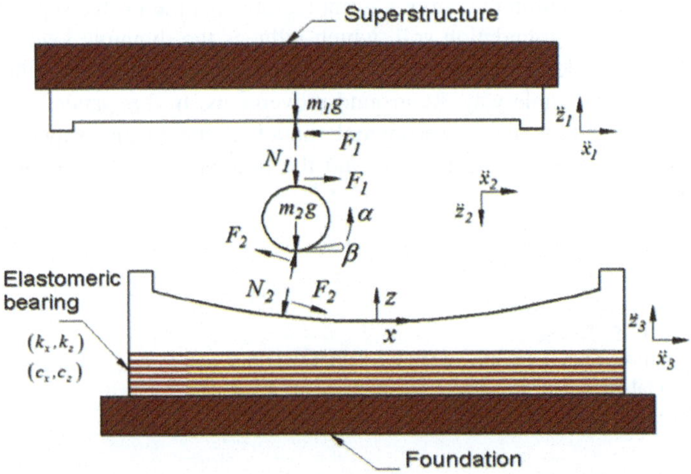

Fig. 47.1 The mathematical model for the dual vibration insulation model

By adding elastomeric bearings between the foundation of a structural system and the bottom plate of the rolling pendulum insulating system in order to optimize the response of the isolated structure to dynamic actions, two equations are added to the equations system obtained for the rolling friction insulating system, which include the elasticity and stiffness characteristics of the elastomeric material [1, 2]:

$$\begin{cases} \ddot{x}_1 = -\cos^2\frac{\beta}{2}\ddot{x}_3 - \frac{1}{2}\sin\beta(\ddot{z}_3 + g)sign(\dot{x}_1) \\ \ddot{z}_1 = \left(-\frac{1}{2}\sin\beta\ddot{x}_3\right)sign(\dot{x}_1) - \sin^2\frac{\beta}{2}(\ddot{z}_3 + g) \\ \ddot{x}_2 = -\frac{1}{2}\cos\beta\ddot{x}_3 + \tan\frac{\beta}{2}(\ddot{z}_3 + g)sign(\dot{x}_1) \\ \ddot{z}_2 = -\frac{1}{2}\sin\beta\ddot{x}_3 + \tan\frac{\beta}{2}(\ddot{z}_3 + g)sign(\dot{x}_1) \\ \alpha = -\frac{1}{2R}\ddot{x}_3 + \tan\frac{\beta}{2}(\ddot{z}_3 + g)sign(\dot{x}_1) \\ F_1 = -F_2 = m_1\left(-\sin^2\frac{\beta}{2}\ddot{x}_3\right) + \frac{1}{2}\sin\beta(\ddot{z}_3 + g)sign(\dot{x}_1) \\ N_1 = N_2 = m_1\left(-\frac{1}{2}\sin\beta\ddot{x}_3\right)sign(\dot{x}_1) + \cos^2\frac{\beta}{2}(\ddot{z}_3 + g) \\ m(\ddot{z}_3 + \ddot{z}_4) + c_z\dot{z}_3 + k_z z_3 = 0 \\ m(\ddot{x}_3 + \ddot{x}_4) + c_x\dot{x}_3 + k_x x_3 = 0 \end{cases} \tag{47.1}$$

47.3 Modeling of Dual Vibration Insulation System for Bridge or Viaduct Structures

The virtual model of an isolated structure assembly endowed with the dual vibration isolation system was realized (Fig. 47.2). Also, an experimental stand representing a small-scale bridge model was built and four vibration isolation systems were installed (Fig. 47.3).

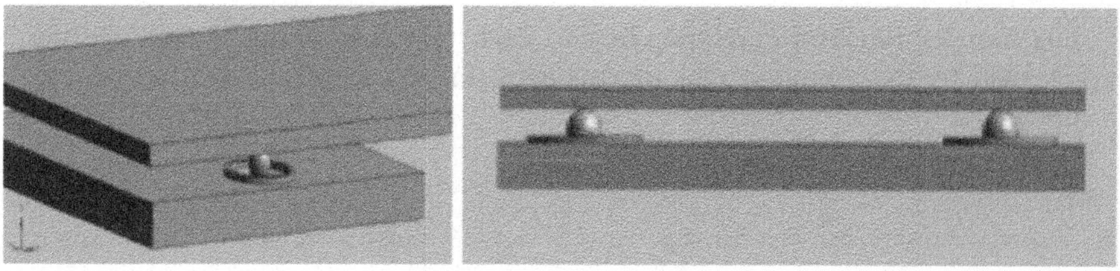

Fig. 47.2 The virtual model for a bridge structure isolated with the dual system

Fig. 47.3 The constructed experimental stand

47.4 Vibration Analysis of the Isolated Structural System with Dual System Mounted and Obtained Results

Applying the excitation to the isolated system favors the movement between the two structural elements through relative accelerations, recorded due to the use of tri-axial accelerometer mounted on the support pier and superstructure.

The recorded different values between the acceleration on identical directions between the superstructure and the support pier are highlighting the insulating and dissipating role of the dual vibration system.

The experimental results obtained are presented in terms of accelerations in time and the acceleration amplitude in relation to the recorded excitation frequency on the three acquisition channels corresponding to the three main orthogonal directions for both structural elements: support pier (inlet) and superstructure (outlet) where the tri-axial accelerometers were mounted.

The obtained results as recorded values at inlet (support pier) and outlet (superstructure) are presented for the analyzed cases regarding the two principal directions of motion (X, Y) as follows [3]:

Case 1—Structural system endowed with rolling pendulum bearing without elastomeric components (Table 47.1);

Case 2—Dual vibration insulation system composed of rolling pendulum bearing with the elastomeric elements, triangular disposed between structural elements (Table 47.2);

Case 3—Dual vibration insulation system composed of rolling pendulum bearing with the elastomeric elements, parallel disposed between structural elements (Table 47.3);

The transfer function corresponding to each vibration propagation direction is represented by the vibration magnitude depending on the excitation frequency.

Table 47.1 The obtained result for case 1

<table>
<tr><th></th><th>Inlet</th><th>Outlet</th></tr>
<tr><td>x</td><td>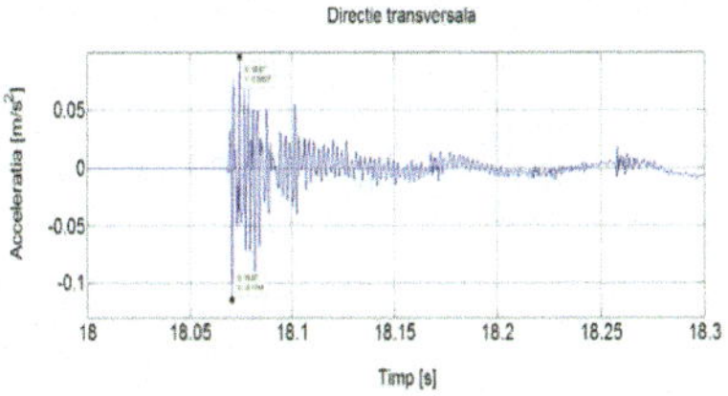
</td><td>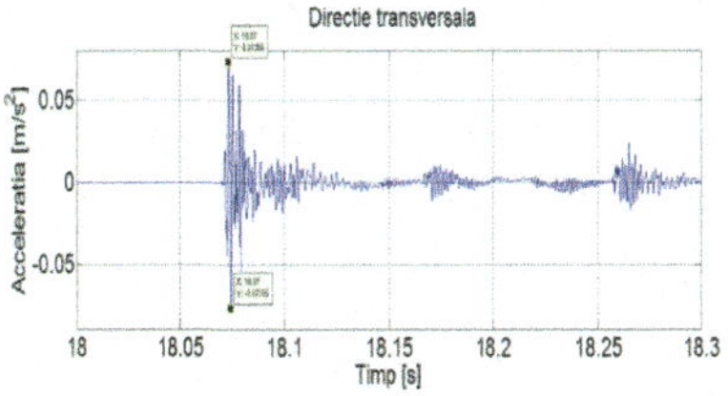
</td></tr>
<tr><td>y</td><td>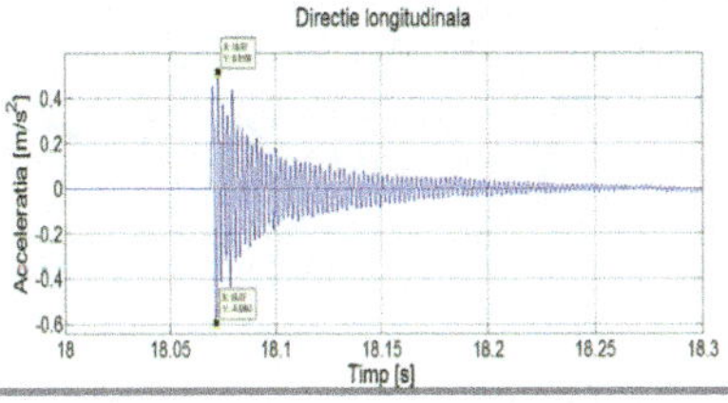
</td><td>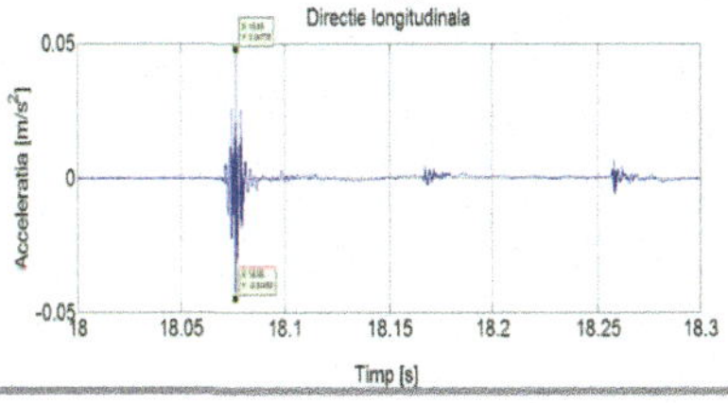
</td></tr>
</table>

Table 47.2 The obtained results for case 2

<table>
<tr><th></th><th>Inlet</th><th>outlet</th></tr>
<tr><td>x</td><td>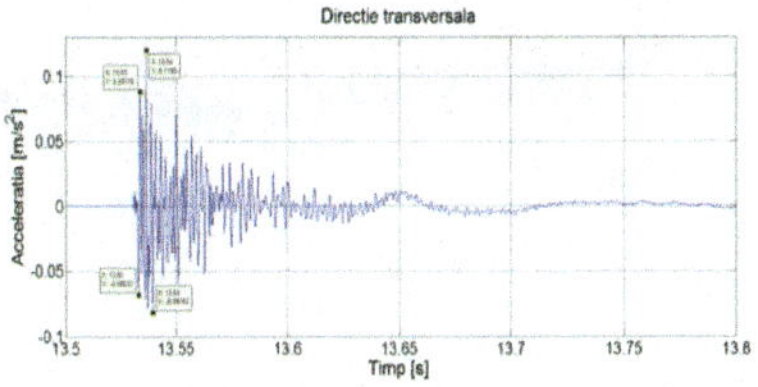
</td><td>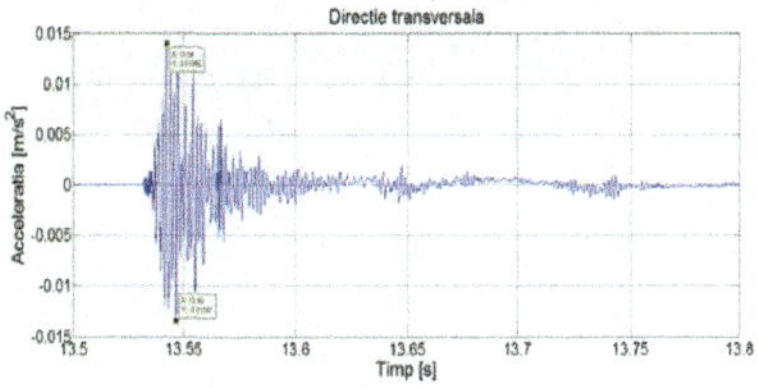
</td></tr>
<tr><td>y</td><td>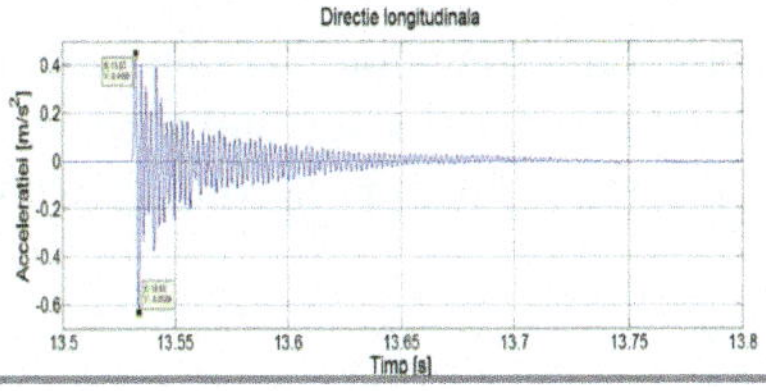
</td><td>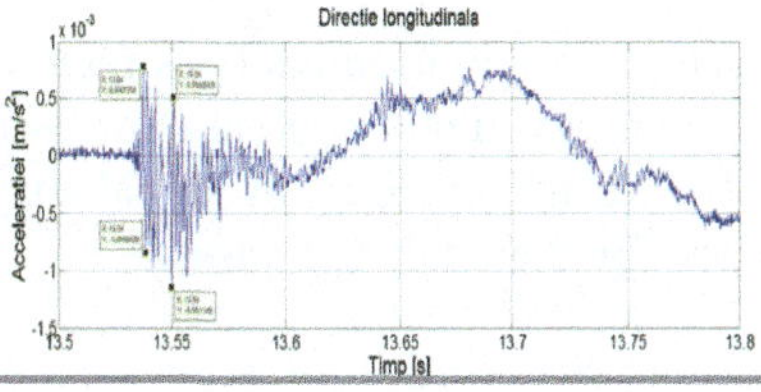
</td></tr>
</table>

Thus, for each analyzed case, the transfer functions for both transversal and longitudinal wave propagation directions were represented, so that the magnitude data can be compared to the excitation frequency (Table 47.4).

Table 47.3 The obtained results for case 3

	Inlet	outlet
x	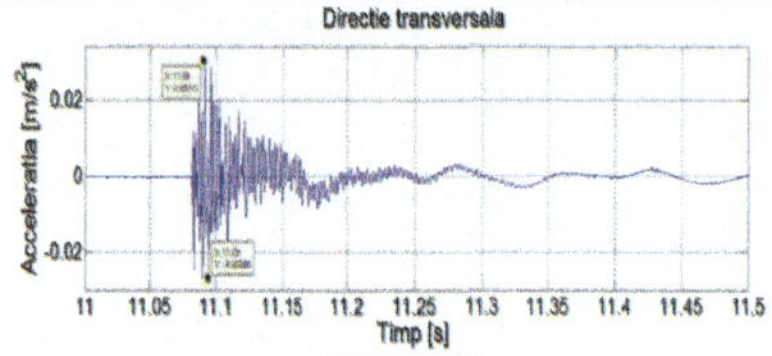	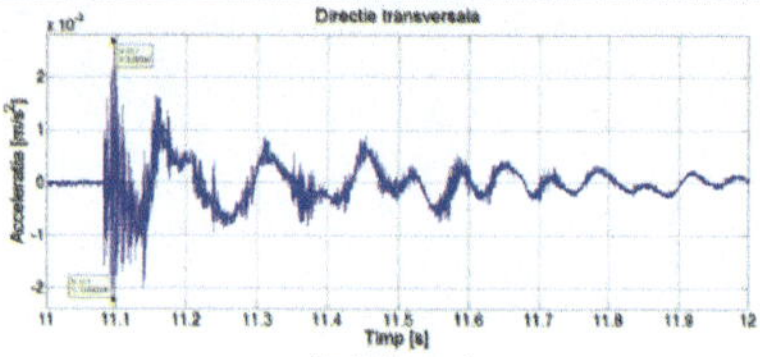
y	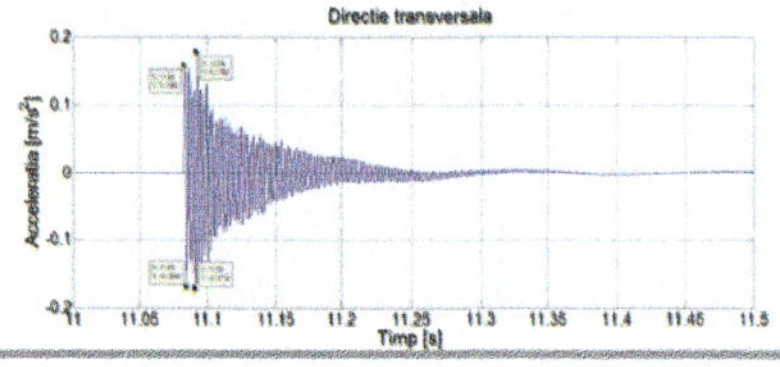	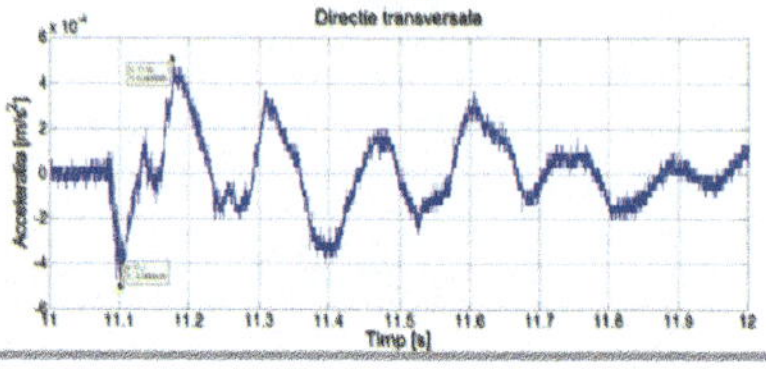

Table 47.4 The transfer functions recorded on the main directions of motion

Transversal direction (Ox)	Longitudinal direction (Oy)
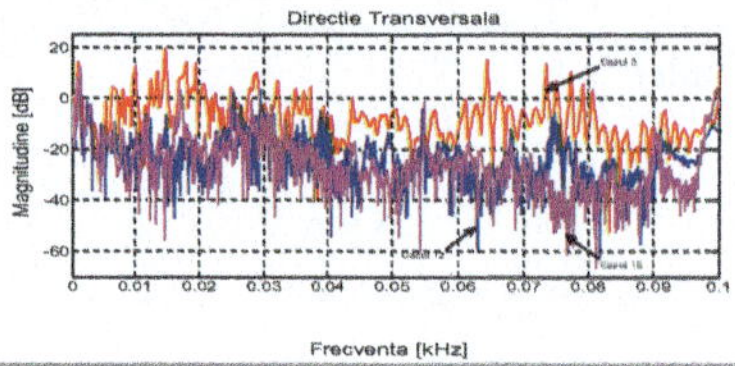	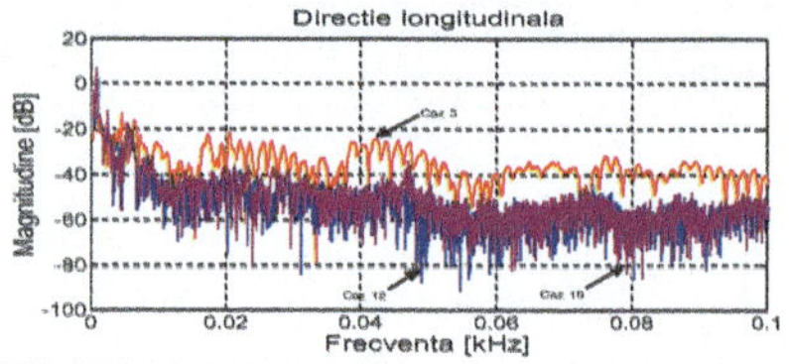

Using the vibration analysis method graphic representation have been achieved for the energy signal content expressed as function of frequency and time, in order to observe the changes of signal magnitude content in time according with frequency for the analyzed cases.

47.5 Conclusions

A dual vibration insulation device based on elastomeric kernel and rolling pendulum dissipation system is presented. This dual system ensures a disconnection between the structural elements where it is mounted.

An experimental model was built as a small scale bridge beam endowed with four vibration isolation devices having different configurations according to the presented cases.

After applying the excitation on the isolated experimental model, the obtained results are confirming the insulation and dissipative role of the dual isolation system according with acceleration amplitude over time values for the two main motion directions.

Also, the representation of transfer functions for the three analyzed cases highlights the magnitude values according to the excitation frequency and the attenuated values are recorded for the cases when the dual anti-vibration insulation system is in use.

References

1. G.C. Lee, Y.C. Ou, J. Song, T. Niu, Z. Liang, A roller seismic isolation bearing for highway bridges, in *The 14th International Conference on Seismic Engineering*, Beijing, China (2008)
2. M.H. Tsai, K.C. Chang, S.Y. Wu, Seismic isolation of a scaled bridge model using rolling-type bearings, in *a-4-a Conferință Internațională în Ingineria Seismică*, Taipei, Taiwan (2006)
3. F.D. Scheaua, Behavior analysis of the dry friction dissipating systems at dynamic actions. PhD thesis, "Dunarea de Jos" University of Galati (2013)

Chapter 48
The Velocity Evolution of an Isolated Element in a Friction Pendulum for Different Sliding Surfaces

Gelu Balan, Tiberiu Manescu, Camelia Jurcau, Ovidiu Milos and Vasile Iancu

Abstract In this paper, the velocity evolution for a structure isolated by friction pendulums (FP) with diverse sliding surfaces is presented. The surfaces are characterized by the curvature; as reference a cylindrical surface is considered and compared against surfaces generated by several polynomials. The position and velocity of the isolated element in different time moments is derived analytically for all surfaces, in order to evaluate the energy loss due to friction. This aspect is important because the materials used in manufacturing of seismic isolation devices can be sensitive to temperature variations, manifested by changing of the physical and mechanical properties. Findings of the study presented in this paper have practical importance, constituting guidelines for friction pendulum design.

48.1 Introduction

Using the state of the art techniques against earthquakes is desired in order to increase the safety and integrity of the structures. The optimal approach is to dissipate the energy generated by the earthquake [1, 2]. Nowadays, the usage of elastomeric devices [3] or FPs is the actual trend. The last one can sustain significant loads and can assure the control of oscillating structures [4]. The classical FP

G. Balan (✉) · T. Manescu · C. Jurcau · O. Milos · V. Iancu
Universitatea "Eftimie Murgu" Resita, P-ta Traian Vuia 1-4, 320085 Resita, Romania
e-mail: gelubalangelu@yahoo.com

T. Manescu
e-mail: manescu.tiberiu@gmail.com

C. Jurcau
e-mail: c.jurcau@uem.ro

O. Milos
e-mail: o.milos@uem.ro

V. Iancu
e-mail: v.iancu@uem.ro

N. Herisanu and V. Marinca (eds.), *Acoustics and Vibration of Mechanical Structures—AVMS-2017*, Springer Proceedings in Physics 198,
https://doi.org/10.1007/978-3-319-69823-6_48

bearing contains one [5] or more [6] cylindrical or spherical sliding surfaces, providing two parameters controlling the dynamics of the isolated structure. The variable curvature FP system was discovered by Tsai et al. [7], and the FP with the sliding surface based on an ellipse was introduced by Pranesh and Sinha [8].

The approach proposed by the authors is a friction pendulum with a surface profile defined by a polynomial function of a superior order. The seismic isolation system would gain flexibility, and the control of the dissipated energy and displacement would be significantly higher.

48.2 Mathematical Model of a Friction Pendulum with the Surface Generated by Polynomials

For the FP with the sliding surface generated by polynomial functions, illustrated in Fig. 48.1a, the equation of the curves generating the sliding surface is

$$h = a \cdot |u|^b \tag{48.1}$$

In (48.1), the parameters a and b have no physical meaning, the single role being to define the curve. By reasons of simplicity, only the positive values of the horizontal displacement u are considered. Norms presented in [6] limit the angle value to $\alpha_{max} \approx 36°$.

For cylindrical sliding surfaces with the radius $R = 5$, $u_{max} = 2.93$ and $h_{max} = 0.948$ m was determined for $\alpha_{max} \approx 36°$.

In Fig. 48.1b we have graphically represented a circular arc with the radius $R = 5$ (with thin continuous line) and the arcs of equations $h = 0.101u^2$ (with dashed line), $h = 0.101|u|^{2.5}$ (with solid line) and $h = 0.101u^4$ (with dotted line).

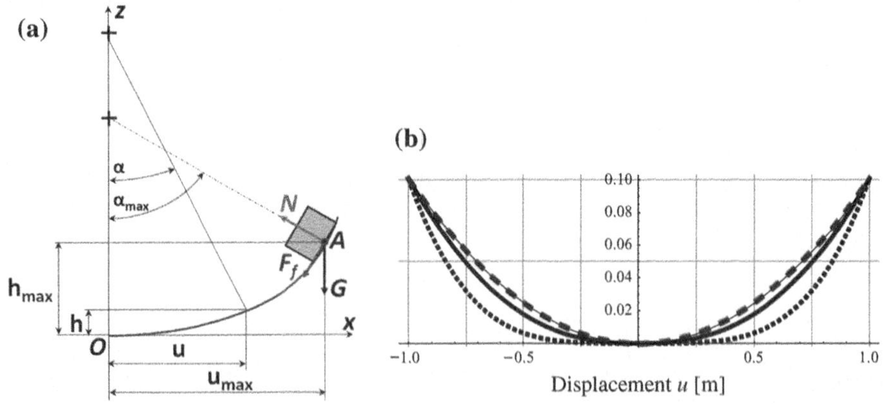

Fig. 48.1 Friction pendulum scheme (**a**) and the polynomial sliding surfaces (**b**)

One observes the similitude between the circular arc and the second order polynomial [9].

The dissipated energy, if the friction force is $F_f = \mu N \cdot sign(\dot{u})$, becomes

$$E_d = \int_0^{u_{\max}} F_f(u)du \tag{48.2}$$

Considering the motion in the positive direction of Ox axis, result

$$E_d = \int_0^{u_{\max}} \mu G \cdot \cos\alpha du = \mu G \int_0^{u_{\max}} \cos\alpha du \tag{48.3}$$

At point A, $tg\alpha = h' = a \cdot b \cdot u^{b-1}$. The instantaneous displacement in α becomes:

$$u = \left(\frac{1}{a \cdot b} tg\alpha\right)^{\frac{1}{b-1}} \text{ thus } du = \left(\frac{1}{a \cdot b}\right)^{\frac{1}{b-1}} \frac{1}{b-1} (tg\alpha)^{\frac{1}{b-1}-1} \frac{1}{\cos^2\alpha} d\alpha \tag{48.4}$$

The integral on $[0,\ u_{max}]$ becomes an integral on $[0, \alpha_{\max}]$, for $\alpha_{\max} = arctg\left(a \cdot b \cdot u_{\max}^{b-1}\right)$, so:

$$E_d = \mu G \left(\frac{1}{a \cdot b}\right)^{\frac{2}{b-1}} \cdot \frac{1}{b-1} \int_0^{\alpha_{\max}} \frac{1}{\cos\alpha} (tg\alpha)^{\frac{2-b}{b-1}} d\alpha \tag{48.5}$$

Denoting $\frac{2-b}{b-1} = d$ and $c = \left(\frac{1}{a \cdot b}\right)^{\frac{2}{b-1}} \cdot \frac{1}{b-1}$ results:

A. for $b \neq 2$:

$$E_d = \mu G \frac{c}{d} \left\{ {}_2F_1\left[-\frac{d}{2}, \frac{1-d}{2}, \frac{2-d}{2}, \cos^2\alpha_{\max}\right] \left(\frac{1}{\cos\alpha_{\max}}\right)^d - \frac{\Gamma\left(1-\frac{d}{2}\right)\Gamma\left(\frac{1}{2}+\frac{d}{2}\right)}{\sqrt{\pi}} \right\} \tag{48.6}$$

and the function ${}_2F_1$ is

$${}_2F_1\left[-\frac{d}{2}, \frac{1-d}{2}, \frac{2-d}{2}, \cos^2\alpha_{\max}\right] = \frac{\Gamma\left(\frac{2-d}{2}\right)}{\Gamma\left(\frac{1-d}{2}\right)\Gamma\left(\frac{2-d}{2}-b\right)} \int_0^1 \frac{t^{\frac{1-d}{2}-1}(1-t)^{\frac{2-d}{2}-\frac{1-d}{2}-1}}{\left(1-t\cos^2\alpha_{\max}\right)^{-\frac{d}{2}}} dt$$

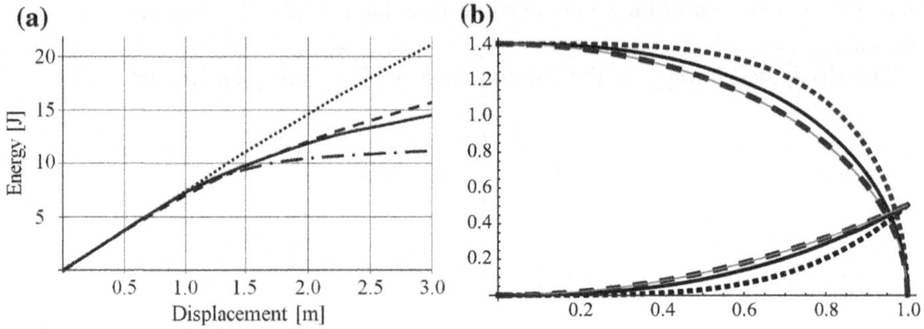

Fig. 48.2 Dissipated energy for different sliding surface profiles (**a**) and evolution of the velocities and heights (**b**)

B. for $b = 2$:

$$E_d = \mu G\left(\frac{1}{2a}\right)\ln\left(\frac{\cos\frac{\alpha_{\max}}{2} + \sin\frac{\alpha_{\max}}{2}}{\cos\frac{\alpha_{\max}}{2} - \sin\frac{\alpha_{\max}}{2}}\right) \tag{48.7}$$

In Fig. 48.2a we have shown the dissipated energies for the different sliding surfaces generated by the curves represented in Fig. 48.1b. For higher values of u there are significant changes in the dissipated energy in the analysed cases.

The kinetic energy of the structure of mass m is given by $E_k = mv^2/2$, where v is the velocity. Denoting $E_{p-\max} = mgh_{\max}$ the maximum potential energy, if friction is not taken into consideration, thus $E_k = E_{p_{\max}} - E_p$, results

$$\frac{mv^2}{2} = mgh_{\max} - mgh \tag{48.8}$$

and thus the maximum velocity, achieved for $u = 0$ is given by

$$v = \sqrt{2g(h_{\max} - h)} \tag{48.9}$$

In Fig. 48.2b we have illustrated the evolution of velocities and the heights during the displacement of the structure from point A to point O (see Fig. 48.1a), on different types of sliding surfaces generated by $y = 0.101u^2$ (with dashed line), $y = 0.101|u|^{2.5}$ (with solid line), $y = 0.101u^4$ (with dotted line) and a circular arc with the radius $R = 5$ (with thin continuous line).

If friction is taken into consideration, the velocity in point O is given by

$$v = \sqrt{2g(h_{\max} - h) - \frac{2E_d}{m}} \tag{48.10}$$

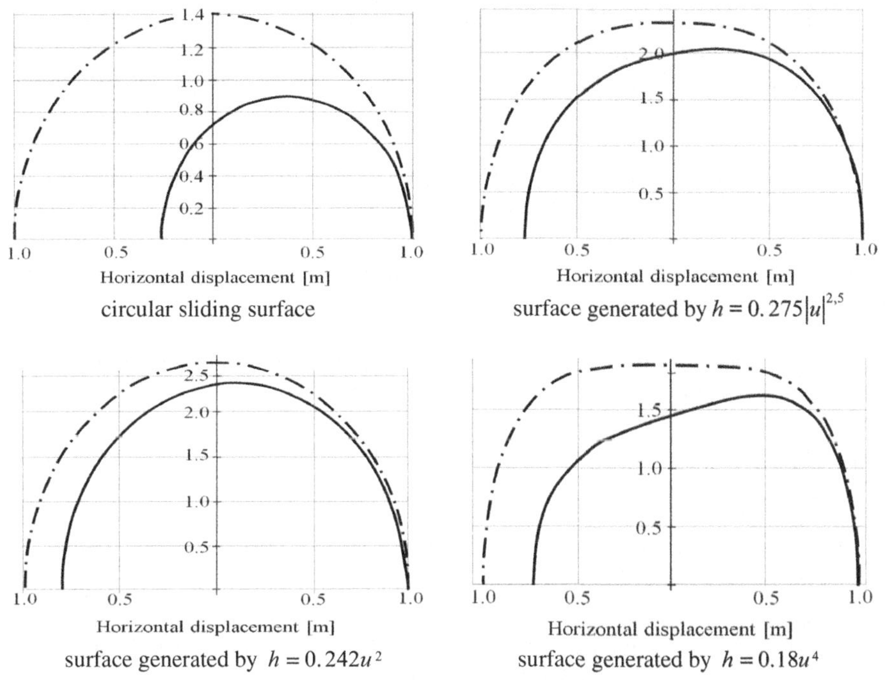

Fig. 48.3 The evolution of velocities for the different sliding surfaces, with and without friction

where E_d is the dissipated energy due to friction when the structure slides from point A to point O.

In Fig. 48.3 we have presented the evolution of velocities, first for the sliding without friction (with dash-dot line), for the sliding with energy dissipation due to friction (with solid line), for the sliding surfaces generated by: a circular arc with the radius $R = 5$ (Fig. 48.3a), a polynomial $h = 0.275|u|^{2.5}$ (Fig. 48.3b), a polynomial $h = 0.242u^2$ (Fig. 48.3c) and a polynomial $h = 0.18u^4$ (Fig. 48.3d). Note that, to have a uniformly distributed contact, the sliding element requests an elastomeric intermediate coated with low-friction layer [10]. These have to resist to high compression forces, but need also special deformation capability [11].

48.3 Conclusion

Classical earthquake isolation systems have limited the dynamic response control to two control parameters. In order to increase the control parameters, the study presented in this paper shows an advanced method to improve the fine tuning capacity starting from the design phase.

Using plates with concave surfaces generated by polynomials ensures an increased speed in the equilibrium position, thus a better circumstance to restore the position achieved prior to the earthquake. Profiles generated by curves of different polynomial functions considered herein proved the capacity to assure improvement in controlling the energy dissipation.

A disadvantage of the proposed sliding surfaces is the fact that, while for the spherical surface the slider has the same shape with the concavity of the plate (i.e. circular), in the case of using sliding surfaces generated by polynomials other than circular, it is obvious that interposing a coated elastomeric element able to follow the changes of the curvature between the slider and sliding surface is compulsory.

Acknowledgements The work has been funded by the Sectoral Operational Programme Human Resources Development 2007–2013 of the Ministry of European Funds through the Financial Agreement POSDRU/159/1.5/S/132395.

References

1. T. Hyakuda, K. Saito, T. Matsushita, N. Tanaka, S. Yoneki, M. Yasuda, M. Miyazaki, A. Suzuki, T. Sawada, The structural design and earthquake observation of a seismic isolation building using Friction Pendulum system, in *Proceedings of the 7th International Seminar on Seismic Isolation, Passive Energy Dissipation and Active Control of Vibrations of Structures*, Assisi, Italy (2001)
2. P. Roussis, M.C. Constantinou, Uplift-restraining friction pendulum seismic isolation system. Earthq. Eng. Struct. **35**, 577–593 (2006)
3. V. Iancu, O. Vasile, G.R. Gillich, Modelling and characterization of hybrid rubber-based earthquake isolation systems. Mater. Plast. **49**(4), 237–241 (2012)
4. Y.B. Yang, T.Y. Lee, I.C. Tsai, Response of multi-degree-of-freedom structure with sliding supports. Earthq. Eng. Struct. **19**, 739–752 (1990)
5. V.A. Zayas, S.S. Low, S.A. Mahin, A simple pendulum technique for achieving seismic isolation. Earthq. Spectra **6**, 317–333 (1990)
6. M.C. Constantinou, Friction pendulum double concave bearing. NEES Report, available at: http://nees.buffalo.edu/docs/dec304/FP-DC%20Report-DEMO.pdf (2004)
7. C.S. Tsai, T.C. Chiang, B.J. Chen, Finite element formulations and theoretical study for variable curvature friction pendulum system. Eng. Struct. **25**, 1719–1730 (2003)
8. M. Pranesh, R. Sinha, VFPI: an isolation device for aseismic design. Earthq. Eng. Struc. **29**, 603–627 (2000)
9. C.S. Jurcau, G.R. Gillich, V. Iancu, D. Amariei, in *Evaluation and Control of Forces Acting on Isolated Friction Pendulum*. International Conference on Engineering Mechanics, Structures, Engineering Geology, International Conference on Geography and Geology—Proceedings (2010), pp. 220–225
10. C.M. Iavornic, Z.I. Praisach, O. Vasile, G.R. Gillich, V. Iancu, Study of Stress and Deformation in Elastomeric Isolation Systems Using the Finite Element Method, Recent Advances in Signal Processing, Computational Geometry and Systems Theory—ISCGAV'11, ISTASC'11 (2011), pp. 239–244
11. G.R. Gillich, P. Bratu, D. Frunzaverde, D. Amariei, V. Iancu, Identifying Mechanical Characteristics of Materials with Non-linear Behavior using Statistical Methods, in *Proceedings of the 4th WSEAS International Conference on Computer Engineering and Applications (CEA 10)* (2010), pp. 96–103

Chapter 49
Analytical and Numerical Study of Critical Speed for Right Shafts

I. Geonea, N. Dumitru and A. Margine

Abstract In this paper it is presented the theoretical, experimental and numerical simulation of critical speed for right shafts. For this purpose, it is designed a virtual model of a test bench for shaft critical speed analysis. The studied shaft is mounted on two bearings and on the middle it has mounted a heavy disc. The system has the possibility to modify the shaft speed with a frontal friction continuous variable transmission. It is presented a mathematical model to calculate the critical speed of the shaft. The result obtained by analytical calculation is verified by experimental way and by numerical simulation. The simulation of shaft flexibility is achieved in ADAMS multibody software and obtained results are presented and discussed. Shaft experimental flexional deformations and the results obtained by numerical simulation are similar. Study of shaft flexibility by numerical simulation in ADAMS is an easy tool and very important for practical application to avoid resonance in elastic mechanical systems.

49.1 Introduction

The traditional approach for dynamic analysis of mechanical system is studied on the assumption that the mechanisms links and machine parts are made up from rigid bodies [1]. This assumption is not always good, because the machine parts are manufactured from steel that is an elastic material and they deflect when are subject to external forces and torques or inertia forces. Particularly at high operation speed a mechanism linkage may support serious elastic deformations due to its inertia, so the operation becomes improper [2].

I. Geonea (✉) · N. Dumitru · A. Margine
Faculty of Mechanics, University of Craiova, Craiova, Romania
e-mail: igeonea@yahoo.com

N. Dumitru
e-mail: nicolae_dtru@yahoo.com

A. Margine
e-mail: fam_margine@yahoo.com

N. Herisanu and V. Marinca (eds.), *Acoustics and Vibration of Mechanical Structures—AVMS-2017*, Springer Proceedings in Physics 198,
https://doi.org/10.1007/978-3-319-69823-6_49

Several authors have recently proposed methods and developed models of mechanisms dynamics, considering flexible links, rather that rigid links [3–5].

The finite elements method it's used by several authors to study the mechanism with flexible links. In this method the geometrically nonlinear relations for the element deformations in terms of the nodal position and orientation coordinates are considered. The equations of motion are formulated in terms of mixed sets of generalized coordinates of the mechanism with rigid links and deformation mode coordinates (characteristic to the elastic links) [6–8].

49.2 Analytical and Experimental Study of Shafts Critical Speed

Any elastic system is characterized by a period of its own oscillations, determined by its rigidity and its own mass. If such a system is subjected to forced oscillations due to the action of variable forces and moments from unbalanced masses or other causes and if the frequency of variation of these forces and disturbing moments is equal to the frequency of the oscillations or with a multiple, the amplitude of the oscillations increases rapidly, and it is occurring resonance.

In the case of shafts, depending on the forces of action and disturbing torsion moments, longitudinal, torsion and transverse vibrations may occur. From the practical point of view, most frequently, only the last two types of vibrations are concerned, because in general the longitudinal restoring forces are very high and the amplitude of the longitudinal vibrations are small. In most cases, transverse vibrations occur as a result of transverse forces, which are repeated periodically.

The mathematical model of the experimental test bench is represented by a shaft with constant section, with horizontal axis, on which a disk is mounted (Fig. 49.1).

Under the action of the weight of the disc, the shaft is deformed by making an arrow f_{st} in the section where the disc is mounted. In this case, the critical speed calculation relation is:

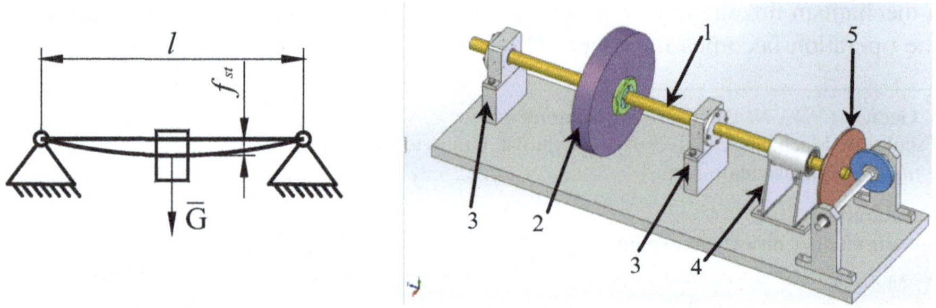

Fig. 49.1 Kinematic scheme and virtual model of a test bench for shafts critical speed

$$n_{cr} = \frac{30}{\pi}\sqrt{\frac{g}{f_{st}}}\ (\text{rpm}) \tag{49.1}$$

A first observation upon (49.1) is that the critical speed is not influenced by the disc eccentricity on the shaft. Considering shaft rigidity (49.1) is written as:

$$n_{cr} = \frac{30}{\pi}\sqrt{\frac{Cg}{G}}\ (\text{rpm}) \tag{49.2}$$

where: n_{cr} is the critical speed of the shaft; *C*—shaft rigidity; *g*—gravity force in (mm/s^2); *G*—disc weight.

For a right shaft with tubular constant section, rigidity is computed with (49.3):

$$C_i = \frac{G_\iota}{f_i}\ (\text{N/mm}) \quad i = 1 \div 3 \tag{49.3}$$

$$f_{st_i = \frac{G_i}{3EI_Z}\frac{a^2(l-a)^2}{l}}\ (\text{mm}) \tag{49.4}$$

$$I_z = \frac{\pi}{64}\left(d_{ext}^4 - d_{int}^4\right)\ (\text{mm}^4) \tag{49.5}$$

where: f_{sti}—shaft flexion; *E*—material Young's module; I_z—cross section inertia moment; *a*—distance between disc center and right bearing, in (mm), *l*—distance from shaft bearings in (mm). For the experimental determination of the rigidity of the shaft, the experimental installation (Fig. 49.2) further comprises, the support 6 on which the special lever 7 is mounted. By mounting the weights 8, the shaft is subjected to bending in the vertical plane. Using the comparator 9, the bending

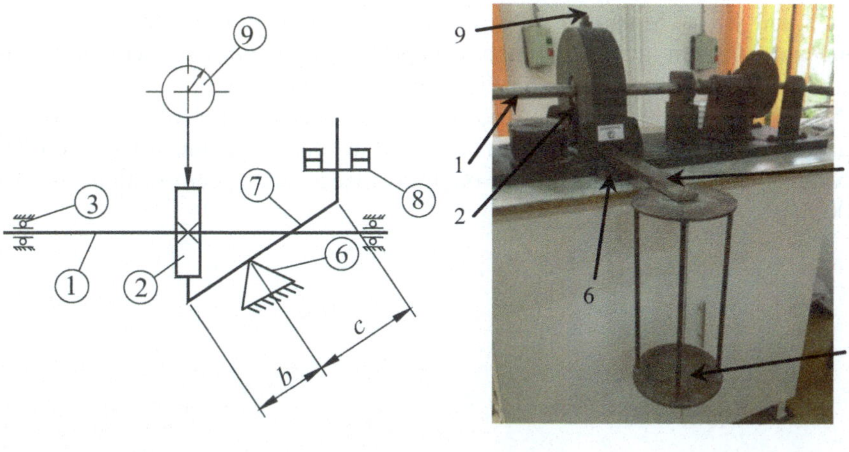

Fig. 49.2 Kinematic scheme and experimental model of a test bench for shafts critical speed

Table 49.1 Experimental shaft bending deformation and computed critical speed

a (mm)	l (mm)	$f_{st.\ exp.}$ (mm)	$f_{st.\ analytic}$ (mm)	d_{int} (mm)	d_{ext} (mm)	G (N)
267	587	0.031	0.096	18	31	196
G_1 (N)	G_2 (N)	G_3 (N)	f_1 (mm)	f_2 mm)	f_3 (mm)	$n_{cr.\ experimental}$ (rpm)
46	93	140	0.026	0.053	0.082	2967
C_1 (N/mm)	C_2 (N/mm)	C_3 (N/mm)	n_{cr1} (rpm)	n_{cr2} (rpm)	n_{cr3} (rpm)	$n_{cr\ analytic}$ (rpm)
1769.23	1754.71	1707.31	5868.66	4110.43	1142.99	3054.15

produced by weights is determined. Experimental data and results are presented in Table 49.1.

49.3 Numerical Simulations in ADAMS of Shaft Vibrations

ADAMS multibody software offers the possibility to study the shafts deformations, considering the deformability of the shaft. For that purpose, the shaft is meshed with tetrahedral finite elements and is achieved a modal dynamic analysis. For that purpose, it is used the option "Create Flexible Body". The shaft speed used for this simulation is 3000 rpm, a value closer of computed critical speed.

In Fig. 49.3 are presented two deformed shapes of the shaft. First one corresponds to the mode 8 of vibration with a frequency of 144.69 Hz. The second deformed shape of the shaft corresponds to the mode 12 of vibration that appears to 765.66 Hz. ADAMS simulations computes also the shaft elastic deformations. In Figs. 49.4 and 49.5 it is presented the elastic deformation of the shaft mass center upon x and y axis. Maximum value of the deformations is 0.02 mm in horizontal plane and 0.03 mm in vertical plane. Computed deformations accelerations of the shaft mass center are presented in Figs. 49.6, 49.7 and 49.8.

Numerical simulation in ADAMS allows computation of shaft deformation in dynamic regime. Elastic deformation computed in dynamic regime is 0.03 mm. The maximum value of shaft center of mass acceleration deformation, reaches 2500 mm/s^2.

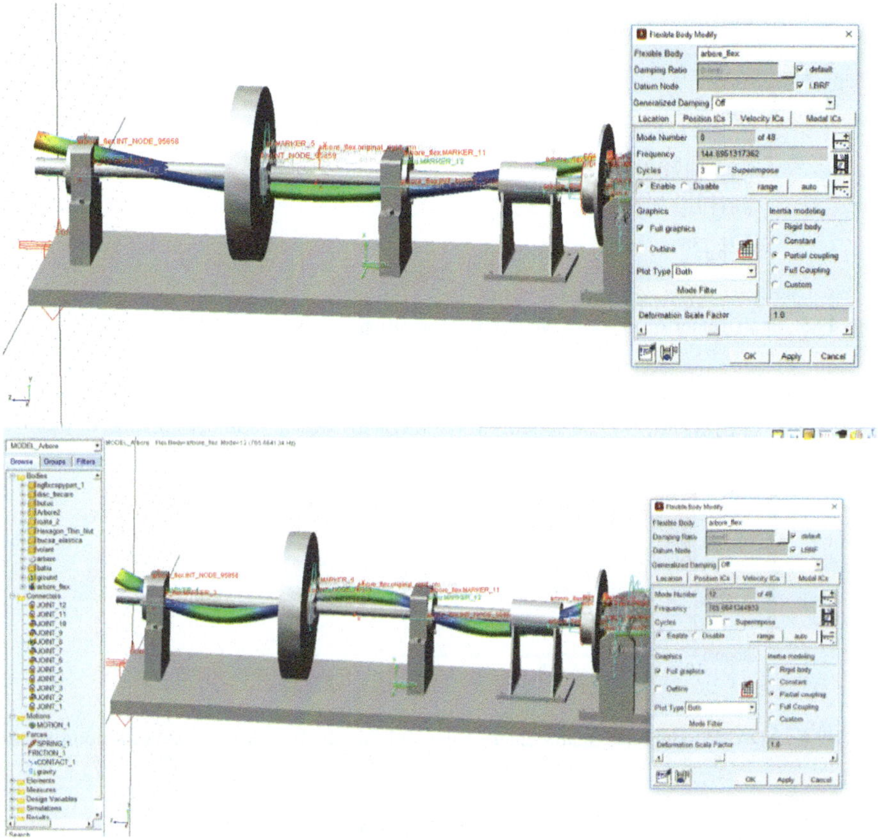

Fig. 49.3 Deformed shaft shape computed in ADAMS

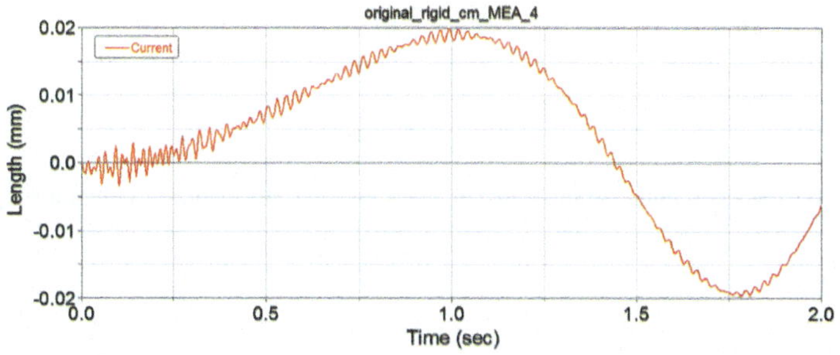

Fig. 49.4 Shaft center of mass elastic deformation upon x axis

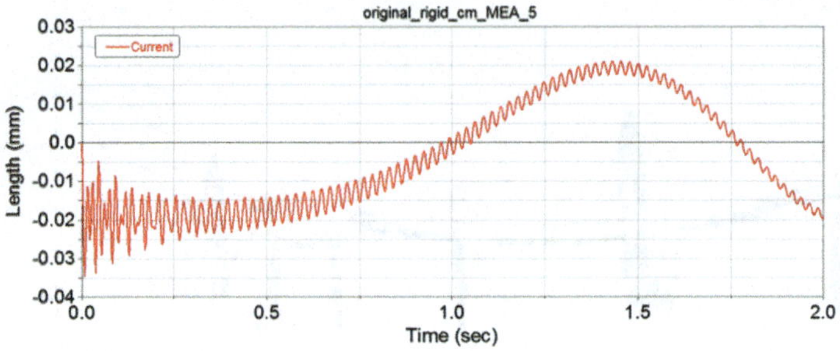

Fig. 49.5 Shaft center of mass elastic deformation upon y axis

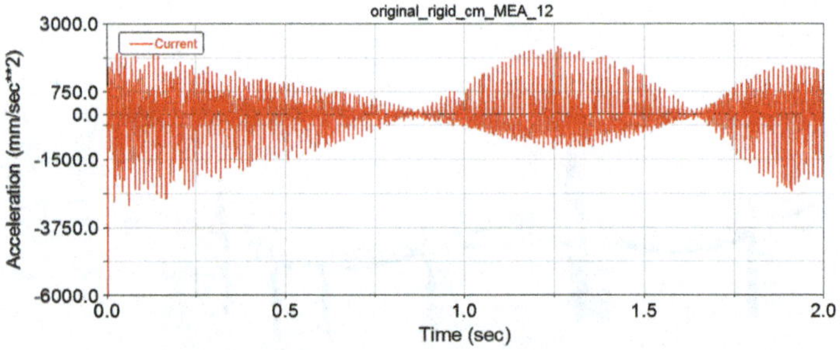

Fig. 49.6 ADAMS computed shaft center of mass elastic deformation acceleration upon x axis

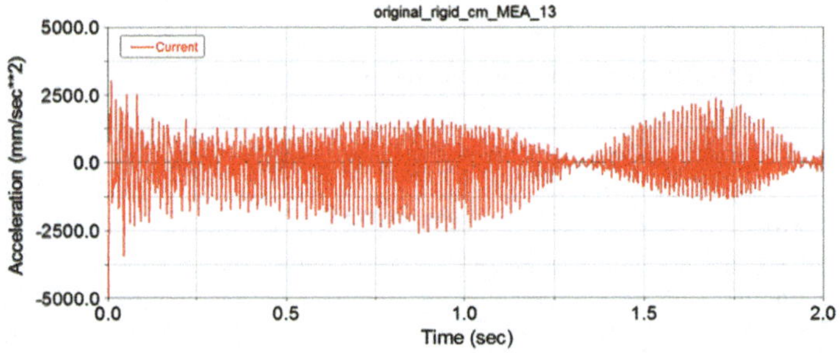

Fig. 49.7 ADAMS computed shaft center of mass elastic deformation acceleration upon y axis

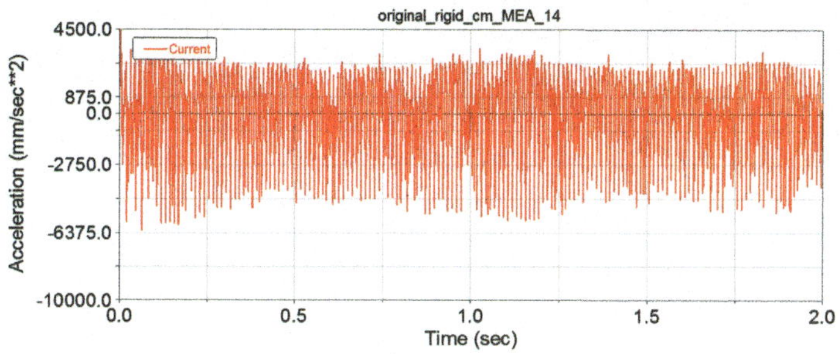

Fig. 49.8 ADAMS computed shaft center of mass elastic deformation acceleration upon z axis

49.4 Conclusion

This paper presents the results obtained for the critical speed of a right shaft. It is presented a mathematical model to establish the shaft critical speed, considering the shaft rigidity. Experimental critical shaft speed is computed in static conditions, considering the shaft rigidity. Numerical simulations are performed in ADAMS, and are obtained the shaft center of mass elastic deformations and accelerations of deformation. The critical speed, determined experimentally will differ by the value computed analytically with (49.1) and (49.2), due to the simplifying assumptions adopted for analytical calculation: by bending of the shaft the disk moves parallel to itself; rigidity is constant in all planes; bearing friction is neglected.

References

1. N. Dumitru et al., Differential transmission for robotizing a powered wheelchair, in *Proceedings of the OPTIROB 2010* (2010), pp. 47–51
2. G. Piras, W.L. Cleghorn, J.K. Mills, Dynamic finite-element analysis of a planar high-speed, high-precision parallel manipulator with flexible links. Mech. Mach. Theory **40**(7), 849–862 (2005)
3. W.L. Cleghorn, G. Fenton, B. Tabarrok, Finite element analysis of high-speed flexible mechanisms. Mech. Mach. Theory **16**(4), 407–424 (1981)
4. B.M. Bahgat, K.D. Willmert, Finite element vibrational analysis of planar mechanisms. Mech. Mach. Theory **11**(1), 47–71 (1976)
5. J.O. Song, J. Haug, Dynamic analysis of planar flexible mechanisms. Comput. Methods Appl. Mech. Eng. **24**(3), 359–381 (1980)
6. B. Jonker, A finite element dynamic analysis of spatial mechanisms with flexible links. Comput. Methods Appl. Mech. Eng. **76**(1), 17–40 (1989)
7. I. Geonea, A. Margine, A. Ungureanu, Study upon the dynamic answer of plane manipulators. Adv. Mater. Res. **463–464**, 1304–1308 (2012)
8. A. Margine et al., *Dynamic Answer and Experimental Research Concerning the Mechanisms of Mowers Machine*. Lecture Notes in Engineering and Computer Science (2010), pp. 1236–1241

Index

N. Herisanu and V. Marinca (eds.), *Acoustics and Vibration of Mechanical Structures—AVMS-2017*, Springer Proceedings in Physics 198,
https://doi.org/10.1007/978-3-319-69823-6

The manufacturer's authorised representative in the EU is Springer Nature Customer Service Centre GmbH, Europaplatz 3, 69115 Heidelberg, Germany. If you have any concerns regarding our products, please contact ProductSafety@springernature.com

Printed and bound by CPI Group (UK) Ltd, Croydon, CR0 4YY
30/09/2026
02225228-0017